PUBLISHED FOR THE
Institute of Early American History and Culture
AT WILLIAMSBURG, VIRGINIA

Philosophical Hall on Independence Square
Erected, 1785-1789

Photograph of the building after its restoration in 1948-1949.

THE
Pursuit of Science
IN
Revolutionary America
1735-1789

‡‡‡‡‡‡‡‡‡‡‡‡‡‡‡‡‡‡‡‡‡‡‡‡‡‡‡‡

BY BROOKE HINDLE

The Norton Library
W·W·NORTON & COMPANY·INC·
NEW YORK

First published in the Norton Library 1974 by
arrangement with The University of North Carolina Press

Published simultaneously in Canada
by George J. McLeod Limited, Toronto

Books That Live
The Norton imprint on a book means that in the publisher's
estimation it is a book not for a single season but for the years.
W. W. Norton & Company, Inc.

Library of Congress Cataloging in Publication Data
Hindle, Brooke.
    The pursuit of science in Revolutionary
America, 1735–1789.
    (The Norton library)
    Reprint of the ed. published by the University
of North Carolina Press, Chapel Hill.
    Bibliography: p.
    1. Science—History—United States. 2. United
States—History—Revolution. I. Title.
[Q127.U6H5   1974]   509'.73   73-18143
ISBN 0-393-00710-3

Printed in the United States of America
1 2 3 4 5 6 7 8 9 0

TO HELEN

✠✠

ACKNOWLEDGMENTS

AMONG MY MANY DEBTS, the greatest is more than can well be described. Carl Bridenbaugh, now of the University of California, has provided constant inspiration and encouragement over a period of many years. Except for him, this book would not have been written.

To the Institute of Early American History and Culture, I am indebted not only for publication but also for a two-year term as Research Associate during which time I was able to complete an important segment of my research. More especially, I owe debts to individuals now and formerly associated with the Institute. Lester J. Cappon, present Director of the Institute, has been a steady guide and a sure reliance. His effort and the effort of James M. Smith, Editor of Publications, have saved me from numerous literary and historical faults. The imaginative insight of Douglass Adair, now of Claremont College, and the understanding of Lyman H. Butterfield, now of the Adams Papers, were important aids.

Upon another small group, I have relied heavily. The doctoral dissertation which introduced me to this field of study was written under Richard H. Shryock, now of the Institute of the History of Medicine, a man with whom I have never had a conversation that did not leave me with a head full of new ideas. I. Bernard Cohen of Harvard University has most generously shared his own extensive knowledge of the field, even to the point of lending me notes prepared for his own use. His careful criticism of the manuscript offered many important correctives. Whitfield J. Bell, Jr., of the Papers of Benjamin Franklin, has befriended me over a long period of time with countless pieces of information and with the opportunity to test many of my concepts against his own rich understanding.

It is a pleasure to acknowledge the indispensible assistance offered by the librarians and keepers of the archives who have so willingly made available their collections. I must content myself with naming the libraries and other institutions where I have been so well served: in Cambridge, Harvard College Library and the Harvard University Archives; in Boston, the American Academy of Arts and Sciences, the Boston Medical Library, and the Boston Public Library; in Providence, Brown University Library; in New Haven, Yale University Library; in New York, the New York Public Library, the New-York Historical Society, the New York Academy of Medicine, New York University Library, and Columbia University Library; in Philadelphia, the American Philosophical Society, the Library Company of Philadelphia, the Historical Society of Pennsylvania, the University of Pennsylvania Library and Archives, the College of Physicians of Philadelphia, and the Pennsylvania Hospital; in Washington, D.C., the Library of Congress; in Richmond, the Virginia Historical Society and the Medical College of Virginia; in Charlottesville, the University of Virginia Library; in Williamsburg, the College of William and Mary Library, the Institute of Early American History and Culture, and Colonial Williamsburg; in Newport News, the Mariner's Museum; in Charleston, the Charleston Library Society; in Chicago, the Newberry Library and the John Crerar Library; and in Evanston, Northwestern University Library.

To those who made it possible for me to use the illustrative material included in this book, I want to express my gratitude. Special thanks should be offered to individuals who aided in the collection of this material. I am grateful, too, to the Princeton University Library for permission to quote copyrighted material.

My wife has helped at every stage in the preparation of the book, but most of all by providing an environment in which the writing of a book was possible.

New York University
March 1956

Contents

[ix]

Contents

PART THREE

The New Nation, 1775-1789

LIST OF ILLUSTRATIONS

[xi]

THE PURSUIT OF SCIENCE
IN REVOLUTIONARY AMERICA, 1735-1789

‡‡‡

Chapter One

INTRODUCTION

As early as 1743, Benjamin Franklin gave voice to exalted aspirations for American achievement in science. He announced: "The first drudgery of settling new colonies which confines the attention of people to mere necessaries is now pretty well over; and there are many in every province in circumstances that set them at ease, and afford leisure to cultivate the finer arts and improve the common stock of knowledge." He went on to sketch a plan for the cooperative promotion of science on an intercolonial basis. In this, he demonstrated a fine appreciation of one of the noblest dreams of the Enlightenment —the dream that man by studied effort could unlock the secrets of nature and apply them to increase his power "over matter, and multiply the conveniences or pleasures of life." [1] Franklin hoped that America was sufficiently mature to contribute to the advancement of science; he was convinced that the promotion of science would yield great practical benefits.

This was almost the first moment at which it would have been reasonable to think in terms of the intercolonial promotion of science. Until sometime after 1735, the idea of voluntary intercolonial activity of almost any sort was chimerical. Cooperation had generally been limited to matters of defense and war and even there the most conspicuous success had been met by small groups of closely related provinces or by the forced efforts of the home government in London.

1. Albert H. Smyth (ed.), *The Writings of Benjamin Franklin* (New York, 1905-7), II, 228, 230.

[1]

Yet, slowly and almost unobserved, a web of ties had drawn the colonists from Massachusetts to Georgia into a closer relationship. Trade not only brought Americans closer ties with Englishmen, but increasingly with other Americans. Social relationships between Americans of different colonies expanded, marked by increasing inter-colonial marriages, correspondence, and organizations. Roads were improved; travel became more frequent and postal service more dependable. Newspapers circulated more widely and contributed much to intercolonial understanding by the news they carried of occurrences in other colonies. Finally, the Great Awakening of the 1730's and 1740's crossed colonial boundaries with impunity and demonstrated the religious and cultural similarity of America. Travellers of the time found the inhabitants "much alike" in essential respects from one colony to another.[2]

War still had the power to call forth the most emotional expression of this growing rapport. The capture of the reputedly impregnable fortress of Louisburg in 1745 by a New England force provided occasion for the most unrestrained celebration of a common patriotism in Maryland as well as in Massachusetts.[3] The ecstatic feelings of the Americans were expressed in the metaphor of English nationality but no Englishman living on the other side of the Atlantic could fully share in this reaction.

Behind overt evidences of a common outlook and a common culture lay the amazingly rapid development of the colonies which had succeeded in pushing the crude ways of the frontier well into the interior. Wealth and a measure of refinement had been won in the seaboard towns and in some of the settled rural regions. Indeed, portions of the American scene looked much like a microcosm of Europe itself for not only were the fine homes, the furnishings, and the clothes of the well-to-do patterned after European models, but political, economic, and religious institutions were similarly European in conception.

On the other hand, no institution and no custom to be found in America was the exact copy of its European prototype. Divergence had begun as soon as the first colonist stepped ashore to discover that

2. Carl Bridenbaugh (ed.), *Gentleman's Progress; The Itinerarium of Dr. Alexander Hamilton, 1744* (Chapel Hill, 1948), 199.
3. Max Savelle, *Seeds of Liberty; the Genesis of the American Mind* (New York, 1948), 571.

the American environment required adaptation on his part. Differentiation had continued despite the strengthening and deepening of ties
with the mother country and with other nations of Europe. Some men
had seized upon the opportunities presented in the colonies to promote
changes that were impossible in the Old World. Others had found
change inescapable, hard as they might try to reproduce familiar conditions. The abundance of land, of fish, and of forest, the location of
rivers, the shortage of labor, the width of the Atlantic, the great distances in America itself, and many more subtle factors had worked
continually to modify the European heritage. The result was the
gradual emergence of a variant of European civilization; America was
a part of Europe, yet still apart from it.

The direction and magnitude of modification and of attainment
were different for every aspect of life. Perhaps the most spectacular
achievement was to be found in self-government where American
experience led to new and successful patterns. The popular assem-
blies were going far toward wresting the essential elements of
power from the governor, the council, and the agencies of the crown.
Economic accomplishments, too, were impressive, particularly in developing a remarkably high standard of living. Religious diversity and
social fluidity were not different in kind but certainly they differed in
degree from the patterns known in the mother country. In working
out many of the alterations and advances, the Americans had enjoyed
certain advantages upon which they had willingly improved.

Not always, however, did the American environment encourage
improvement and advance. For the growth of some institutions and
capacities, the American soil proved thin and weak with the result
that flourishing imports atrophied and barely survived their transplanting. With respect to science, there was serious question whether
America could support scientific work of the sort Franklin demanded
in 1743. Nowhere in the colonies could be found the rich libraries, the
ancient universities, or the conversation of the learned that graced the
centers of Old World culture. Great endowed institutions were lacking. So, for the most part, was the patronage of the king and of
enlightened nobles. Stimulus and support from Europe could supply
only a part of the deficiency. Communication was slow and uncertain
and the encouragement offered was not balanced but biased by the
particular needs of the individuals who offered it. Development of a

[3]

fruitful scientific life demanded far more of the people and the society that sought it than did the successful practice of self-government. As Dr. Thomas Bond later remarked, "[Science] is a child of a thousand Years,—Approaches slowly to Maturity, and is long in dying." [4]

As far as the past record went, there had been some notable accomplishments. Contributions were made to scientific knowledge by Europeans who had visited the colonies in order to study the natural history of the New World. Some Americans, too, had sent back valuable information. A taste for learning had been demonstrated by a handful of planters, merchants, and professional men who studied abroad or assimilated otherwise the values of the European gentleman.[5] Still more had been accomplished in the intellectual atmosphere of New England from which region a great majority of the colonial fellows of the Royal Society of London had been chosen. It was a New Englander who had supplied the key American astronomical observations used by Sir Isaac Newton in his great *Principia Mathematica*. It was in New England that the first great test in the Western world had been made of smallpox inoculation. New Englanders had contributed most of the American articles in the *Philosophical Transactions* including some of first-rate importance.[6]

The greatest achievements, centered in Boston and in Cambridge, were a result of the cultivation of matters of the mind by clergy and laymen who had readily accepted the "new science" of the seventeenth century. However, they had their limitations. To the erudite clergy, the study of nature was "not sufficient for salvation, nor for saving faith." [7] A certain indifference, therefore, lingered on and added to the other obstacles which beset the pursuit of science in America. The great complaint of Thomas Brattle in 1705 and of Paul

4. Thomas Bond, *Anniversary Oration Delivered May 21st* (Philadelphia, 1782), 30.

5. See Louis B. Wright, *First Gentlemen of Virginia* (San Marino, California, 1940).

6. Raymond P. Stearns, "Colonial Fellows of the Royal Society of London, 1661-1788," *William and Mary Quarterly*, 3d ser., 3 (1946), 208-68; Otho T. Beall, Jr., and Richard H. Shryock, "Cotton Mather: First Significant Figure in American Medicine," American Antiquarian Society, *Proceedings*, 63 (1953), 133-66.

7. Richard Mather quoted by Perry Miller, *The New England Mind, from Colony to Province* (Cambridge, 1953), 438.

Dudley a few years later was that even in Boston they could find none of their neighbors interested in experimental science.[8]

The new enthusiasm for scientific attainment voiced by Benjamin Franklin was a reflection of the Enlightenment flooding into the American colonies. In New England, it provoked further adaptation on the part of the clergy but, at the same time, leadership in science passed to nonclerical hands. Dynamic efforts to advance science were also made in the Southern and Middle colonies—especially at Philadelphia. Growing rapidly in wealth and power, the younger city endeavored with an earnest dedication to fulfill the ideals of the Enlightenment. Perhaps because Boston already taught science to its youth with a greater competence than any other American city and because it had behind it an unmatched record of intellectual attainment, the Massachusetts city was a shade less eager to achieve. Perhaps it was the closer commercial, religious, and personal ties of Philadelphians to the dynamic intellectual circles of England and Scotland. Enthusiasm was to be found at many points in the colonies; actual accomplishment was still a relatively rare commodity.

Certainly more fundamental than enthusiasm was the question of the ability and willingness of the Americans to support scientific work. On the one hand, the material well-being of the Americans was remarkable and it was clear that the average American had more leisure in his day than the average European.[9] On the other hand, both wealth and leisure were so widely distributed that there were only a few to act in the capacity of patrons even if they would. It was easy enough to ask for scientific work from the Harvard professor of mathematics and natural philosophy but where else were the men to be found to fulfill the scientific ideals of the Enlightenment?

As John Bartram, American seedsman and botanist, pondered that question in 1745, he arrived at some rather discouraging conclusions. "I sometimes observe," he wrote, "that the major part of our inhabitants may be ranked in three Classes, the first Class are those whose thought and study is intirely upon getting and laying up large estates

8. Frederick G. Kilgour, "The Rise of Scientific Activity in Colonial New England," *Yale Journal of Biology and Medicine*, 22 (1949), 126; Samuel E. Morison, *The Puritan Pronaos* (New York, 1936), 263.

9. The Marquis de Chastellux asserted, "Every American has twice as much leisure in the day as an European." *Travels in North-America in the Years 1780-81-82* (New York, 1827), 381.

and any other attainment that dont turn immediately upon that hinge thay think is not worth thair notice, the second class are those that are for spending in Luxury all they can come at and are often the children of avaritious Parents, the third class are those that nessesity obliges to hard labour and Cares for a moderate and happy maintenance of thair family and these are many times the most curious tho deprived mostly of time and matereals to pursue thair natural inclinations." [10]

Why, after all, should an American turn his leisure time to the pursuit of science? It was not possible to rely upon the desire to learn God's ways through the study of nature for the clergy had always maintained that there were more direct routes to that sort of knowledge. No one asserted that scientific investigation was a "way to wealth." Europe did offer some encouragement; not wealth, but honor. Perhaps more reliance might be placed upon the natural curiosity of man in the broad eighteenth-century sense in which John Bartram used the term, but it was not clear how many "curious" Americans there were or how deep-running their curiosity. Most promising was the American acceptance of the pronouncement of Francis Bacon that "knowledge and power are synonomous," even though there was still no very clear indication that science had had practical benefits. [11] The nature of the American response to the call for scientific attainment could not well be predicted, or the direction in which American effort would run.

When Franklin wrote in 1743, the gathering together of Americans into new patterns for the advancement of science had already begun, although the outlines were far from clear. To offset partially the many obvious disadvantages afflicting all efforts to promote science in America there were some advantages. Certain investigations could be carried on in America and nowhere else in the world. Three reservoirs existed from which Americans with scientific knowledge might be drawn: the naturalists, the physicians, and the college teachers. As Franklin perceived, there was a direct connection between growing social maturity and the capacity to advance science. No one could yet see the manner in which patriotism might spur a cultural nationalism that could in turn

10. John Bartram to [Cadwallader Colden], April 7, 1745, Boston Public Library.
11. Francis Bacon, *Novum Organum, or True Suggestions for the Interpretation of Nature* [1620] (London, 1850), 10.

provide still more encouragement for the advancement of science. For some time, it would be accurate merely to say that in America "arts and sciences" were "just dawning." [12]

12. Andrew Burnaby, *Travels through the Middle Settlements in North America* [1759] (London, 1775), 103.

PART ONE

‡‡‡‡‡‡‡‡‡‡‡‡‡‡‡‡‡‡‡‡‡‡‡‡‡‡‡‡‡‡‡

Colonial Circles and Ties

1735-1765

Chapter Two

THE NATURAL HISTORY CIRCLE

America was settled in a wonderful age of expanding knowledge and hope. As Europeans became familiar with Asia, Africa, and America, a new picture of the physical world began to take shape; their imagination soared; and the limits of the possible receded before their dreams. Those who stayed at home were fed upon tales of strange new lands and unknown peoples—those who crossed the seas paid with their hardships for the sight of these strange scenes. Decades and centuries after the initial discoveries, the wonder still remained. New explorations and new knowledge seemed to expand the bounds of the unknown even faster than the limits of the known.

The curiosity aroused in the days of Columbus and Sir Walter Raleigh could not be satisfied despite the traffic and the books that followed. Princes, merchants, and scholars, as well as the barely literate succumbed. Even Shakespeare lamented, "When they will not give a doit to relieve a lame beggar, they will lay out ten to see a dead Indian." [1] Fantasies bred easily in such a soil. An American magazine article argued the plausibility of mermaids from the similarity of the orangutan to man. [2] A book appeared in Constantinople with a picture of the American Wakwak tree—a tree which bore women rather than

1. *The Tempest* quoted by George Browne Goode, "The Beginnings of Natural History in America," Smithsonian Institution, *Annual Report for the Year Ending June 30, 1897*, in the United States National Museum, *Report*, pt. 2 (Washington, 1901), 378.
2. *The American Magazine and Monthly Chronicle for the British Colonies*, 1 (1757-58), 28-29.

the more usual type of fruit.[3] Travel accounts abounded in similar efforts to satisfy the appetite for the bizarre, resorting to fiction and plagiarism as well as to fact.

This insatiable curiosity about the expanding physical world provided a sure support for the encouragement of natural history. In France, Buffon's *Histoire Naturelle* actually received more attention than the literary classics of Voltaire and Rousseau.[4] Across the channel, Peter Collinson reported, "We are very fond of all branches of Natural History; they sell the best of any books in England." [5]

The eighteenth-century scholar shared the popular curiosity and, in addition, was influenced by a philosophical outlook which placed heavy emphasis upon description and classification. Both the Newtonian world machine, running according to law, and the empiricism of John Locke stressed the need for first discovering the external facts and then arranging them in an orderly manner so that the structure of the natural law underlying them might be revealed. In natural history, attention to classification was made more imperative by the rapid multiplication of known species which had outmoded previous classification schemes. The result was that few naturalists would have disputed the great Swedish systematist, Carl Linnaeus, when he declared, "the true botanists are of two sorts, collectors or methodical writers"; that is, collectors or classifiers.[6] Some careful experimental and physiological work was done, but it was a minor interest. Even Dr. John Huxham who objected to the overattention being paid to botanical nomenclature sought only to stimulate more interest in the medical properties of plants.[7] In all branches of natural history, general attention was given only to taxonomy and to use.

3. William and Mabel S. C. Smallwood, *Natural History and the American Mind* (New York, 1941), 24.

4. Daniel Mornet, *Les Sciences de la nature en France au XVIII° siècle* (Paris, 1911), 179, 248-49.

5. Peter Collinson to Carl Linnaeus, April 16, 1747, James Edward Smith (ed.), *A Selection of the Correspondence of Linnaeus and Other Naturalists* (London, 1821), I, 18-19. Linnaeus' name was more consistently Carolus Linnaeus or later Carl von Linné but he was usually referred to merely as Linnaeus; the form used here follows the usage of Knut Hagberg, *Carl Linnaeus*, trans. by Alan Blair (New York, 1953).

6. Linnaeus, *The Elements of Botany*, trans. by Hugh Ross (London, 1775), 5.

7. Alexander Garden to Cadwallader Colden, Jan. 14, 1755, *The Letters and Papers of Cadwallader Colden*, V, New-York Historical Society, *Collections for 1921*, 54 (1923), 2. Hereafter cited as *Colden Papers*.

Some nations had clear advantages in developing a descriptive natural history which fed upon materials garnered from the newer parts of the world. The colonizing countries had the most direct contacts with America, and in the early period Spain and France produced several works dealing with American natural history, but by the eighteenth century England was enjoying unparalleled advantages from her commerce and from the wealth it brought to some of her people. As her American colonies matured, too, political, religious, and commercial channels all became increasingly effective in the service of natural history. The commerce and colonies of all the nations grew, but none so rapidly as those of England.

As western commerce and wealth increased, the more fortunate Europeans were enabled to enjoy the exhilaration of confronting the New World more directly than by reading about it and more pleasantly than by actually visiting it. These men could revel in the taste of an American bear, realizing that it was an exotic dish beyond the reach of most of their fellow men.[8] They could import dried flowers, or bottled bugs, or chips of stone, and they did all of these things to the gratification of scientists as well as the satisfaction of their own collecting urge. They could also grow gardens of living plants, and these proved to be the most conspicuous means of enjoying the natural riches of the far corners of the earth. The Old World interest in gardens gave a decidedly botanical tinge to the development of American natural history.

Gardens were cultivated throughout western Europe: useful botanical gardens in connection with the medical faculties of the universities and ornamental gardens at the seats of royalty, nobility, and wealthy commoners. Germany lagged in this development because of her poverty but the Netherlands could boast not only of the collections of the University of Leyden but of excellent private gardens such as that of the wealthy banker, Thomas Clifford.[9] Most impressive in France was the Jardin Royal. In England, there were many types of gardens. The interest of the royal family, particularly Frederick, Prince of Wales, and his widow, had led to the establishment of the Kew Gardens. Oxford proclaimed its Physick Garden, the oldest in

8. Collinson to Linnaeus, Oct. 26, 1747, Smith (ed.), *Correspondence of Linnaeus*, I, 20-21.

9. Albrecht von Haller to Linnaeus, Aug. 25, 1740, *ibid.*, II, 345; John Frederick Gronovius to Colden, Aug. 6, 1743, *Colden Papers*, III, 31.

England, and Linnaeus considered it superior to those of other European universities.[10] The garden of the Apothecaries Company at Chelsea was headed for many years by Philip Miller who distinguished himself by introducing over two hundred American plants.[11] Most English gardens were cultivated by private citizens: Lord Petre assembled the staggering total of 10,000 American plants and Dr. John Fothergill owned close to 3,400 different species of exotics from various continents.[12] Gardens were cultivated so seriously that the intervention of Parliament was required to curb the continuing theft of rare plants. Indeed, an act was passed punishing men who stole American plants by transporting the thieves to America![13]

The English advantages of trade and patronage contrasted markedly with the less fortunate position enjoyed by Linnaeus whose system of biological classification supplanted all competing schemes and brought its author an acknowledged primacy in the field. Linnaeus worked and studied for a time in the Netherlands, but in 1739 he gave up the advantages of the rich and bustling Atlantic community to return to his native Sweden. There, "in the fagg End of the World," he could only envy the English their "free and frequent Intercourse" with distant lands.[14] This deficiency was remedied in some measure by Linnaeus' extraordinary ability to stimulate exploration in the most remote parts of the earth; at one time, he reported botanical field work at the Cape of Good Hope by Sparrmann, in Japan by Thunberg, in Persia by Gmelin, in Tartary by Falk, at Tranquebar by Koenig, in Surinam by Rolander, and in Arabia by Forshall.[15] John James Dillenius at Oxford felt that the "few seeds sent hither from America" were a poor match for the diligent exertions of the learned pupils and

10. W. J. Bean, *The Royal Botanic Gardens, Kew* (London, 1908), 12-16; R. T. Gunther, *Oxford Gardens* (Oxford, 1912), 17.

11. John Barnhart, "The Significance of John Bartram's Work to Botanical and Horticultural Knowledge," Reprint from *Bartonia* (1931), 28; Philip Miller, *The Gardener's Dictionary* (London, 1731), xiii; Humphrey Sibthorp to John Bartram, April 30, 1762, William Darlington, *Memorials of John Bartram and Humphry Marshall* (Philadelphia, 1843), 428.

12. Benjamin Waterhouse, *The Botanist* (Boston, 1811), 115; Collinson to Bartram, Sept. 1, 1741, Aug. 21, 1766, Darlington, *Memorials*, 145, 282.

13. Collinson to Bartram, March 20, 1766, Aug. 21, 1766, Darlington, *Memorials*, 276, 282.

14. Collinson to Colden, March 9, 1743/44, *Colden Papers*, III, 51.

15. Linnaeus to John Ellis, Dec. 20, 1771, Smith (ed.), *Correspondence of Linnaeus*, I, 275.

friends of Linnaeus, but Linnaeus was not convinced.[16] He continued
to praise the English for their rich gardens, their annual importation
of seeds and specimens, and their liberal patronage which he considered unexampled throughout the world.[17]

Most of England's advantages in the cultivation of natural history
were shared by her American colonists. In addition, they possessed
an advantage that no Englishmen could match—they dwelt in the
midst of the flowers, trees, animals, birds, and minerals that were so
highly valued in Europe. American specimens and seeds could be
collected only in America, whether by Europeans who had made the
difficult and costly trip across the Atlantic or by Americans. The great
advantage of having Americans collect the desired objects was obvious.
Whenever intelligent, interested Americans could be found to perform
this task, they were used, but the colonies had to reach a certain level
of maturity before American collectors predominated over the European visitors.

It was not until the late seventeenth century that groups of men
in England began to make continuing efforts to exploit the natural
history of the world, taking advantage for the first time of the remarkable growth of English commerce, colonies, and wealth. The
Royal Society of London was the principal focus of this effort but
associated clubs and informal circles dedicated to the cause of natural
history also became important.[18] Such men as Sir Hans Sloane, well-
to-do physician and president of the Royal Society; James Petiver,
apothecary; and William Sherard, one time English consul, were
writers and collectors, but they were primarily important as patrons
who stood behind much of the collecting that was done.[19] They made
use of sea captains and itinerants, and whenever a resident of an unexplored region showed interest, they immediately encouraged him to
send them curiosities.

These English promoters had a hand in most of the natural history
of America that was written in the late seventeenth and early eight-

16. Dillenius to Linnaeus, Sept. 23, 1742, *ibid.*, II, 123.

17. Collinson to Colden, March 9, 1743/44, *Colden Papers*, III, 51; Linnaeus
to Ellis, Aug. 8, 1771, Smith (ed.), *Correspondence of Linnaeus*, I, 266.

18. Raymond P. Stearns, "James Petiver," American Antiquarian Society,
Proceedings, 62 (1952), 252.

19. See George Pasti, Jr., Consul Sherard (Doctoral Dissertation, University
of Illinois, 1950).

eenth centuries. English subscriptions in part supported the Reverend John Banister in his voyage to Virginia where he compiled a catalogue of plants, published by John Ray in his influential *Historia Plantarum* of 1686.[20] Four of his letters appeared in the *Philosophical Transactions* of the Royal Society.[21] Shortly after Banister's death, the Reverend Hugh Jones was appointed to a vacant Maryland parish because of the strenuous efforts of a group of Englishmen to secure a naturalist in that post.[22] Even the careful observations of John Lawson, surveyor-general of North Carolina who wrote *A New Voyage to Carolina*, were in some measure owing to the encouragement of James Petiver.[23] Mark Catesby's more influential *Natural History of Carolina, Florida, and the Bahama Islands* was the result of a period of study in America from 1722 to 1726 made possible by the financial backing of William Sherard, Sir Hans Sloane, and other gentlemen.[24] All of these men were members of the English natural history circle; the intellectual roots of those who lived in America were just as thoroughly English as those of their patrons who remained in England.

The New Englanders who interested themselves in natural history at this time were never so thoroughly integrated into the circle or so much supported by it, although the stimulus for their activity was just as clearly English. In 1712, Richard Waller as secretary of the Royal Society sought to expand an overseas correspondence upon natural history but it was Dr. John Woodward who met with the most voluminous response. Woodward merely suggested to Cotton Mather that he send from Boston "such subterraneous curiosities, as may have been in these parts of America mett withal," and the New Englander replied with a total of eighty-two letters.[25] Much of the

20. Richard Pulteney, *Historical and Biographical Sketches of the Progress of Botany in England* (London, 1790), II, 55; *Philadelphia Medical and Physical Journal*, 2 (1805), pt. 2, 134.
21. *Philosophical Transactions*, 17 (1693), 667-72; 22 (1700-1), 807-14; Stearns, "Petiver," 330.
22. Stearns, "Petiver," 292-95.
23. *Ibid.*, 335-42; John Lawson, *A New Voyage to Carolina* (London, 1709).
24. Mark Catesby, *Natural History of Carolina, Florida, and the Bahama Islands* (London, 1754), I, ii; Pulteney, *Progress of Botany*, II, 219; Elsa G. Allen, "The History of American Ornithology before Audubon," American Philosophical Society, *Transactions*, new ser., 41 (1951), 465-74.
25. Quoted by Frederick G. Kilgour, "The Rise of Scientific Activity in Colonial New England," *Yale Journal of Biology and Medicine*, 22 (1949), 127;

contents was useless hearsay, but some of the letters were published in the *Philosophical Transactions,* and along with the dross, appeared two gleaming ideas. The eager clergyman gave the first recorded account of plant hybridization and suggested the large scale use of inoculation to combat epidemic smallpox.[26] Also from Boston, Benjamin Bullivant sent more that was useful in the way of specimens and Paul Dudley wrote careful, critical letters on deer, whales, bees, and earthquakes, twelve of which were published in the *Philosophical Transactions.*[27] Dudley and Mather were elected to fellowship in the Royal Society but thereafter, New England interest in natural history abated. At the very time that activity in the Southern colonies was increasing to a new peak and even the Middle colonies for the first time were beginning to take a leading part in the development of American natural history, New England lost interest.

By the middle of the eighteenth century, an international circle devoted to the cultivation of natural history was one of the most dynamic intellectual forces in Europe and in America. Naturalists in England, France, Holland, Sweden, Germany, and Italy kept in frequent correspondence, visited each other, and accepted posts in foreign countries. England was a particularly active member of the community. John James Dillenius was brought from Germany by William Sherard and placed years later in the Sherardian chair at Oxford. Linnaeus, shortly after the publication of his epoch-making *Systema Naturae* in 1735, visited England where he established lasting ties with the English members of the circle. A little later, one of Linnaeus' best pupils, Daniel Charles Solander, went to England from Sweden to become associated with the British Museum and to accompany Sir Joseph Banks on his famous exploration. The relationship of English and French members of the circle was made strikingly clear during the frequent Anglo-French conflicts when it was usual to provide that communications between the American colonies and England

Cf. Raymond P. Stearns, "Remarks upon the Introduction of Inoculation for Smallpox in England," *Bulletin of the History of Medicine,* 24 (1950), 108.

26. Conway Zirkle, *The Beginnings of Plant Hybridization* (Philadelphia, 1935), 104-6; Kilgour, "Rise of Scientific Activity," 130.

27. *Philosophical Transactions,* 31 (1720-21), 27-28, 145-46, 148-50, 165-68; 32 (1722-23), 69-72, 231-32, 292-95; 33 (1724-25), 129-32, 194-200, 256-69; 34 (1726-27), 261-62; 39 (1735-36), 63-73.

THE PURSUIT OF SCIENCE

be sent to Jussieu, Buffon, or Duhamel or to the Académie des Sciences in case of capture by a French ship.[28]

In the years after 1735, a group of Americans became equal partners in the international effort to advance natural history, their acceptance following in remarkable degree from the initiative of one man—Peter Collinson, the London Quaker merchant. Collinson stimulated the initial interest of some of the Americans, but more important, he made sure that their efforts were used and rewarded. He gave coherence and direction to the American components of the circle. His great role was made possible by his own success as a merchant and by his far-flung trade which brought him into contact with Carolina, Virginia, Maryland, Pennsylvania, and New England as well as many other quarters of the world.[29] He was able to make of his counting house something of a clearing house from which specimens and communications received from America were redirected to Linnaeus in Sweden, Dillenius at Oxford, Gronovius at Leyden, and other members of the circle. Because of his position, all of the important doors in London were open to him including those of cabinet members and at least one prime minister. He was on terms of intimacy with several of the nobility, more of the gentry, and most of the naturalists.

Throughout his life, Peter Collinson remained an enthusiastic Quaker with close ties to the liberal, dissenting circles of England. In common with the dissenting tradition, his own faith favored the empirical and utilitarian rather than the dialectical and humanistic. Of natural history in particular, the Quaker leaders, Fox and Penn, had spoken with special warmth.[30] This favorable attitude toward science was also a characteristic of Collinson's dissenting friends: the Reverend Joseph Priestley, the Reverend Richard Price, and Sir John Pringle with whom he was associated in the Club of Honest Whigs.[31]

Peter Collinson became a leading member of the Royal Society but,

28. Garden to Ellis, Jan. 26, 1771, Smith (ed.), *Correspondence of Linnaeus,* I, 587; Colden to Collinson, Aug. 23, 1758, *Colden Papers,* V, 254; Collinson to Colden, March 30, 1745, *ibid.,* III, 109; Collinson to Bartram, Feb. 10, 1756, Darlington, *Memorials,* 203.
29. Norman G. Brett-James, *The Life of Peter Collinson* [London, 1928], 52.
30. Frederick B. Tolles, *Meeting House and Counting House* (Chapel Hill, 1948), 208.
31. Carl B. Cone, *Torchbearer of Freedom* (Lexington, Ky., 1952), 54.

[18]

despite his attachment to science and to the English scientific society, he was not himself a scientist. In his day, the nonscientific members of the society outnumbered the scientists by more than two to one, yet few of the members in either category contributed as much to the advance of science as did the Quaker merchant.[32] Many important papers came before the society through his hands, most of them written by his correspondents. The papers he himself wrote all demonstrated a notable clarity of thinking.[33] In one, for example, he entered into that debate which had been raging from the days of Albertus Magnus over the submersion of swallows. Even Linnaeus accepted the possibility that swallows could spend the winter beneath the mud bottom of ponds, and innumerable citations could be found to people who had discovered birds in such a condition. Although a state approaching torpidity is now recognized in swallows, Collinson quite properly scoffed at the idea of a bird being able to sustain life under water. In a private letter, he called upon Linnaeus to reconcile his view of the matter with the anatomy of the swallow.[34]

Collinson consistently maintained that he was an amateur in science. He spoke for a whole class of gentlemen gardeners when he protested against the complexity of the Linnaean system of classification with the remark, "The Science of Botany is too much perplex'd already." [35] "I profess myself no Botanist," Collinson declared, "my Skill is Slender, and I have not time to make any proficiency in the New Method." [36] He complained that far from being able to do any scientific writing, he could barely keep up with his correspondence by snatching time from his counting house. Cadwallader Colden was pursuing a vain hope when he urged Collinson to publish a catalog of the plants in his garden.[37] Collinson loved his flowers and trees; he gained endless pleasure from his correspondence; he admired those who advanced the frontiers of knowledge; but it was not his ambition to be numbered

32. Sir Henry Lyons, *The Royal Society, 1660-1940* (London, 1944), 126.
33. For example: *Philosophical Transactions*, 44 (1746), 70-74; 54 (1764), 65-68; 57 (1767), 464-69.
34. Collinson to Linnaeus, May 20, 1762, Smith (ed.), *Correspondence of Linnaeus*, I, 54; W. L. McAtee, "Torpidity in Birds," *American Midland Naturalist*, 38 (1947), 195.
35. Collinson to Colden, Sept. 4, 1743, *Colden Papers*, III, 28.
36. *Ibid*.
37. Colden to Collinson, Nov. 13, 1742, *Colden Papers*, II, 281.

among them. He wanted to be remembered only for the many plants he had introduced into England—at least 171 different species.[38]

His love of rare plants led to the importation of seed from Maryland and New England as early as 1725 but his business correspondents in America were generally unsatisfactory seedsmen. They were more ready to give "fair promises" than to fulfill them.[39] Almost inevitably, Collinson found some of the Quaker merchants of Philadelphia among his consignees and one of them turned out to have interests that were not all kept within a ledger.

Joseph Breintnall was a good-natured man with wide-ranging interests, a member of Benjamin Franklin's famous Junto. He supplied some of Collinson's botanical requests and he sent him experiments on grafting, but his scientific interests were not primarily botanical. Through Collinson, he submitted two papers to the Royal Society which were published in the *Philosophical Transactions;* one on the aurora borealis and the other on the effects of a rattlesnake bite.[40] A more important paper recounting experiments upon heat absorption as a function of color remained in manuscript form.[41] Breintnall's greatest service to Collinson and to science generally was his recommendation of John Bartram "as a very proper person" to furnish the seeds and plants which the Londoner was constantly demanding.[42] This may have been, as Collinson later stated, a contrivance "to get rid of my importunities" but it had the most far reaching consequences.[43]

John Bartram, in the vernacular of his time, was an "original." Born and bred, but not very fully educated, in America, he was properly described as a "down right plain Country Man." He was "a Quaker too Into the Bargain" until he was read out of meeting for rejecting the divinity of Christ, but that did not prevent him from continuing to attend the Darby meeting or cause him to abandon Friendly ways.[44]

38. Brett-James, *Collinson*, 50.
39. *Ibid.*, 52.
40. *Philosophical Transactions*, 41 (1739-41), 359-60; 44 (1746), 147-50.
41. Tolles, *Meeting House*, 217-18.
42. Quoted by Carl and Jessica Bridenbaugh, *Rebels and Gentlemen* (Philadelphia, 1942), 309.
43. Quoted by Brett-James, *Collinson*, 52.
44. Collinson to John Custis, Dec. 24, 1737, E. G. Swem (ed.), "Brothers of the Spade," American Antiquarian Society, *Proceedings*, 58 (1948), 66; Ernest Earnest, *John and William Bartram* (Philadelphia, 1940), 66-67; Tolles, *Meeting House*, 221n.

This independence was typical of the man. He was no more amenable to the artificial restrictions of society than he was to the discipline of the Friends' meeting. Bartram stood in no awe of a governor, a nobleman, or a Virginia aristocrat; instead, he venerated such men "of innocency, integrity, ingenuity, and Humanity" as Carl Linnaeus, Benjamin Franklin, Dr. John Fothergill, and his great friend, Peter Collinson.[45] He was an individualist, unaffected and unbowed, but possessed of a bluntness that must have hurt the sensitive. He did not hesitate to tell one English friend who had sent him a gift of three books that he had read two of them and the third was of no value.[46] When finally honored by a pension from the King, his reaction was that it was too small in amount.[47] He worked hard to repair his shortcomings and lack of advantages and he enjoyed his attainments. He remarked that the letters of his European correspondents afforded him "the secret pleasure of modestly informing them of some of their mistakes." [48]

Before John Bartram was called to the attention of Peter Collinson, he had already attracted the interest of James Logan, wealthy Quaker merchant and powerful Pennsylvania politician. James Logan was one of the most impressive intellectual figures in the colonies at this time; his library was probably unexcelled in scientific titles among private colonial collections. He is credited with importing the first copy of Sir Isaac Newton's *Principia Mathematica* into the colony, and although largely self-taught in mathematics, he was able to comprehend it.[49] More than that, he went on to suggest various alterations in the theory of the moon's motion and to publish a treatise upon optics in which he suggested improvements in some of Huygens' methods of treating lenses.[50] In the *Philosophical Transactions*, he published several papers

45. Bartram to Benjamin Franklin, Nov. 24, 1770, Franklin Letters, III, 34, American Philosophical Society.
46. Bartram to Alexander Catcot, Nov. 24, 1743, Darlington, *Memorials*, 325-26.
47. Collinson to Bartram, Nov. 13, 1765, Darlington, *Memorials*, 273.
48. Bartram to Colden, April 29, 1744, Gratz Collection, Historical Society of Pennsylvania.
49. Frederick E. Brasch, "James Logan," American Philosophical Society, *Proceedings*, 86 (1942), 6.
50. Jacobo Logan, *Canonem pro Inveniendis Refractionum* (Leyden, 1739) cited in Tolles, *Meeting House*, 214-15; Jacobo Logan, *Demonstrationes de Radiorum Lucis* (Leyden, 1741), called to the author's attention by Albert E. Lownes of Providence, R.I.

including one on astronomy, another on lightning, and a third in which he supported the claim of Thomas Godfrey of Philadelphia to the invention of an improved quadrant.[51] In the case of Godfrey, an unlettered glazier, Logan had given aid in his initial studies and had successfully sustained his claims against those of the rival English inventor, John Hadley.[52]

It was in botany that James Logan made his greatest contribution to science, a contribution which revealed his close relationship to the English natural history circle. He carried out a series of carefully controlled experiments which demonstrated the function of pollen in fertilizing maize. The sexuality of plants had been stated long before, and Cotton Mather as well as Paul Dudley had recorded instances of plant hybridization, but Logan was the first to demonstrate experimentally the function of the various organs.[53] It was Peter Collinson who presented to the Royal Society the first results of this influential work in the form of a letter Logan had written him. It appeared in the *Philosophical Transactions* of 1736 as "Experiments Concerning the Impregnation of the Seeds of Plants." In 1739, an enlarged version was published in Latin under the eyes of Gronovius at Leyden and, still later, Dr. John Fothergill translated it into English and saw it through publication in London.[54] Peter Collinson made sure that Linnaeus and other naturalists received copies.[55] James Logan was well and favorably known to the natural history circle; in fact, Linnaeus named a genus for him.

No one in America was better fitted to help John Bartram along his way than Logan. He began by giving him the first three botanical books Bartram ever read, all of them herbals.[56] For a time, he continued to give and lend books, he explained the Linnaean system as

51. *Philosophical Transactions*, 39 (1735-36), 404-5, 240; 38 (1733-34), 441-50.
52. *American Magazine and Historical Chronicle*, 1 (1757-58), 475-80, 525-34; Harold E. Gillingham, "Some Early Philadelphia Instrument Makers," *Pennsylvania Magazine of History and Biography*, 51 (1927), 291.
53. Tolles, *Meeting House*, 216.
54. Logan, *Experimenta et Meletemata de Plantarum Generatione* (Leyden, 1739).
55. Collinson to Linnaeus, May 13, 1739, March 10, 1747, Oct. 26, 1747, Smith (ed.), *Correspondence of Linnaeus*, I, 6, 17, 19.
56. Bartram to Sir Hans Sloane, Sept. 23, 1743, Darlington, *Memorials*, 304; Earnest, *Bartram*, 22.

early as 1736 from a set of Latin tables, and he advised on the use of the microscope.[57] Peter Collinson considered the relationship so close that he called John Bartram "his pupil." [58] Then suddenly, Logan's interest cooled. In 1743, he refused to encourage a projected subscription to subsidize Bartram's botanical expeditions and in the following year he refused to support the formation of the American Philosophical Society in which Bartram was a prime mover. Thereafter, John Bartram and James Logan seem to have had little contact to the undoubted injury of both.

Without support from overseas John Bartram's activities would necessarily have been greatly circumscribed, and they would not have become so widely known. The scope of Bartram's botanical interest was limited by the necessity of working his farm on the Schuylkill just above Philadelphia in order to support his numerous family. Occasionally he also practiced medicine among his poorly served neighbors, a pursuit which gave his botany a medical twist but did not provide wealth. His first voluntary shipments of seed to Peter Collinson were repaid in kind and by little gifts of clothes but it was John Bartram who proposed that he be granted an annual allowance to permit him to exercise his talents more freely. The fulfillment of this idea was the work of Peter Collinson who each year undertook the business of soliciting among the gentlemen gardeners subscriptions to boxes of seeds, distributing them when they arrived, collecting the charges, and even providing planting instructions.[59] At the outset, this yielded Bartram twenty guineas annually which was expected to free him from other pursuits for two or three months a year. In later years, boxes of seeds cost each subscriber £5 5s each, and in 1764 twenty-two such boxes were sent overseas.[60] To Bartram's complaint of the inadequacy of the £50 annual pension assigned by the King, Collinson replied, "Thou knows the length of the chain of fifty links; go as far as that goes, and when that's at an end, cease to go any farther." [61]

By these means John Bartram was enabled to range over the colonies

57. Logan to Bartram, June 19, 1736, Darlington, *Memorials*, 307-8.
58. Collinson to Bartram, Dec. 20, 1737, Darlington, *Memorials*, 107.
59. Quoted by Brett-James, *Collinson*, 53.
60. *Ibid.*, 54; Bartram to Collinson, Nov. 22, 1764, Darlington, *Memorials*, 268.
61. Collinson to Bartram, Nov. 13, 1765, Darlington, *Memorials*, 273.

from New England to Florida and from the coast to the lakes, seeking seeds for his patrons but seeking knowledge as well. From the beginning, he responded to the desire of his English correspondents that he observe animals and minerals in addition to plants; that he send back dried specimens of plants, insects preserved in bottles, and anything else related to American natural history. From his experiences he wrote several papers that were published in the *Philosophical Transactions*: four on insects and others on the aurora borealis, snake teeth, and mollusks.[62] Surprisingly, he published nothing in this series upon botanical matters. His two published journals contained comments upon plants but his only specific publication in the field was an annotation to Thomas Short's *Medicina Britannica*, in which he indicated the American habitat of many of the plants mentioned and provided a new list of American plants with their medical virtues.[63] John Bartram depended upon botany for a part of his support but he never became a specialist. In many respects, his most notable proposal was his plan for a geological survey of the colonies which, by borings, would uncover the mineral wealth of the country.[64]

The Pennsylvania naturalist was speedily accepted by the European natural history circle. Sir Hans Sloane, Professor John J. Dillenius, and Mark Catesby sent him copies of their own works; he was introduced to a correspondence with Carl Linnaeus, John Frederick Gronovius, Philip Miller, and John Hope, among others. As these relationships expanded his horizon, John Bartram became more conscious of his own shortcomings and of the deficiencies of his intellectual environment. To Peter Collinson, in 1739, he proposed that a learned society or academy be established in Philadelphia with a staff of academic instructors. This would have created an intellectual circle adequate to nourish John Bartram's rapid development, but Collinson felt that it was beyond the present capacities of the colonies.[65] Collinson did see the problem, however, and it was through his efforts that Bartram was introduced to an important group of men throughout the colonies. Peter Collinson requested the members of the Library Company of

62. *Philosophical Transactions*, 41 (1742), 358-59; 43 (1744-45), 157-59, 363-66; 46 (1750-51), 278-79, 323-25, 400-2; 52 (1762), 474; 53 (1764), 37-38.
63. Thomas Short, *Medicina Britannica* (3d ed., Philadelphia, 1751), especially the preface and appendix.
64. Bartram to Garden, March 14, 1756, Darlington, *Memorials*, 393.
65. Collinson to Bartram, July 10, 1739, Darlington, *Memorials*, 132.

Philadelphia to extend to Bartram borrowing privileges free of any of the usual charges. This institution had been created by Benjamin Franklin and a group of his friends because of the same kind of intellectual needs that had led to Bartram's suggestion of a learned society. They had created a library that was fast becoming the finest in the colonies and they were more than ready to accede to a request from the man who from the beginning had acted as their purchasing agent in London, frequently adding little gifts of his own.[66] By mail, Peter Collinson also introduced Bartram to interested men throughout the colonies: to Thomas Penn, proprietor of Pennsylvania, to Cadwallader Colden in New York, and to John Custis in Virginia.[67]

After John Bartram had been introduced in this manner and when he received increasing recognition from Europe, the Americans began to value him more highly. Benjamin Franklin became particularly interested. When English orders for seed from Bartram fell off badly in 1740, Franklin made a valiant effort to make good the defections by setting on foot a subscription to finance a botanical expedition. He pushed the program through the channel of his newspaper where he soon announced that the original plan for a single expedition would be expanded to a series of annual expeditions.[68] It did not succeed. John Bartram reported bitterly, "Some people lay the blame upon James Logan and not without cause," but it may be doubted that the failure could be attributed entirely to Logan.[69] The province was not sufficiently developed or urbane enough to provide an annual living for a botanist.

Throughout his long life, John Bartram remained basically a noble nurseryman. He was elected to membership in the Royal Academy of Sciences at Stockholm and given a gold medal by the Edinburgh Society of Arts and Sciences. All of the many men of learning who knew him valued his knowledge of North American flora and en-

66. Bartram to Collinson, May 27, 1743, Darlington, *Memorials*, 162; Franklin to Michael Collinson, [1768?], Franklin Papers, XLVI, 15, American Philosophical Society; Minutes of the Directors of the Library Company of Philadelphia, April 28, 1743, I, 131, Library Company of Philadelphia.
67. Collinson to Custis, Dec. 24, 1737, Swem (ed.), "Brothers of the Spade," 66; Collinson to Bartram, Feb. 25, 1740/41, Darlington, *Memorials*, 142; Collinson to Bartram, April 7, 1741, Gilbert Papers, I, 343, College of Physicians of Philadelphia.
68. *Pennsylvania Gazette*, March 10, 1741/42, March 17, 1741/42.
69. Bartram to Collinson, June 11, 1743, Darlington, *Memorials*, 164.

joyed his infectious enthusiasm. When the first of his two rambling journals was published, Peter Kalm remarked that it did not display "a thousandth part of the great knowledge which he has acquired." [70] There was much to Dr. John Fothergill's assertion that "the eminent naturalist, John Bartram," had been created by his friend Peter Collinson, but neither Fothergill nor anyone who knew Bartram would have denied that the American was "a wonderful Natural Genius." [71] The most capable botanists in America who knew John Bartram well, Cadwallader Colden and Alexander Garden, delighted in his company and considered him a "wonderful observer" but they agreed that his knowledge of the principles of botany was limited. [72] He never mastered Latin and while he was pretty well able to make out a Latin description, he could not characterize properly new plants that he found or determine well between species and varieties. [73] Both Garden and Colden may have been somewhat biased by their own aristocratic condescension toward old John and by their eagerness to promote their own interests. Nevertheless, Bartram was not a good systematic botanist. He was one member of a team, and as an assiduous collector, he contributed more to the advance of knowledge than any of the more learned botanists in his country.

Johann David Schöpf was right when he described him as "more collector than student," and Colden and Franklin were wrong when they sought to spur him into writing a natural history of America. [74] With European stimulus, he did some experimental work on the generation of plants and he delighted in the results he obtained from crossbreeding to produce flowers unknown to nature. [75] He enjoyed all sorts of philosophical speculations; he was eager to learn theory; but

70. Adolph B. Benson (ed.), *Peter Kalm's Travels in North America* (New York, 1937), I, 61.

71. John Fothergill, *Some Account of the Late Peter Collinson* (London, 1770), 11; Collinson to Colden, March 7, 1741/42, *Colden Papers*, II, 247.

72. Colden to Collinson, [June 1744], *Colden Papers*, III, 61; Garden to Linnaeus, June 2, 1763, Smith (ed.), *Correspondence of Linnaeus*, I, 316.

73. Garden to Ellis, July 15, 1765, *Correspondence of Linnaeus*, I, 536-37.

74. Johann David Schöpf, *Travels in the Confederation, 1783-1784*, trans. by Alfred J. Morrison (Philadelphia, 1911), I, 90; Colden to Bartram, [Dec. 1744], *Colden Papers*, III, 94-95; Franklin to Bartram, Jan. 9, 1769, Darlington, *Memorials*, 403.

75. Bartram to William Byrd, [1739], Darlington, *Memorials*, 315; Collinson to Bartram, July 22, 1740, *ibid.*, 136.

it was not reasonable to expect systematic studies of natural history from him. John Bartram would send his annual shipment of seeds and specimens to England and then wait expectantly until Daniel Solander could classify them and report his findings. He did not always agree with the classification assigned because he knew the plants better than anyone else and dried specimens were always an uncertain dependence, but he never attempted to take over the task himself.[76] The number of American plants cultivated in England doubled during Bartram's activity and in large part because of that activity.[77] It was not neces-sary or even desirable that he be able to classify the new plants he found, for this could be done much more satisfactorily by the more erudite Europeans who were able to integrate Bartram's discoveries into the general body of Western science.

John Bartram contributed to the general advance of knowledge but in the American scene he was more important for the stimulus he provided to the development of an intellectual community and for focusing the attention of the Europeans upon those close to him. Even his own family was drawn into the natural history circle—particularly Billy—one of his seven sons. Billy produced excellent maps, drew plants and birds which attracted the attention of George Edwards, the English naturalist, and in 1769 he did a series of drawings of turtles patronized by Dr. John Fothergill and the Duchess of Portland.[78] He accompanied his father on his expedition to Florida in 1765. William's greatest deficiency, his lack of dependability, was clearly revealed in the small initial yield he returned upon Dr. John Fothergill's later investment in an expedition of his own.[79] John Jr. proved much more reliable when he took temporary charge of his father's business, but he lacked his brother's genius and all of the boys lacked their father's range and industry.

There were others in John Bartram's neighborhood who competed with him as seedsmen. James Alexander, the proprietor's gardener at Pennsbury, developed this trade as a sideline, adding a new technique.

76. Bartram to Collinson, Aug. 15, 1762, Darlington, *Memorials*, 240.
77. Barnhart, "Significance of Bartram's Work," 27-28.
78. Collinson to Bartram, July 6, 1768, Darlington, *Memorials*, 300; Collinson to William Bartram, July 18, 1768, Darlington, *Memorials*, 301.
79. Fothergill to Bartram, July 8, 1774, Darlington, *Memorials*, 347-48.

He paid the country people to collect seeds and roots for him.[80] Alexander's seeds reached the English market but Alexander himself never enjoyed a close relationship with the English naturalists. William Young went further. He was a local boy of German descent who learned a good bit from John Bartram and then tried to follow in his footsteps. For a time, he operated a nursery at Charleston, South Carolina, and with the enthusiastic aid of Dr. Alexander Garden, he was soon supplying seed for the local and for the European market. He went to England and there succeeded so well that he was appointed Queen's botanist at the reported salary of £300. Dr. John Fothergill and Peter Collinson tried to aid him without expecting him to achieve very much.[81] There were other Pennsylvanians who occasionally sent over new specimens but most of them gained little credit in Europe.

Pennsylvania very clearly outdistanced the other colonies in developing naturalists and seedsmen at this time, but the best systematic treatise on American botany was the result of the industry of a Virginian. This was the *Flora Virginica,* a book that had to be used for many years by all who dealt with American plants. It was the outstanding example of fruitful collaboration between an American collector and a European "methodical writer." The first edition of this work appeared at Leyden in 1739 and 1741 under the name of John Frederick Gronovius but it was based upon the collections of John Clayton of Virginia.

John Clayton was a devoted, retiring collector who snatched hours and days from his petty post as clerk of Gloucester County to spend them more agreeably in the pursuit of nature. He was well known to the English natural history circle and he corresponded with the American naturalists—occasionally receiving visits from such men as John Bartram, though he complained that he often had to rely upon Peter Collinson for news of other Americans.[82] He ransacked his own neigh-

80. Bartram to Collinson, June 8, 1756, June 12, 1756, Darlington, *Memorials,* 209, 210.

81. Garden to Ellis, July 25, 1761, Nov. 19, 1764, Smith (ed.), *Correspondence of Linnaeus,* I, 512, 522; Ellis to the Duchess of Norfolk, Oct. 11, 1768, *ibid.,* II, 73; Bartram to Collinson, Dec. 5, 1766, Darlington, *Memorials,* 285; Fothergill to Bartram [prob. 1772], *ibid.,* 344; Collinson to Bartram, May 28, 1766, *ibid.,* 279.

82. Clayton to Bartram, July 23, 1760, Feb. 23, 1761, March 16, 1763, Feb. 25, 1764, Darlington, *Memorials,* 407, 409, 410, 411.

borhood for specimens but did not venture much farther afield than the eastern slope of the Blue Ridge Mountains.[83] After the praise accorded him in *Flora Virginica* and the honor of election to the Royal Academy of Sciences at Stockholm, his neighbors, too, came to hold him in high esteem. They elected him first president of the Virginian Society for the Promotion of Usefull Knowledge when it was founded in 1773.

The first edition of *Flora Virginica* was compiled by Gronovius from specimens collected by John Clayton. In this task the Dutch botanist was assisted by Linnaeus who was then working in the Netherlands and whose sexual system of classification alone permitted Gronovius to reduce several of the plants to their proper genus.[84] Probably Clayton was introduced to Gronovius through London, for his primary relationships were always with England. Gronovius gave Clayton's duplicate specimens to Linnaeus but a large collection of them found its way back to England where it was catalogued in the Banksian Museum as the specimens from which *Flora Virginica* had been compiled.[85]

The second edition (1762) was a much superior production, utilizing the binomial method of classification which Linnaeus had not worked out at the time of the first edition. Its genesis was with John Clayton himself who wrote out an enlarged and improved manuscript and sent it with dried specimens, not to Gronovius or his son, but to Peter Collinson in 1758.[86] Collinson referred it to the naturalist, John Ellis, for criticism, and by the time it appeared, again under the name of Gronovius, it had been worked over by many hands and had received the benefit of comparison with the writing done by a group of American naturalists in the past twenty years. In it, older writings by Sloane, Plumier, Petiver, Banister, and even Josselyn were used, but they were less important than such works as John Bartram's

83. *Philadelphia Medical and Physical Journal*, 2 (1805), 140n.
84. "Diary of Linnaeus," in Richard Pulteney, *General View of the Writings of Linnaeus* (London, 1805), 530; Gronovius to Richard Richardson, Sept. 2, 1738, Smith (ed.), *Correspondence of Linnaeus*, II, 179; Linnaeus to Haller, Jan. 23, 1738, *ibid.*, 314-15.
85. Frederick Pursch, *Flora Americae Septentrionalis* (London, 1814), I, xvi; "Diary of Linnaeus," 530.
86. Ellis to Linnaeus, April 25, 1758, Smith (ed.), *Correspondence of Linnaeus*, I, 93; Collinson to Linnaeus, Dec. 25, 1757, *ibid.*, 42.

Journal of 1751, Dr. Cadwallader Colden's "Plantae Coldenghamiae" of 1743, and Dr. John Mitchell's "Nova Genera Plantarum" of 1748. For its perfection, Peter Collinson made available plants collected by John Bartram and a manuscript by Paul Dudley which has never yet been published.[87] *Flora Virginica* was a tribute to the collaboration of Clayton and Gronovius but it was more than that; it was a product of the whole natural history circle and a tribute to its effectiveness.

Before John Clayton died, he prepared two manuscript volumes for the press accompanied by a collection of dried specimens and notes for the engraver who would prepare the plates designed to be included in the projected publication. All of this material was destroyed by fire shortly after the Revolution and its nature has become a mystery.[88] However, Clayton's whole history makes it clear that it was either another volume on local plants or a third edition of the *Flora Virginica*. Whatever has been lost, Clayton's greatest work had already been published.

When Peter Collinson was called upon to name the competent Linnaean botanists in America, he cited Clayton, Colden, and Mitchell, but behind them and behind John Bartram there was a large number of men and women who occasionally made contributions to the field of botany.[89] The colonies were overwhelmingly agricultural and there was extensive interest in plants especially among such gentlemen as Abraham Redwood of Rhode Island, Henry Laurens of South Carolina, and William Byrd of Virginia who kept gardens of rarities following the English fashion. William Byrd's garden was reputed to be the finest in Virginia, replete with greenhouses and many rare plants. In the same colony, John Custis cultivated his own garden with a deeper understanding of natural history, and in South Carolina, Martha Logan actually published a treatise on gardening.[90] Charles

87. John Frederick Gronovius, *Flora Virginica* (2d ed., Leyden, 1762).

88. *Philadelphia Medical and Physical Journal*, 2 (1805), 140.

89. Collinson to Linnaeus, Jan. 18, 1743/44, Smith (ed.), *Correspondence of Linnaeus*, I, 9.

90. Donald Wyman, "The Arboretums and Botanical Gardens of North America," *Chronica Botanica*, 10 (1947), 399; Bartram to Collinson, July 19, 1761, Darlington, *Memorials*, 229; Collinson to Bartram [1738], *ibid.*, 113; Albert E. Benson, *History of the Massachusetts Horticultural Society* ([Norwood, Mass.], 1929), 13. Charles Evans, *American Bibliography* (Chicago, 1903-34), IV, 321, lists Martha Logan, *A Treatise on Gardening* (Charleston, 1772); no copy is now known.

Read in New Jersey and William Allen in Pennsylvania appear to have done more than the southerners in agricultural experimenting.[91] Many of these people exchanged seed and occasionally contributed unknown plants to more active members of the natural history circle.

Officers of the colonial governments, particularly governors, offered great assistance in collecting materials for natural history. Governors Morris of New York, Ellis of Georgia, Fauquier of Virginia, Sharpe of Maryland, Eliot of West Florida, and Pownall of Massachusetts were outstanding in their willingness to cooperate with requests from Europe.[92] Francis Fauquier, Henry Ellis, and Robert Hunter Morris were members of the Royal Society, and one of Fauquier's papers on a hailstorm in Virginia appeared in the *Philosophical Transactions*. When exploring expeditions came under the cognizance of such army officers as General Henry Bouquet, they, too, extended a full measure of support to discovery of the unknown.[93]

Government activity was most conspicuous in the more recently settled colonies. In Georgia, shortly after its founding, the Trustees established a botanical garden for the primary purpose of aiding development of productive agriculture in the colony. As director, they hired a competent botanist, Dr. William Houston, whose work was valued by the natural history circle and used.[94] Even more activity followed the ceding of the Floridas to Great Britain in 1763. In his new post as King's agent to West Florida, John Ellis, the naturalist, hoped to stimulate the exploration of the region.[95] Attempts were made to map, chart, and in some measure explore Florida by William Gerard De Brahm, surveyor-general of the Southern Department, and Bernard

91. See Carl R. Woodward, *Ploughs and Politics* (New Brunswick, 1941); Bridenbaugh, *Rebels and Gentlemen*, 215-16.

92. Collinson to Bartram, Feb. 3, 1741/42, Darlington, *Memorials*, 149; Bartram to his wife, Sept. 4, 1765, *ibid.*, 426; Ellis to the Duchess of Norfolk, Aug. 7, 1769, Smith (ed.), *Correspondence of Linnaeus*, II, 75; Francis Fauquier to Horatio Sharpe, July 22, 1765, Emmet Collection, I, 40, New York Public Library; *Philosophical Transactions*, 50 (1757-58), 746-47.

93. Henry Bouquet to Bartram, Feb. 3, 1762, Darlington, *Memorials*, 427.

94. James W. Holland, "The Trustees Garden in Georgia," *Agricultural History*, 12 (1938), 271-77; Winthrop Tilley, The Literature of Natural and Physical Science in the American Colonies from the Beginnings to 1765 (Doctoral Dissertation, Brown University, 1933), 51; Gronovius, *Flora Virginica* (1762), not paged.

95. Ellis to Linnaeus, Jan. 1, 1765, Smith (ed.), *Correspondence of Linnaeus*, I, 163.

Romans, his deputy. These men as well as John Bartram, who was enabled to explore Florida by his appointment as King's botanist, made important contributions to knowledge though the fruit of most of their work did not come until the years just before the Revolution.

Cartography owed much to the support of government as well as to the demands of private individuals. Surveying was a skill in constant demand in the colonies because of the perennial sale of public and private lands and the need to establish intercolonial boundaries. Yet, surveyors were not scarce and the most satisfactory security in this kind of work lay in appointment as surveyor-general of one of the provinces. In this capacity, Cadwallader Colden served in New York, James Alexander in New Jersey, Nicholas Scull and William Parsons in Pennsylvania, and John Leeds in Maryland, each of them contributing something to the advance of knowledge. One of the most active map makers of this period was Lewis Evans who did not hold such a post though he was occasionally employed by the government of Pennsylvania on various missions. In 1743 he accompanied John Bartram and Conrad Weiser on a journey to Onondaga; in 1750 plans fell through for sending him to explore the Ohio region even to the point of making soil tests.[96] In 1755, Evans published his *General Map of the Middle British Colonies in America* accompanied by the valuable *Geographical Essays*. This work brought him to the attention of the natural history circle on both sides of the Atlantic and made him the center of political controversy because his boundaries did not please everyone. At about the same time, Peter Jefferson and Joshua Fry published their *Map of the Inhabited Parts of Virginia* which for that region was as important a work as Evans'.[97]

Another figure who came within the cognizance of those who sought to extend their knowledge of American natural history was the Indian agent, trader, and land speculator. By profession he was an explorer and discoverer of the unknown in the West. All of the Indian agents and scouts learned things of value. Christopher Gist and George Croghan were as communicative as any. Both were involved, for in-

96. See John Bartram, *Observations ... in his Travels from Pensilvania to Onondago* (London, 1751); Paul A. W. Wallace, *Conrad Weiser* (Philadelphia, 1945), 155-59.
97. Lewis Evans to Colden, March 13, 1748/49, *Colden Papers*, IV, 107; Lawrence H. Gipson, *Lewis Evans* (Philadelphia, 1939), 18, 63; Lawrence C. Wroth, *An American Bookshelf* (Philadelphia, 1934), 49-52.

stance, in making known the fabulous remains of ancient bones at Big Bone Lick on the Ohio River. Such bones had been reported by Banister, Mather, Dudley, and others but Christopher Gist was probably the first Englishman to bring back bones from the Lick. It was after George Croghan first visited it in 1765 that bones in any number were placed before the natural history circle in America and England.[98]

In such an overwhelmingly agricultural country with growing scientific interests it was odd that no more concentrated attention was given to attempts to improve agriculture. Some gentlemen farmers showed interest but the only man to write anything extensive on the subject was Jared Eliot of Killingworth, Connecticut.[99] Eliot was a clergyman and a medical practitioner who did considerable travelling near his home where he could observe agricultural practice. He read English agricultural publications and even manuscripts forwarded to him by Peter Collinson and he tried to interpret them in terms of his own experience. When he became interested in the seed drill of Jethro Tull, he went on to develop an improvement upon it with the help of Benjamin Gale, his son-in-law, and Thomas Clap. Gale continued the experiments after the death of Eliot. More important were Eliot's observations upon cultivation, plowing, use of fertilizers, and other techniques. They were all included in his *Essays upon Field Husbandry* which he published first in a series of newspaper articles and later in book form. In this work Eliot succeeded not merely in making known the developing English techniques but in many cases in modifying them to suit American conditions. He was always very conscious of the peculiarities of the American environment. Eliot became a member of the natural history circle with a small but significant correspondence outside New England including John Bartram, William Logan, and Charles Read.[100]

Although field work in American natural history was generally con-

98. See George G. Simpson, "The Beginnings of Vertebrate Paleontology in North America," American Philosophical Society, *Proceedings*, 86 (1942), 134-35.

99. The only publications sought to encourage new crops and they were often reprints. Lionel Slator, *Instructions for the Cultivating and Raising of Flax and Hemp* (Boston, 1733); *Observations Made by Richard Hall of the City of Dublin, Hemp and Flax Dresser* (Boston, 1735).

100. Jared Eliot, *Essays upon Field Husbandry in New England and Other Papers, 1748-1762* [First collection of the *Essays* published 1760], ed. by Harry J. Carman and Rexford G. Tugwell (New York, 1934).

fined to Americans and Britishers, continental naturalists occasionally visited the English colonies in North America. Linnaeus, who had been so successful in sending his followers to every other part of the world, introduced one of his students to America as well. Peter Kalm, newly appointed professor at Abô, Finland, was sent on an expedition to America financed by the Royal Academy of Sciences at Stockholm and other groups in an effort to find plants and trees that would be useful in Sweden.[101] Kalm interpreted his mission in broader terms; he investigated not only botany but all of natural history and astronomy as well. It was universally agreed that he was competent to perform this task. Cadwallader Colden even felt that he was so much more competent than the American naturalists that Linnaeus would no longer benefit from them.[102] Nevertheless, some of the English naturalists were irked at seeing a foreigner invade this field.[103]

Linnaeus had considered Canada the primary object of Kalm's interest because of the climate and Kalm himself thought in terms of New England and Canada but it did not turn out that way.[104] The Canadians were more than hospitable, providing all his needs without charge in return for similar favors previously granted to French scientists who had gone to Sweden to measure the length of a degree of latitude.[105] Yet Kalm stayed in Canada but briefly and avoided New England altogether to spend most of his time in the Middle colonies, particularly in Philadelphia at the home of Benjamin Franklin. There he found a society of men with intellectual interests and individuals with knowledge of natural history. John Bartram knew more about American natural history than Kalm could have found in years and he was very willing to share his information.

Peter Kalm's Travels in North America was published in Swedish in 1753 and in English in 1770. It was this book by which his contemporaries evaluated his science although it was not a scientific

101. Benson (ed.), *Kalm's Travels*, I, vii; Pulteney, *Linnaeus*, 121; Per Axel Rydberg, "Linnaeus and American Botany," *Science*, new ser., 26 (1907), 70.

102. Colden to Linnaeus, Feb. 1, 1750/51, *Colden Papers*, IV, 257; Kalm to Colden, Sept. 29, 1748, *ibid.*, IV, 77.

103. Garden to Colden, Oct. 27, 1755, *Colden Papers*, V, 33; Garden to James Parsons, May 5, 1755, John Nichols, *Literary Anecdotes of the Eighteenth Century* (London, 1812-15), V, 484n.

104. Linnaeus to Haller, Oct. 23, 1747, Smith (ed.), *Correspondence of Linnaeus*, II, 417.

105. Kalm to Bartram, Aug. 6, 1749, Darlington, *Memorials*, 369.

treatise but a chronological account of travels. It contained many pene-trating observations but, like most travel books of this time, it also contained much that was second or third hand—and consequently inaccurate. He gave credit for his facts to the wrong people or not at all, injuring some of the Americans to such an extent that one Swede felt he had to apologize for him.[106] Benjamin Franklin concluded that one should beware of "Strangers that keep Journals" but the whole fault was not Kalm's. The best of his science was published in Swedish journals which were not known to the Americans of his day and which are only now being translated into English.[107] His work resulted in adding new species and even new genera of American plants.

Kalm's visit in 1748 revealed the existence of an Anglo-American circle dedicated to the advance of natural history. It was a part of the larger Western community of science but it had an identity of its own. By 1765 the Americans were playing an essential role in this international community and their work was applauded and used in Sweden, France, and the Netherlands but most of it reached the Continent through England—much of it through one man, Peter Collinson. The American naturalists enjoyed the closest ties with England but they were also developing direct contacts with one another. A heterogeneous group, they represented the beginning of the first scientific community in America.

106. Charles M. Wrangel to Bartram, July 2, 1769, Darlington, *Memorials*, 445.

107. Franklin to David Colden, March 5, 1773, *Colden Papers*, VII, 185. Esther Louise Larsen has translated a series of Kalm's papers; see *Agricultural History*, 9 (1935), 98-117; 13 (1939), 33-64, 149-56; 16 (1942), 149-57; 19 (1945), 58-64; 21 (1947), 75-78; 22 (1948), 142-43.

Chapter Three

THE DOCTORS: NATURALISTS AND PHYSICIANS

ONE OF THE MOST remarkable things about the pursuit of natural history in the eighteenth century was the prominence of physicians in the effort. On both sides of the Atlantic, they formed so vital a part of the natural history circle that they might be said to have been its core. In England, the surgeons and physicians formed the largest single group among the scientific fellows of the Royal Society and it was they who accounted for much of the attention accorded natural history within the organization.[1] Most of the leading naturalists on the continent were physicians, Boerhaave, Haller, and Linnaeus among them; while in England both naturalists and patrons were drawn in large numbers from this group. To have removed Sir Hans Sloane, Dr. John Fothergill, John Hunter, and John James Dillenius from the circle in England would have injured it irreparably.

This prominence of physicians in natural history was no accident but resulted from the medical training of that day which afforded an introduction to the study of the sciences of nature. At the progressive schools of Padua, Leyden, and Edinburgh and in the great hospitals of the city of London, medical students took formal courses in botany, in chemistry, and in anatomy—sometimes including comparative anatomy. The eyes of the medical student were particularly directed toward botany because of the predominance of vegetable remedies and this bias served to increase the great attention that was lavished upon

1. Sir Henry Lyons, *The Royal Society, 1660-1940* (London, 1944), 341-42.

that branch of natural history. His studies were seldom deep or profound but they did provide the best academic foundation available in natural history and often stimulated continuing interest and study.[2]

In the early years of the eighteenth century, it was usual for Englishmen and Scots to resort to the continent for their medical education, a pattern which was reflected in the colonies. Many of the trained physicians who emigrated to the colonies, went there after having studied medicine on the continent. Thomas Dale in Charleston, Isaac DuBois in New York, William Douglass in Boston, and many others had come to the colonies with Leyden degrees. Indeed, except in Boston, the leading colonial towns were served by a medical faculty, a majority of which had been trained in Europe.[3] When Americans sought a formal medical training, they too followed the established pattern of the mother country and studied on the continent. Beginning with William Bull of Charleston who obtained a doctorate in medicine from Leyden in 1734, a small group of active Americans went abroad to study and often to obtain medical degrees. The breadth of the training they received and the personal ties they established usually served them after their return from Europe.

Many of the Americans who studied on the continent were introduced to the European natural history circle and remained in some measure attached to it. William Bull continued to be held in high esteem by Baron Gerhard Van Swieten who had been a fellow student under the great Hermann Boerhaave, and as late as 1755 Alexander Garden described Bull as the only man in Charleston who had any acquaintance with natural history.[4] From Philadelphia, John Redman and Benjamin Morris also took degrees at Leyden, but more important

2. John D. Comrie, *History of Scottish Medicine* (London, 1932), I, 289, 300; Alexander Bower, *History of the University of Edinburgh* (Edinburgh, 1817-30), II, 125; Sir Alexander Grant, *The Story of the University of Edinburgh* (London, 1884), II, 378-85; T. B. Johnston, "The Medical School," in H. A. Ripman (ed.), *Guy's Hospital, 1725-1948* (London, 1951), 59-60; F. G. Parsons, *History of St. Thomas's Hospital* (London, [1934]), II, 166-232; Stewart C. Thomson, "The Great Windmill Street School," *Bulletin of the History of Medicine*, 12 (1942), 377-91.

3. Carl Bridenbaugh, *Cities in the Wilderness* (New York, 1938), 404.

4. Garden to Dr. James Parsons, May 5, 1755, Nichols, *Literary Anecdotes*, V, 484n; Eleanor W. Townshend, "William Bull, M.D. (1710-1791)," *Annals of Medical History*, new ser., 7 (1935), 315.

were the Bond brothers.[5] Thomas Bond was introduced to Professor
Bernard de Jussieu of the Jardin Royal in Paris by Peter Collinson in
a visit made memorable by the specimens and queries he carried from
John Bartram.[6] Phineas Bond, who took his degree at Rheims, became
acquainted with Gronovius during the course of his European studies
and retained the Dutch biologist's regard long after he had returned to
America.[7] These men developed an interest in natural history though
none of them pursued it with any vigor.

Strikingly, all of the physicians who did make contributions to the
advance of natural history in America had experienced a strong Scottish influence; either they were Scots themselves or they had received
training in that country. The most important were the Scottish emi-
grants some of whom came to America while others went to England
and other parts of the world in the search of better opportunity. At this
very time, Scotland was giving to England some of her leading physicians and her most prominent naturalists. Sir John Pringle, president of
the Royal Society, was a Scot; so were William and John Hunter, the
famous surgeons. Even John Fothergill who was an English Quaker,
took his medical training at the University of Edinburgh. Among the
men of real capacity who left Scotland were several physicians who
came to America. John Lining in Charleston and Alexander Hamilton in Annapolis played an important role on the colonial scene but
did not pay much attention to natural history. Three others—Cadwallader Colden in New York, William Douglass in Boston, and Alexander Garden in Charleston—became essential elements in the international natural history circle. The only other colonial physician to
rank with these men in accomplishment was John Mitchell, who was
apparently not a Scot but who did take an essential part of his education at the University of Edinburgh.[8]

The most fascinating and least understood of this group was Cad-

5. Whitfield J. Bell, Jr., "Philadelphia Medical Students in Europe, 1750-
1800," *Pennsylvania Magazine of History and Biography*, 67 (1943), 2.
6. Thomas Bond to Bartram, Feb. 20, 1738/39, William Darlington, *Memorials of John Bartram and Humphry Marshall* (Philadelphia, 1943), 316.
7. Gronovius to Bartram, June 2, 1746, Darlington, *Memorials*, 356.
8. Mitchell's birthplace remains an open question. Herbert Thatcher, "Dr.
Mitchell, M.D., F.R.S., of Virginia," *Virginia Magazine of History and Biography*, 39 (1931), 126-35, gives all the evidence but places too much reliance
upon Kalm's statement that Mitchell was born in America.

wallader Colden. Colden was born and educated in Scotland to the level of a master of arts degree which he received from the University of Edinburgh.[9] In the course of this work, he attended the lectures of Dr. Charles Preston on botany but since Edinburgh did not then maintain a faculty of medicine, he went to London for his medical education.[10] There, he took courses with some of the outstanding private teachers of the time. When he came to Philadelphia for the first time in 1710, he brought with him an excellent education, a few ties to the mother country, and an inquiring mind fond of "disputation."[11]

By means of careful planning and continued effort he succeeded in overcoming one of the greatest obstacles in the way of scientific or intellectual attainment in the colonies; he sought and found the necessary leisure. While still in Philadelphia, he obtained the support of James Logan, his wife's cousin, for a bill to allow him a salary as physician to the poor and to establish a public medical lecture but this endeavor failed.[12] In 1718 he removed to New York to accept appointment by Governor Robert Hunter, a fellow Scot, to two minor government posts. In 1720, he was appointed surveyor-general of that province, a position he continued to hold until in his old age he was able to turn it over to his son. During that time it always yielded an income without demanding much from him because he was able to fulfill his obligations very largely by means of deputies.[13] Two years later he was made a member of the council and later still he became lieutenant-governor of New York. Periodically, public duties demanded his full attention, but in 1739 he withdrew from the city to his farm called Coldengham, which he left largely to the management of his wife while he worked in his study. In 1750 he announced a retirement from public life to "such kind of philosophical amusement in which

9. Lewis L. Gitin, "Cadwallader Colden, as Scientist and Philosopher," *New York History*, 16 (1935), 169.

10. Colden to Kalm [no date], Asa Gray (ed.), "Selections from the Correspondence of Cadwallader Colden," *American Journal of Science*, 1st ser., 44 (1843), 108; Grant, *University of Edinburgh*, II, 379.

11. William Smith, Memoirs, II, not paged, N.Y. Public Library.

12. Francis R. Packard, *History of Medicine in the United States* (New York, 1932), I, 286.

13. Smith, Memoirs, II, n.p.; Agreement of David Colden and Gerard Bancker, Jan. 1, 1775, *Colden Papers*, VII, 258, shows how Colden's son continued to deputize his duties when he became surveyor-general.

I take most pleasure," although he never fully attained this goal.[14] He continued to press his friends in England in the quest of remunerative appointments for himself and for his family.[15] Few men in colonial America were so successful in finding sinecures as Cadwallader Colden and still fewer were interested in applying the leisure thus gained to intellectual pursuits.

Colden's most important scientific contributions were made in natural history—particularly in botany—in the period after his withdrawal to Coldengham. He had begun an investigation of American plants shortly after his arrival in America, but the lack of encouragement and the press of other business had caused him to abandon this beginning.[16] Just about the time he left the city, he happened to run across a copy of Linnaeus' *Genera Plantarum* which so impressed him that he began to collect and study the plants in the neighborhood of Coldengham. It may have been his friend James Alexander, another Scottish immigrant, who introduced him to a 'correspondence with Peter Collinson. Alexander was surveyor-general of New Jersey as Colden was of New York and both of them were members of the Council of New York at the same time. Alexander, too, had scientific interests that were somewhat less extensive than Colden's but just as lasting. Collinson's first service to Colden was in encouraging him to continue his *History of the Five Indian Nations* of which he had published the first segment in 1727.[17] Upon Collinson's urging, Colden extended his account to 1697 and after a few years the London merchant succeeded in having it published. Though based admittedly upon secondary accounts in very large measure, this history attained great esteem and was used as an authority on both sides of the Atlantic.[18] Colden's botanical interests seem to have led him to write Gronovius

14. Colden to James Alexander, Dec. 15, 1750, *Colden Papers*, IV, 243.
15. Colden to Mitchell, July 18, 1751, *Colden Papers*, IX, 101; Colden to Collinson, July 28, 1752, *ibid.*, IX, 117; Collinson to Colden, March 27, 1746/47, Jan. 15, 1752, *ibid.*, III, 369, IX, 110; Colden to Thomas Osborne [1748?], *ibid.*, VII, 345.
16. Colden to Linnaeus, Feb. 9, 1748/49, Smith (ed.), *Correspondence of Linnaeus*, II, 452.
17. Collinson to Colden, March 5, 1740/41, March 4, 1743, Aug. 3, 1747, Smith (ed.), *Correspondence of Linnaeus*, II, 207, III, 27, 410; Colden to Collinson, April 9, 1742, *ibid.*, II, 251.
18. Alice M. Keys, *Cadwallader Colden* (New York, 1906), 5; Lawrence C. Wroth, *An American Bookshelf* (Philadelphia, 1934), 94.

on his own initiative, although it is not impossible that Peter Collinson brought the work of Gronovius to Colden's attention.[19]

Colden sent his descriptions of the plants he had collected to Gronovius because he valued the *Flora Virginica* as the greatest contribution to American botany and felt that Americans owed Gronovius any help they could give him. Gronovius wrote Linnaeus about Colden's work and both Collinson and Gronovius continued to feed information about him to the Swede. Colden developed an irregular correspondence with Linnaeus himself, sending him occasional specimens and descriptions as he did to Gronovius.

The reception Colden's activities met in Europe must have overwhelmed him. Linnaeus called him "Summus Perfectus" and Peter Collinson was surprised at his demonstration of "what Leisure and Application assisted with a Great Genius can attain too." [20] Even Gronovius had to confess, "I hath a great deal light in your Characters about somme plants, mentioned in the flora Virginica." [21] Linnaeus was sufficiently impressed with Colden's classification and description of the plants near Coldengham that he published the first part of this work under the title "Plantae Coldenghamiae" in *Acta Upsaliensis*. When Linnaeus then applied the name Coldenia to a plant in his *Flora Zeylanica*, the master of Coldengham had already won a measure of immortality.

This recognition had been won with relative ease. Colden had attained proficiency in the Linnaean system at a time when only a handful of men in England were as well versed in it.[22] He had done more than John Bartram in that his descriptions of plants were written in the standard Latin form and classified with some success. Moreover, he perceived the weaknesses of the Linnaean system of sexual classification and the desirability of a "natural" system.[23] He was able to talk intelligently with the collectors of America and with the classifiers of Europe. Yet, he was honored for making known new plants rather

19. Gronovius to Colden, Aug. 6, 1743, *Colden Papers*, III, 31; Colden to Gronovius, Oct. 29, 1745, Gray (ed.), "Selections from Colden," 98.

20. Colden to Linnaeus, Feb. 9, 1748/49, *Colden Papers*, IV, 99; Collinson to Colden, March 9, 1743/44, *ibid.*, III, 50.

21. Gronovius to Colden, Oct. 3, 1743, *Colden Papers*, III, 33.

22. Collinson to Linnaeus, Jan. 18, 1743/44, April 10, 1755, Smith (ed.), *Correspondence of Linnaeus*, I, 9, 33.

23. Colden to Gronovius [December, 1744], *Colden Papers*, III, 84.

than for his criticism of schemes of classification. He was esteemed in Europe for the same basic reason that other American collectors were valued. The rewards he received were very generous but somewhat deceptive. They did not mark him as a peer of Linnaeus, Gronovius, or Jussieu.

It was his ambition, however, to attain that distinction. In common with so many men of the eighteenth century, he set the highest premium upon intellectual activities. In approach and aspiration he was not a good field worker, for he was essentially rational and speculative in all of his science rather than experimental or clinical. On the other hand, he had nothing but contempt for the "mere Scholar" who knew only what he read in books and made no discovery of his own. His own formula was to read and think "by turns" filling in the intervals with conversation.[24] He even warned against excessive industry lest it fatigue the imagination, since he valued creative imagination most highly as the source of such truly great discoveries as the synthesis of Sir Isaac Newton. In all of his writings, Colden was more notable for his ability to indulge in untrammeled speculation than for a meticulous knowledge derived either from close observation or experimentation. His political enemy, William Smith, was probably not far from the truth when he described Colden as "rather a Man of Genius than Erudition."[25] It is an epitaph that would have pleased him, at any rate.

Almost inevitably, Colden soon became weary of the demands made upon his time by members of the natural history circle who sought from him a constant flow of seeds and specimens. His complaints that field trips were too much for him and that his eyes were too weak to examine the parts of the flower accurately suggested an unusual experiment.[26] Colden prepared an English guide to Linnaeus for his daughter, Jane, and taught her enough Latin to enable her to make out botanical descriptions. When she exhibited interest as well as talent in sketching plants, her father sent the drawings to Europe with much satisfaction. Linnaeus, Gronovius, Ellis, and Bartram were all impressed. Even Alexander Garden described her work as "ex-

24. Colden, Introduction to the Study of Natural Philosophy, Colden Manuscripts, New-York Historical Society.
25. Smith, Memoirs, II, not paged.
26. Colden to Gronovius, Oct. 1, 1755, *Colden Papers*, V, 29.

tremely accurate" and Collinson thought it "marvellous." [27] Indeed, Peter Collinson did what he could to bring together Colden's botanically-minded daughter and William Bartram—unimpressed by the social gulf that lay between them.[28] Cadwallader Colden soon reached a point where he was able to shift some of the burden of European botanical requests to his daughter.[29]

It is true, Colden was sixty-six years old when he began to groom his daughter as a botanist but it was not simply his advanced age which caused him to restrict his botanical activities. He had already turned his attention to a very much more important object—one in which imagination rather than tiring field work would be the important factor. He had found an object worthy of the most exalted genius, one which would place him in the company not only of Linnaeus but of Newton and Aristotle if he could but gain it. Indeed, he sought to pick up the baton where Newton himself had faltered and confessed himself unable to go on. He "pretend[ed] to have discover'd the Cause of Gravitation." [30]

Now this was a very different thing from discovering new plants. This was not within the realm of natural history but rather belonged to the "abstruser" parts of natural philosophy. Here, nearly all of the advantages which had smoothed the way to achievement and recognition in his botanical work were converted into disadvantages. It was no longer an advantage to be located in America, especially on a remote farm, far from the rich libraries and the stimulating conversations of learned men. Colden's situation in this respect was inferior not only to that of European scholars but even of other Americans who enjoyed the richer intellectual atmosphere of Boston or Philadelphia. The deficiencies of Coldengham were not remedied by an irregular correspondence with scholars, by occasional visits from John Bartram, by extended arguments with such third-rate thinkers as Captain John Rutherfurd, or by trips to New York City which was itself poor in

27. Bartram, "A Journey to the Katskill Mountains, ... 1753," Darlington, *Memorials*, 195; Collinson to Colden, April 6, 1757, *Colden Papers*, V, 139; Garden to Colden, Dec. 17, 1754, *ibid.*, IV, 475; Collinson to Linnaeus, May 12, 1756, April 30, 1758, Smith (ed.), *Correspondence of Linnaeus*, I, 39, 45.

28. Collinson to Colden, Oct. 5, 1757, *Colden Papers*, V, 190.

29. Colden to Gronovius, Oct. 1, 1753, Gray (ed.), "Selections from Colden," *American Journal of Science*, 1st ser., 44 (1843), 104.

30. Colden to Collinson, [1745], Colden Mss., New-York Historical Society.

matters of the mind.[31] Colden was aware of the drawbacks of his intellectual isolation but they did not deter him from his great effort. His approach to this basic problem in theoretical physics was rational, non-experimental, and uninformed. He had apparently never heard of Leonard Euler, the greatest mathematical physicist of that day; he did not know the pertinent work of Huygens, Leibnitz, or the Bernoullis; and even his understanding of Newton's laws of motion was defective.[32] His method reflected a disposition of the age for synthesis based upon inadequate data in the quest of all-embracing laws comparable to those of Sir Isaac Newton. To explain gravity, he first divided the material world into three substances: matter, light, and ether. Gravity was the result of the force exerted by the particles of ether upon all planets and stars. Since there were fewer particles between the sun and the earth than there were surrounding them, he mistakenly assumed that there would be less force exerted by the ether between the two bodies and that the resultant force would cause them to move together.[33] He asserted that they were prevented from colliding by the force exerted by light emanating from the sun which in addition imparted a tangential motion to the earth in some unexplained manner.

Colden believed his synthesis to be of fundamental importance to the understanding of the whole material world and, at points, his metaphysical approach yielded striking insights. He seemed almost to equate energy and matter when he insisted that action was the basis of all matter. Colden was constantly aware of the sweeping significance of his assertions and of their permanent value if he could but get them accepted. His great goal was made pathetically clear in the dedication to his first edition published in New York in 1745. There he expressed the hope that the pamphlet might be preserved "from Oblivion" and the name of his friend, James Alexander, transmitted to posterity along with his own.[34] Colden was more interested in at-

31. Rutherford was a young Scottish officer in command of the garrison at Albany. See Rutherford to Colden, March 2, 1742/43, April 19, 1743, *Colden Papers*, III, 6-9, 17.
32. Colden to Franklin, Nov. 19, 1753, *Colden Papers*, IV, 414.
33. Colden, *The Principles of Action in Matter* (London, 1751), [vi].
34. [Colden], *An Explication of the First Causes of Action in Matter* (New York, 1745), v-vi.

taining immortality through his work than in understanding the cause of gravity.

The initial reaction to Colden's theory was one of bewilderment. Colden sent a pre-publication draft to Philadelphia where James Logan, the man most competent to judge, could make no sense out of it although at the same time he praised Colden "as the ablest Thinker ... in this part of the World." [35] After publication, Lewis Evans declared he could not understand it and John Bartram decided it was too much to consider until after harvest. Seven other "ingenious" Philadelphians admitted it was beyond their comprehension but Logan then pointed out that the concept of gravity being the effect of elasticity was originally Jacob Bernoulli's.[36] Logan and the others read whatever they could find that was pertinent in their attempts to understand, but to no avail. Benjamin Franklin would not give up so quickly—or perhaps he was merely more kind to Colden. He wrote, "I imagine, that if I had an Opportunity of reading it with you, and proposing to you my Difficulties for Explanation as they rise, I might possibly soon Succeed." [37]

When the pamphlet reached England, it met just as much bewilderment but with somewhat different results. A London bookseller, deciding that anything so difficult must be valuable, printed a pirated edition in 1746 under the same title: *An Explication of the First Causes of Action in Matter; and of the Cause of Gravity*. From the English men of science Colden had a hard time getting any evaluation. Collinson sent copies to scientists in France, Russia, Germany, Holland, Sweden, and Scotland as well as in England where one copy was presented to the Royal Society.[38] He was told that it "was no trifleing affair" and although it was much read, no one would commit himself on it.[39] One mathematician did charge that it was a plagiarism which had originated in Europe—a kind of backhanded compliment. Collinson could only console Colden by expressing his conviction that "some

35. Franklin to Colden, Oct. 25, 1744, *Colden Papers*, III, 77.
36. Franklin to Colden, July 10, 1746, Oct. 16, 1746, *Colden Papers*, III, 227, 273-75.
37. *Ibid.*, 274.
38. Collinson to Colden, March 27, 1746/47, April 12, 1747, *Colden Papers*, III, 368, 371.
39. *Ibid.*

or other will at Last Saye Something for it or against It." [40] Occasional criticisms did not end the period of bewilderment which saw a French translation, a German translation, and a revised and enlarged English edition which appeared in 1751.[41]

When the criticisms finally were added up, the effect was devastating. In America, Samuel Johnson pointed out, two bodies in ether would be pressed as much from the side between them as from any other side.[42] One Englishman simply declared, "Mr. Colden is Mistaken in every part of his Conjectures." [43] Dr. John Bevis, the only English astronomer Collinson could persuade to examine Colden's explanations of his original work, was friendly but not very hopeful.[44] Abraham Gotthelf Kästner, who translated the work into German, was very critical, feeling that Colden's definitions of his three forms of matter were "too light to be right" and that Colden was deficient in mathematical understanding.[45] The final stroke was delivered from Berlin by Leonard Euler who found so much absurdity and lack of "knowledge of the principles of Motion" that he declared the author incapable of "Establishing the True Forces requisite to the Motion of the Planets, from whatever Cause He may attempt to Derive them." [46]

Cadwallader Colden never understood. He admitted that he had applied "himself to Physics and Mathematics, only by way of amusement, to fill up a vacant hour" but he continued to think that he could salvage his synthesis by a little patching and revising.[47] When he was unable to get his third edition published in England, he printed some corrections and replies in magazines and turned to Scotland for encouragement. Learning that his correspondent at Edinburgh, Dr. Rob-

40. Collinson to Colden, Aug. 3, 1747, *Colden Papers*, III, 411.
41. Colden, *Explication des premières causes de l'action dans la matière* (Paris, 1751); [Colden], *Erklärung der ersten wirkenden Ursache in der Materie* (Hamburg, 1748); Colden, *The Principles of Action in Matter* (London, 1751).
42. Samuel Johnson to Colden, Jan. 12, 1746/47, *Colden Papers*, III, 331.
43. Collinson to Colden, Aug. 3, 1747, *Colden Papers*, III, 412.
44. Alexander Colden to Cadwallader Colden, Sept. 20, 1756, *Colden Papers*, V, 92; Dr. John Bevis to Collinson, Aug. 10, 1755, *ibid.*, 23; Bevis to Alexander Colden, June 9, 1759, *ibid.*, 302.
45. A translation of Kästner's remarks is among the Colden Mss. at the New-York Historical Society without date or title.
46. Collinson to Colden, March 7, 1753, *Colden Papers*, IV, 356.
47. Comments by Colden on criticism of his *Principles of Action*, Colden, Mss.

ert Whytt, was unable to find a publisher, Colden deposited his final revision in the library of the University of Edinburgh in the hope that posterity, at least, would vindicate him.[48] He never realized that his training and environment, which had been more than adequate to permit him to add to the world's knowledge in botany, were quite inadequate to permit him even to understand such a basic problem in theoretical physics.

In fact, he continued to view his *Principles of Action in Matter* as a beginning which could be expanded profitably to explain the entire material world. He asserted that upon this foundation he could erect a new theory of the moon's motion which would not require the use of conic sections and would be of great advantage to navigation. He stood at the elbow of his crippled son, David, while he wrote out "A Supplement to the Principles of Action in Matter" in which he tried to apply his father's explanation of the cause of gravity to the cause of electrical attraction and other electrical phenomena. When Colden came back again into the realm of natural history and medicine, he sought to explain "The Animal OEconomy ... Mechanically according to the Laws of Matter in Motion." Here, however, Colden was much better informed about the facts so that his ninety-five-page manuscript turned out to have very little relation to his great theory but to be generally acceptable in the context of his time. In a paper he called "An Inquiry into the Principles of Vital Motion" he did bring his theory into closer touch with medicine by relating it to the motion of microscopic life in fermentation and in sperm. Though metaphysical, this again was more reasonable than his explanation of gravitation.[49]

Colden's failures were only partly a result of inadequate preparation and of an impoverished environment; they also owed much to the speculative quality of his thinking which sometimes led him far astray and sometimes to significant insights. His analysis of waterspouts was

48. *Monthly Review*, 1st ser., 21 (1759), 397-403; 23 (1760), 380-86; Colden to Porterfield, [n.d.], Colden Mss.; Colden to Robert Whytt, March 7, 1763, Colden Mss.; Whytt to Colden, May 16, 1763, *Colden Papers*, VI, 218.

49. Betts to Alexander, Sept. 7, 1749, *Colden Papers*, IV, 137; Colden to Bevis, May 23, 1757, *ibid.*, V, 146; David Colden to Franklin, [Sept. 18, 1757], *ibid.*, 184; David Colden's "Supplement to Principles of Action in Matter" and his father's "Animal OEconomy" and "Vital Motion" are among the Colden Mss.

THE PURSUIT OF SCIENCE

correct and his suggestion of stereotype printing was ahead of his
time, both having been derived from rational considerations altogether. His medical writings were similarly rational and almost scholastic. The best example of this approach was probably his support
of Bishop George Berkeley's metaphysical attempt to revive tar water
with more metaphysics and a few cases.[50] His pamphlet on the *Iliac
Passion*, his essay on yellow fever, and his articles on poke weed and
cancer were a fair reflection of the interests of medical men of his
time.[51] They all contained some speculation but also some fact and
understanding. On one occasion he was brought up sharply after writing a paper which displayed uninformed speculation upon the subject
of yellow fever. When he sent a disquisition on the topic to Dr. John
Mitchell of Virginia, who had attended many cases, Mitchell completely demolished Colden's theories. Colden replied very meekly that
he had not seen a case of yellow fever and appreciated Mitchell's comments.[52]

Among the several American physicians with whom Colden was
friendly, his most important early correspondent was Dr. William
Douglass of Boston. Douglass had a background similar in many respects to that of Colden himself. He was a Scot who had studied at
the University of Edinburgh before it had a medical faculty, but he
had completed his medical training at Leyden and at Paris. He came
to Boston in 1718, having been induced to take that step by the prospect of preferment under a man who had erroneously anticipated
being appointed governor of Massachusetts Bay. Douglass immediately
found himself the only physician in Boston with a medical degree, an
advantage which permitted him to live "handsomely" by his practice
and soon to become the leader of the profession.[53] With his superiority
in learning and his own positive personality, Douglass was able to set
the tone of medical practice in Boston in accord with his own empiri-

50. *An Abstract from Dr. Berkley's* [sic] *Treatise on Tar-Water with Some
Reflexions Thereon, Adapted to Diseases Frequent in America* (New York,
1745); Alexander to Colden, Feb. 10, 1744/45, *Colden Papers*, III, 102, identifies
Colden as the author.

51. *Gentleman's Magazine*, 21 (1751), 305-8; 22 (1752), 302; Packard, *History of Medicine*, I, 493, 500, 505.

52. Colden to Mitchell, 1745, [draft], Colden Mss.

53. Douglass to Colden, Feb. 20, 1720/21, "Letters from Dr. William Douglass to Cadwallader Colden of New York," Massachusetts Historical Society,
Collections, 4th ser., 2 (1854), 164.

cal, nonrational approach. He established a dominance that the equally learned Dr. Alexander Hamilton of Maryland considered unfortunate.[54]

Douglass was certainly in as favorable a position to accomplish great things in natural history as in medicine, and he gave a very early indication of interest in that direction. In 1720, he recorded the state of the weather in Boston as well as he could without a thermometer or barometer. By early 1721, he had collected seven hundred plants within four or five miles of Boston and had made a beginning on a collection of minerals in the area.[55] In his meteorological observations he followed the lead of Cadwallader Colden but he preceded Colden in his botanical work. It was the outbreak of the famous smallpox epidemic of 1721-1722 that deflected Douglass from this fine beginning; after that he never recovered a driving interest in natural history.[56] He did collect more plants, classifying them according to the system of Joseph de Tournefort as he had the others. John Bartram waited expectantly for this description of 1,100 plants to appear but although Douglass mentioned them once again in his best known publication, *A Summary, Historical and Political . . . of the British Settlements in North America*, he never got them into print.[57] In his *Summary*, Douglass included a few notes on natural history which attracted the attention of Peter Collinson.[58] He also wrote privately on an earthquake, on astronomy, and on cartography but he never really entered the natural history circle or offered much of value to its members.[59]

This failure to fulfill his early promise in natural history may have been related to the peculiar environment of Boston for it was just after the smallpox controversy that Massachusetts' contributions to natural history declined to practically nothing. It seems quite probable

54. Carl Bridenbaugh (ed.), *Gentleman's Progress* (Chapel Hill, 1948), 116, 131.
55. Douglass to Colden, Feb. 20, 1720/21, "Letters from Douglass," Massachusetts Historical Society, *Collections*, 4th ser., 2 (1854), 164.
56. Douglass to Colden, July 28, 1721, May 1, 1722, *ibid.*, 166, 169.
57. Bartram to Colden, Nov. 2, 1744, *Colden Papers*, III, 78.
58. Collinson to Bartram, July 20, 1756, Darlington, *Memorials*, 210.
59. Douglass to Colden, Nov. 20, 1727, Sept. 14, 1729, "Letters from Douglass," 172, 185; William Douglass, *A Summary, Historical and Political of the First Planting, Progressive Improvements, and Present State of the British Settlements in North America* (London, 1760), I, 404.

that such men as Colden, Mitchell, and Garden would have become involved in the great inoculation controversy, had they lived in Boston. Indeed, Colden from New York did write opposing inoculation in much the same vein as Douglass. In the highly rational society of Boston, Douglass found ample scope for his intellectual energies in trying to lead the Boston medical profession down the proper path, in combating Cotton Mather, and in entering occasional political disputes.[60] His most important scientific writing was his *Practical History of a New Epidemical Eruptive Miliary Fever* which provided an accurate description of scarlet fever. The "sore throat distemper" had infected a very large area between 1735 and 1740, providing the occasion for this treatise as it did for numerous pamphlets by other observers of the epidemic.[61] The extent of concern about the problem practically demanded a superior performance from Douglass if he were to justify the pre-eminence he sought to maintain in the medical profession. In the stimulating environment of Boston, he had many more claims upon his time and energy than other Scottish physicians in America who did make important contributions to American natural history but who lived in quieter, more rural surroundings.

The Charleston environment was always closer to the soil and less intellectual than that of Boston, but it had hazards of its own for the pursuit of natural history. By mid-century, it had attracted a surprising number of physicians few of whom were impressed with their environment. The most usual complaint was the unbearable summer heat, but the lack of intellectual interests on the part of the residents was just as debilitating.[62] Alexander Garden wondered "how there should be one place abounding with so many marks of the divine wisdom and power, and not one rational eye to contemplate them." When his friend, Dr. John Lining, gave up medicine and "turned a planter," Garden considered him a lost soul for he asserted that the planters of South Carolina were "absolutely above every occu-

60. See William Douglass, *A Discourse Concerning the Currencies of the British Plantations in America* (Boston, 1740).

61. Ernest Caulfield, *A True History of the Terrible Epidemic Vulgarly Called the Throat Distemper* (New Haven, n.d.).

62. David Ramsay, *The History of South-Carolina* (Charleston, 1809), II, 472, indicates that Garden did not complain of the weather interfering with his studies and that this was unusual, but see Garden to Colden, Aug. 14, 1756, *Colden Papers*, V, 89.

pation but eating, drinking, lolling, smoking, and sleeping, which five modes of action constitute the essence of their life and existence." [63]

Despite Garden's opinion, Lining was the first of the Scottish doctors in Charleston to do important scientific work. His primary interest was and remained medicine although he wrote upon botany as a phase of medicine. A paper he published in the *Edinburgh Essays* was instructive in this respect for it presented an account of the medical properties of Indian pink drawn in part from his own experience without offering the careful botanical description which Garden included in his account of the same plant.[64] His most valued work in his own day was in the field of meteorology where he attempted, as so many of his contemporaries were doing, to relate weather to disease. Of more permanent value were his famous statical experiments in which he weighed the intake and outgo of his own body for a period of one year.[65] This was a study of metabolism but it was designed as an inquiry into the causes of seasonal epidemic diseases. His treatise on yellow fever, in the form of a letter to Robert Whytt at Edinburgh, gave careful clinical descriptions of the symptoms, progress, and chronology of the disease and at the same time provided accurate weather reports for comparative use.[66] It was at Whytt's request that this account was sent to Edinburgh and through his efforts that it appeared in print.[67] Lining's strongest ties were with Edinburgh although his statical experiments were published in the *Philosophical Transactions* and Garden set him on some experiments with vegetable dyes for the Society of Arts in London. A little later, he corresponded with Benjamin Franklin.[68] Lining was never a naturalist although some of his work was of value to the natural history circle.

63. Garden to Ellis, May 6, 1757, March 21, 1760, Nov. 19, 1764, Smith (ed.), *Correspondence of Linnaeus*, I, 407, 477, 520.

64. John Lining, "Of the Anthelmintic Virtues of the Root of the Indian Pink," *Essays and Observations* (Edinburgh, 1771), I, 436-39; Alexander Garden, "An Account of the Indian Pink," *ibid.*, III, 145-53.

65. *Philosophical Transactions*, 42 (1742-43), 491-509; 43 (1744-45), 318-30; Frederick P. Bowes, *The Culture of Early Charleston* (Chapel Hill, 1942), 81-82.

66. John Lining, "A Description of the American Yellow Fever," *Essays and Observations* (Edinburgh, [1771]), II, 404-32.

67. *Ibid.*, 404.

68. Garden to Ellis, May 6, 1757, Smith (ed.), *Correspondence of Linnaeus*, I, 407; *Philosophical Transactions*, 48 (1753-54), 757-64; I. Bernard Cohen (ed.), *Benjamin Franklin's Experiments* (Cambridge, 1941), 331-45.

When Alexander Garden arrived in Charleston in 1752, he was welcomed by Lining, who, on his retirement three years later, went to the extent of turning over his medical practice to the younger man. Garden had gained some botanical knowledge before coming to America, particularly at the garden presided over by Dr. Charles Alston, professor of botany at the University of Edinburgh.[69] It was to Alston that he very naturally addressed his first botanical queries from America in January 1753. He may very well have introduced Lining to the correspondence with Alston which he developed shortly afterward, for Lining had left Scotland too early to have studied under that professor.[70] Garden's early desultory excursions into the country to study plants yielded very little until in 1754 two events gave him a new sense of direction. When William Bull put Linnaeus' *Fundamenta Botanica* in his hands for the first time, Garden was quickly inspired to write the Swedish master, proclaiming himself his disciple and offering his services.[71] In the summer of 1754, Garden took a trip for his health into the northern provinces where he was delighted to find American botanists who were much more advanced in the science than he and who had already won the respect of the European botanists. Cadwallader Colden, a fellow Scot, had the greatest influence upon Garden, introducing him to his daughter and to John Bartram who happened to be visiting at the same time. He also placed before him additional works of Linnaeus, his own "Plantae Coldenghamiae," and evidence of his extensive European correspondence.[72] On his return to Charleston, Garden stopped at Philadelphia to visit John Bartram, and while he was there, he met Benjamin Franklin. He probably talked with John Clayton in Virginia too, for by 1755, he was corresponding with him.[73] The trip opened a new world, leading him to exclaim, "How happy should I be to pass my life with men so distinguished by

69. Grant, *University of Edinburgh*, II, 380; Garden to Ellis, March 25, 1755, Smith (ed.), *Correspondence of Linnaeus*, I, 345.

70. William M. and Mabel S. C. Smallwood, *Natural History and the American Mind* (New York, 1941), 94, 95.

71. Garden to Linnaeus, March 15, 1755, Smith (ed.), *Correspondence of Linnaeus*, I, 284, 288.

72. Garden to Ellis, March 25, 1755, Smith (ed.), *Correspondence of Linnaeus*, I, 343; Garden to Parsons, May 5, 1755, Nichols (ed.), *Literary Anecdotes*, V, 484n.

73. Garden to Ellis, March 25, 1755, Smith (ed.), *Correspondence of Linnaeus*, I, 345.

genius, acuteness, and liberality, as well as by eminent botanical learning and experience," but it made him more than ever dissatisfied with
South Carolina, "a horrid country, where there is not a living soul
who knows the least *iota* of Natural History." [74]

Driven by strong ambition, Garden used all of his spare hours to
earn a place for himself in the natural history circle. He developed a
correspondence with Colden and Clayton and with Bartram, who
visited him on two occasions. He initiated a correspondence with
Peter Collinson as he had with Linnaeus—without having had any previous acquaintance with him. Collinson judged him an "Ingenious
Man" and endeavored to help him but Collinson never played the part
in Garden's life that he did in that of Bartram, Colden, or Clayton. [75]
Garden relied directly upon Linnaeus, upon several Edinburgh acquaintances, and in England, upon his friend, John Ellis. Ellis was one
of the leading English naturalists who had just been admitted to the
Royal Society himself in 1754 and was consequently full of zeal. He
was best known for his *Corals and Corallines* which had settled a much
disputed point, demonstrating that the organisms which construct and
inhabit corals are animal in nature. [76]

After Garden had expressed an early interest in being elected to
fellowship in the Royal Society, he cooled abruptly upon receiving a
real or fancied slight, and for several years refused even to submit
papers for publication in the *Philosophical Transactions*. In this period
he preferred to direct his collections and his writings to Edinburgh and
Upsala rather than to London, although he often forwarded copies to
his friend John Ellis. [77] He did make an early effort to work with the
London Society of Arts or the Premium Society of which Ellis was
also a member. This society was founded in 1755 for the purpose of
granting premiums in Britain and the colonies for the encouragement
of commerce, manufactures, and agriculture. Garden was elected to
membership after he had sent the society a list of products he felt
should be encouraged in South Carolina, and Cadwallader Colden was

74. Garden to Linnaeus, March 15, 1755, Smith (ed.), *Correspondence of
Linnaeus*, I, 286; Garden to Ellis, [about Jan. 1761], *ibid.*, 502.
75. Collinson to Colden, May 19, 1756, *Colden Papers*, V, 81.
76. Garden to Ellis, March 25, 1755, Smith (ed.), *Correspondence of Linnaeus*, I, 342, 345.
77. Garden to Ellis, July 14, 1759, Smith (ed.), *Correspondence of Linnaeus*,
I, 454.

elected a member after Garden had mentioned him to William Shipley, founder of the society.[78]

In 1757 Garden responded energetically to a project proposed to him by a London member of the Society of Arts, Charles Whitworth. Whitworth suggested the establishment of a society dedicated to a study of vegetable dyes with a corresponding unit in London and experimental units in the colonies. The aim would be to seek plants which would yield commercially profitable dyes. Since indigo production was even then sustained by the British bounty on the crop, the thinking behind this plan was meaningful to a resident of South Carolina. Garden immediately conferred with "some of the most public spirited gentlemen" in the colony about the project, put an advertisement in the paper, and tried to encourage the planters to make experiments with vegetable dyes. Almost immediately, he saw a nobler vision, too, suggesting that societies be established in Georgia, South Carolina, Philadelphia, and New York in order to point out "proper articles for your Premium Society to exercise their benevolence on" and to enrich "Natural History and Philosophy" in general. Not only did this larger plan fail to materialize, but his efforts to encourage dye experiments among the planters of South Carolina never produced a single trial.[79]

Garden's own most important contributions were made in botany. He maintained a small garden and explored an area near his home with great care, but except for one 250-mile trip into the Carolina back country with Governor James Glen, he did none of the wide travelling that John Bartram undertook.[80] When Ellis suggested that the garden-fanciers in England would be happy to pay him for any seeds he might send over, Garden bruskly rejected the suggestion. Whatever he could collect would be solely for his friends, he declared.[81] Garden wanted to become a "philosophic botanist" not a mere "observator" and in

78. William Shipley to Franklin, Sept. 13, 1755, Franklin Papers, I, not paged, American Philosophical Society.

79. Garden to Charles Whitworth, April 27, 1757, Smith (ed.), *Correspondence of Linnaeus*, I, 385-87; Garden to Ellis, May 6, 1757, Aug. 11, 1758, *ibid.*, 402, 419; *South-Carolina Gazette*, April 1, 1757.

80. Garden to Colden, Aug. 14, 1756, *Colden Papers*, V, 90; Garden to Whitworth, April 27, 1757, Smith (ed.), *Correspondence of Linnaeus*, I, 387.

81. Garden to Ellis, July 6, 1757, Smith (ed.), *Correspondence of Linnaeus*, I, 415.

the end, he certainly succeeded.[82] His work, including the discovery of many new genera, was mostly assimilated in Linnaeus' *Systema Naturae* rather than in a separately published collection of Carolina plants. His ability to classify plants on his own was particularly notable. In several lengthy disputes he conducted with Linnaeus over the classification of the plants he had found, he often as not "came off conqueror." [83] In one case, although his contention was never accepted by Linnaeus, it has been confirmed by more recent botanists.[84]

More than most of the American botanizers, Garden concerned himself with zoology—by request rather than by inclination. After waiting four years for Linnaeus to answer his letters, Garden was confronted with a request that he collect the fish, reptiles, and insects of Carolina. To this task he immediately set himself.[85] He sent back not only descriptions but specimens which found their way into the *Systema Naturae* in such great number that Garden's name was mentioned more often than that of any other American in the twelfth edition. His most spectacular discovery was the amphibious animal which he called the siren, considered by Linnaeus not merely a new genus but a new class or order.[86] Accounts of it were published in Linnaeus' *Amoenitates Academicae*, in the *Philosophical Transactions*, and in the *Gentleman's Magazine*.[87] John Ellis, too, helped to turn Garden's eye in the direction of zoology when he asked him to study and send specimens of the cochineal insect so much valued for the red dye that could be prepared from it.[88]

Garden did not discover as many new plants as Bartram nor did he publish systematic catalogues as did Colden and Clayton, but in a sense, he operated at a higher level than any of them. He was still not a peer of Linnaeus although he could debate with him on terms of

82. Garden to Ellis, Jan. 13, 1756, Smith (ed.), *Correspondence of Linnaeus*, I, 369.
83. Garden to Ellis, May 15, 1773, Smith (ed.), *Correspondence of Linnaeus*, I, 599.
84. Margaret Denny, "Linnaeus and his Disciple in Carolina: Alexander Garden," *Isis*, 38 (1948), 173.
85. Garden to Linnaeus, Jan. 2, 1760, Smith (ed.), *Correspondence of Linnaeus*, I, 297-302.
86. Garden to Colden, [received June, 1768], *Colden Papers*, VII, 141.
87. *Amoenitates Academicae*, 7 (1769), 311; *Philosophical Transactions*, 56 (1766), 189-92; *Gentleman's Magazine*, 37 (1767), 637-38.
88. *Philosophical Transactions*, 52 (1761-62), 661-67.

equality within the sphere of his knowledge. He received full recognition for his accomplishments, being elected to fellowship in the societies at Edinburgh and Upsala, and finally in 1773 to the Royal Society itself. Linnaeus named the familiar Gardenia for him, although it was not one of his discoveries. When he returned to England after the onset of the American Revolution, he easily took his place among the London virtuosi as a vice-president of the Royal Society.

In training, capacity, and the consuming ambition to achieve, Dr. John Mitchell of Urbana, Virginia, was a match for Garden. The two men never met because Mitchell had left America in 1746, before Garden arrived. Mitchell may have been born in America, but it is certain that he studied at Edinburgh and at Leyden, and like Garden, he received the foundation of his botanical knowledge from the lectures and garden of Dr. Charles Alston at Edinburgh. When he began to study botany in America, he too was disturbed by the obstacles of debilitating climate, absence of books, and lack of "the conversation of the learned." In some measure, he was able to remedy these defects by a European correspondence and by acquaintance with the same circle of American naturalists in which Garden later took his place.[89]

Mitchell's interest in botany was probably set on a productive course by Peter Collinson for he once told the London merchant, "Whatever I have written on Plants is at your request and exhortation."[90] In 1738, he sent Collinson a paper, addressed to Sir Hans Sloane, entitled "Dissertatio Brevis de Principiis Botanicum et Zoologorum" in which he tangled with the problem of classification, and three years later, he sent a list of Virginia plants, "Nova Genera Plantarum." Collinson forwarded both papers to a German friend, Dr. Christian J. Trew, who published them in the "Nuremburg Transactions" of 1748.[91] Mitchell's classification scheme recognized the advantages of convenience in the sexual system but, instead of basing it upon the organs of reproduction, as Linnaeus did, he tried to make the method of reproduction the distinguishing factor. When Linnaeus found his

89. Mitchell to Dr. Charles Alston, Oct. 4, 1738, Herbert Thatcher, "Dr. Mitchell, M.D., F.R.S., of Virginia," *Virginia Magazine of History and Biography,* 40 (1932), 51.

90. Mitchell to Collinson, March 11, 1741, *ibid.,* 103.

91. Though popularly known as the "Nuremburg Transactions," the title of the journal was *Acta Physico-Medica Academiae Caesareae ... Ephemerides,* 8 (1748), 187-224; Theodore Hornberger, "The Scientific Ideas of John Mitchell," *Huntington Library Quarterly,* 10 (1946-47), 281.

writing "very difficult to make out," Mitchell cheerily explained to the master, "If I mistake not, our systems support each other." [92] He took pleasure in pointing out Linnaeus' failure to recognize the classical allusions he had made in the names he assigned to his new genera, and he did not hesitate to point out specific errors made by Linnaeus.[93] Far from being offended by these criticisms Linnaeus wrote of Mitchell to his correspondents in the highest terms. When Mitchell returned to England just about the time of the death of Dillenius, the leading academic botanist in the country, it seemed the logical thing for Collinson to ask that Mitchell answer a letter Linnaeus had written Dillenius.[94] Mitchell did make lasting contributions to botany. Of the twenty-five genera he claimed to have discovered, a recent analysis reveals that twenty-one were new at the time he wrote though by the time his paper was published, eleven of those had been described by others.[95]

Dr. John Mitchell's interests were never confined to botany. Of his zoological excursions, the most important was his study of the opossum which had been consistently misinterpreted by earlier investigators. Through Collinson, he sent two papers on the opossum to the Royal Society in which he carefully described the life cycle of the animal, assigning the proper role to its pouch.[96] Just before he left America, he sent Collinson his "Essay upon the Causes of the Different Colours of People in Different Climates," another very popular question of the moment. That paper had been written in answer to a prize question proposed by the Bordeaux Academy but, because it was too late to be submitted in that contest, it was published in the *Philosophical Transactions*. Mitchell's treatment was based upon reason and speculation although he did describe a few experiments upon living subjects. He accepted environmental modification as the underlying cause of differentiation.[97]

When Mitchell returned to England, he was drawn from botany

92. Linnaeus to Haller, Sept. 13, 1748, Smith (ed.), *Correspondence of Linnaeus*, II, 429; Mitchell to Linnaeus, Sept. 20, 1748, *ibid.*, 448.

93. Collinson to Linnaeus, Sept. 1, 1745, Smith (ed.), *Correspondence of Linnaeus*, I, 13.

94. Collinson to Linnaeus, Sept. 1, 1745, April 16, 1747, Smith (ed.), *Correspondence of Linnaeus*, I, 13, 18.

95. Thatcher, "Dr. Mitchell of Virginia," *Virginia Magazine of History and Biography*, 40 (1932), 269.

96. *Ibid.*, 337, 338-46.

97. *Philosophical Transactions*, 43 (1744-45), 102-50.

and natural history to other pursuits, most of which were related to America. He published three books on America, all of them largely political in character, and his famous *Map of the British and French Dominions in North America* which was political in purpose.[98] The map was prepared at the request of the Board of Trade and issued first in 1755. It was based upon an unmatched collection of original surveys held by the Board of Trade and by the Admiralty, upon Mitchell's own knowledge, and upon correspondence.[99] The first really satisfactory British map of the region, it provided an important counter to French claims. The use of William Douglass' map of New England and other local maps would have improved it, particularly as far as topography was concerned.[100] Nevertheless, the map was a great achievement. It remained a standard until long after its use at the peace conference of 1783 and went through at least nineteen editions by 1792. Mitchell never did get around to writing the monumental natural and medical history of North America which he planned, because his ambition drew him to more promising fields.

From 1740 to the end of the colonial period, the natural history circle was a thriving, creative group. The Americans, whether such assiduous collectors as Bartram or competent classifiers of the capacity of Garden and Mitchell, depended upon the intellectual stimulus they received from Europe. Their raw material was found in America but it was only from Europe that encouragement to exploit it was derived. It was inevitable that the knowledge to understand and apply American raw material must come from Europe. What was remarkable was the extent to which Americans made permanent contributions to the advance of knowledge in natural history.

98. *The Contest in America Between Great Britain and France* (London, 1757) to accompany the map; *A New and Complete History of the British Empire in America* (London, 1756); *The Present State of Great Britain and North America* (London, 1767).

99. Lyman Carrier, "Dr. John Mitchell, Naturalist, Cartographer, and Historian," American Historical Association, *Annual Report for 1918*, I, 207; Mitchell to Colden, April 5, 1751, *Colden Papers*, IX, 89; Hunter Miller, "Mitchell's Map," *Treaties and Other International Acts* (Washington, 1933), III, 330; Wroth, *American Bookshelf*, 32; Erwin Raisz, "Outline History of American Cartography," *Isis*, 26 (1937), 377.

100. Mitchell to Colden, March 25, 1749, Theodore Hornberger (ed.), "A Letter from John Mitchell to Cadwallader Colden," *Huntington Library Quarterly*, 10 (1946-47), 414.

Chapter Four

SHATTERED DREAMS AND UNEXPECTED
ACCOMPLISHMENT

EVERY ONE OF THE active American naturalists complained of the
thinness of the intellectual environment he met in America. Each of
them missed the libraries, the publications, and, most of all, "the conversation of the learned" that would have been open to him in Europe.
The gradual process of maturing might have been expected to bring all
of those things in its train but it was not necessary to wait passively
for this enrichment. Colden, Bartram, Mitchell, Garden, and Clayton
all took part in efforts to create a community that would encourage
scientific activity.

Cadwallader Colden was the first to suggest the formation of an
intercolonial scientific community in the form of a society or academy.
This was the most logical starting point in the eighteenth century
when the academy was the primary focus of scientific activity. Despite
the much heralded decline of the Royal Society of London, it remained the point at which most British science was centered; while
for the French world, the same function was performed by the
Académie des Sciences. Despite its faults, the Royal Society's *Philosophical Transactions* was the medium of regular, scientific intelligence
in all the areas where English was spoken and in many where it was
not. Throughout Europe, the formation of academies was still in full
flood in the first half of the eighteenth century with the establishment
of such societies at Madrid, Vienna, Upsala, Bordeaux, Marseilles, St.
Petersburg, and Edinburgh—all of them patterned after the Royal So-

ciety or the Académie des Sciences. Colden was making the proper approach when he wrote his friend William Douglass in 1728 suggesting the formation of "a Voluntary Society for the advancing of Knowledge." [1]

The society would include members in each province but, he suggested, "because the greatest number of proper persons are like to be found in your Colony . . . the Members residing in or near Boston [ought to] have the chief Direction." [2] In 1728, it was reasonable to look to Boston for leadership in a project of this sort. Boston then was and for many years had been the principal cultural center in the British colonies. Its attainments in science were particularly notable if fellowship in the Royal Society could be used as an index, for of the eight residents of the continental colonies elected to the Royal Society before 1728, seven were Bostonians.[3] Moreover, real contributions to the advance of science had been made by Paul Dudley, Thomas Robie, and Zabdiel Boylston as well as by William Douglass, all of them still living in the area when Colden made his suggestion. Colden may also have known of Increase Mather's attempt in Boston to establish a philosophical society as early as 1683.[4]

Boston could be proud of its citizens who had contributed to science and of its many cultural accomplishments but it is clear that it had not developed a scientific community. Colden, of course, had sought to create one through the formation of his proposed organization; yet organizations are seldom the starting point in the formation of any kind of community. More often they are based upon previously existing, though informal, relationships. Such had been the experience of the two greatest academies in the West: the Royal Society of London had grown out of both the "invisible college" and the Oxford Philosophical Society; the Académie des Sciences had developed when Colbert provided official recognition of the long continued but irregular meetings held by a group of French scientists.[5] Lacking such a

1. Colden to Douglass, [1728?], *The Letters and Papers of Cadwallader Colden*, I, New-York Historical Society, *Collections for 1917*, 50 (1918), 272.
2. *Ibid.*
3. Raymond P. Stearns, "Colonial Fellows of the Royal Society of London, 1661-1788," *William and Mary Quarterly*, 3d ser., 3 (1946), 222-32.
4. Kenneth B. Murdock, *Increase Mather* (Cambridge, 1926), 147-48.
5. C. R. Weld, *History of the Royal Society* (London, 1848), I, 30-71; Martha Ornstein, *The Role of Scientific Societies in the Seventeenth Century* (Chicago, 1928), 144-45.

background, Douglass was well-advised in not doing anything to implement Colden's suggestion.

Eight years later, Douglass wrote to Colden to announce the formation of a medical society in Boston, recalling as he did so the New Yorker's earlier suggestion of "a sort of Virt[u]oso Society or rather correspondence." [6] On the face of it, Douglass' medical society had little relationship to Colden's project. His was not a scientific but a medical society which planned to publish short pieces on medical subjects from time to time. Even as he wrote, Douglass announced that the first number of these "Medical Memoirs" was ready for the press. Somehow it failed to be published and the four medical essays it was to contain have been lost. The society itself continued for some years; indeed, Dr. Alexander Hamilton found it still functioning under the leadership of Douglass in 1744.[7]

This relationship of a medical society in being to a philosophical society that remained an aspiration was closer than it might seem. Even as late as 1736, there was no scientific community anywhere in America except for the medical fraternity which flourished in each of the larger colonial towns. These medical communities had intercolonial and international relationships. They were the most coherent groups of men who had had some kind of scientific training to be found in the colonies. In addition, the physicians had a more immediate motive of expediency which encouraged association and joint effort. Their early organizations were always in some measure dedicated toward improving the conditions of their own profession, toward providing such things as the licensing of practitioners and the regulation of rates. The physicians were associated by material interests as well as by intellectual curiosity.

Douglass's society was only the first such organization of which traces remain. In 1749, a society was meeting weekly in New York and in 1755, the "faculty of physic" in Charleston organized under the presidency of Dr. John Moultrie.[8] The Charleston organization

6. Douglass to Colden, Feb. 17, 1735/36, *Colden Papers*, II, 146.

7. Carl Bridenbaugh (ed.), *Gentleman's Progress* (Chapel Hill, 1948), 137; see also Bartram to Colden, Nov. 2, 1744, *Colden Papers*, III, 78.

8. John Bard, An Essay on the nature of the malignant Pleurisy, 1749, Academy of Medicine of New York; *South-Carolina Gazette*, June 5, 1755, cited by Joseph I. Waring, "An Incident in Early South Carolina Medicine," *Annals of Medical History*, new ser., 1 (1929), 608.

[61]

was primarily concerned with the "better Support of the Dignity, the Privileges, and Emoluments of their Humane Art," but its agreement that physicians should collect fees at each visit aroused a furore in the press.[9] The twofold aim of these societies was best demonstrated by the Boston organization which in addition to its interest in publishing memoirs and observing surgical operations, sought to obtain a system of licensing practitioners in Massachusetts.[10]

All of these societies were in a sense surface manifestations of an American medical community which was already a part of the much larger European medical community. The Americans were related to the Europeans not only through medical studies undertaken by some in Europe and through continuing personal correspondence, but also most effectively through publications. The Royal Society itself had always devoted attention to medical subjects and some Americans had access to its *Philosophical Transactions* both as readers and as contributors.[11] The Americans were also served by two more specialized periodicals published by medical societies: from Edinburgh, the *Medical Essays and Observations*, often called simply the *Edinburgh Essays* which began to appear in 1733, and from London, the *Medical Observations and Inquiries* of Dr. John Fothergill's society beginning in 1757. The *Medical Observations and Inquiries* particularly welcomed American papers from its foundation, recognizing that there were "physicians of great experience and abilities" in the colonies.[12] The general magazines in England and Scotland published medical papers too, the *Gentleman's Magazine* being particularly notable for the number of American essays it saw fit to print.[13] In lieu of medical periodicals in America, the newspapers and the short-lived magazines frequently published medical essays of a type that belonged in a journal.[14] When to all of these facilities was added the publication of medical pamphlets and books on both sides of the Atlantic it is

9. *Ibid.*

10. *Weekly News-Letter*, Boston, Jan. 5, 1738, Nov. 13, 1741.

11. *Philosophical Transactions*, 32 (1722-23), 33-35, 223-27; 33 (1724-25), 67-70; 54 (1764), 386-88; 55 (1765), 244-45.

12. *Medical Observations and Inquiries*, 1 (1757), iv, 67-80, 83-86, 87-110; 2 (1762), 265-68, 369-72.

13. *Gentleman's Magazine*, 20 (1750), 343; 21 (1751), 305-8; 23 (1753), 413-15.

14. *American Magazine and Historical Chronicle* (1743-44), 31-32, 68-70, 588-92; *American Magazine and Monthly Chronicle*, 1 (1756-57), 374-82; *New American Magazine*, 1 (1758-59), 189-90, 565-69, 674-78.

clear that physicians in America were furnished with intellectual outlets as well as means of professional enlightenment.

A large proportion of their writing was intended to influence the lay public rather than to advance knowledge. The great debate over inoculation brought a flood of articles designed to win support for or condemnation of the practice. Inoculation was fought out before the public and before the colonial legislatures in terms of absolute prohibition, regulation, establishment of inoculation hospitals, and specific technique to be used.[15] Every epidemic of smallpox as well as other diseases brought newspaper articles on treatment. Questions of autopsy, quarantine, and care of the poor were similarly of public interest. The publication, however, most certain to sell was the home medical guide for which there was a constant demand.[16]

Americans also made studies of specific diseases often resulting in scholarly medical papers which conformed to the standards of the British periodicals and books they read. Most frequently, they wrote such case reports as John Bard's account of an extra-uterine foetus or Thomas Bond's account of a worm that had been lodged in a patient's liver.[17] Some studies were made in the colonies to provide specific information that seemed important in the light of prevailing medical theory. Such were the numerous efforts to correlate the incidence of sickness with weather conditions; the meteorological studies of Lionel Chalmers, a Scottish physician of Charleston, even approached a climatological determinism as far as disease was concerned.[18] Other papers attempted to support the virtues of some new specific which had been discovered in America. Occasionally these studies brought fame and fortune to individuals although not one of the drugs so re-

15. James Killpatrick, *A Full and Clear Reply to Dr. Thomas Dale* (Charles Town, 1739); Adam Thomson, *A Discourse on the Preparation of the Body for the Small-pox* (Philadelphia, 1750); *Maryland Gazette*, March 4, 1765; Henry R. Viets, *A Brief History of Medicine in Massachusetts* (Boston, 1930), 74-75; Winthrop Tilley, The Literature of Natural and Physical Science in the American Colonies from the Beginnings to 1765 (Doctoral Dissertation, Brown University, 1933), 77, 80.

16. Short's *Medicina Britannica* was this kind of book as was *Every Man His Own Doctor* (Williamsburg, 1734), which was probably the work of John Tennent.

17. *Medical Observations and Inquiries*, 1 (1757), 68-76; 2 (1762), 369-72.

18. Lionel Chalmers, *An Account of the Weather and Diseases of South-Carolina* (London, 1776), I, 168-78.

warded lived up to the claims of its advocate. One of the most dramatic examples of the promotion of a new medicine was the case of
John Tennent of Virginia, who wrote much in support of the virtues
of Seneca snake root, even journeying to England in support of his
claims. Although it had been originally presented as a rattlesnake cure,
Tennent claimed its utility in a wide variety of disorders. Before the
defects of his own character disgraced him, he won election to the
Royal Society and a £100 grant from the Virginia Assembly.[19] The
legislature of South Carolina showed still more gullibility when it
granted upwards of £500 to a man who claimed to have discovered a
sovereign cure for syphilis.[20]

With all the vitality the medical community could boast, it was not
a substitute for a scientific community and it did not satisfy the needs
that had led Colden to suggest an intercolonial scientific society. These
needs came to seem still more imperative to John Bartram as he extended his study of natural history, leading him to address his own suggestion of a scientific academy to Peter Collinson in 1739. While
Collinson was right in asserting that an academy with paid professors
was beyond the capacity of Pennsylvania, Philadelphia was in the midst
of an impressive development. The Library Company of Philadelphia,
into which Collinson succeeded in introducing Bartram, was in 1739
only the most striking evidence of a rapidly developing capacity to
sustain cultural activity of a varied character.

Collinson made a penetrating observation when he told Bartram,
"Your Library Company I take to be an essay towards...a society
[such as you suggested]." The Library Company was a remarkable
institution.[21] Because it was operated on the subscription principle,
regular planned book purchases became possible with the result that
its collections grew rapidly and rationally. Because its purchases were
directed by men seeking general enlightenment, it differed consider-

19. John Tennent, *An Essay on the Pleurisy* (Williamsburg, 1736); John
Tennent, *An Epistle to Dr. Richard Mead* (Edinburgh, 1738); Wyndham B.
Blanton, *Medicine in Virginia in the Eighteenth Century* (Richmond, 1931),
121-29.

20. Blanton, *Medicine in Virginia in the Eighteenth Century*, 126; Garden
to Ellis, May 11, 1759, Smith (ed.), *Correspondence of Linnaeus*, I, 447; see also
Gentleman's Magazine, 20 (1750), 343.

21. Colden to Bartram, July 7, 1739, William Darlington, *Memorials of John
Bartram and Humphry Marshall* (Philadelphia, 1811), 132.

ably from the large libraries collected by colleges for the use of graduate students in theology.[22] It soon held the finest collection of scientific books in the country. It became a good nucleus for the formation of a scientific community because it served not only the members of the Library Company, but also the general public, particularly the city artisans who made much use of the reading privileges freely extended to all.

In addition to serving as the most effective library in the colonies, the Library Company presented the appearance of a museum or a scientific society because of the natural history specimens and scientific apparatus which were acquired. In several cases, shares in the company were given in return for curiosities: for some stuffed snakes, for a dead pelican, for the robes of Indian chiefs, for an old sword, for a set of fossils.[23] In 1738, John Penn, one of the proprietors of Pennsylvania, recognized the general importance of the organization when he presented a "costly Air-Pump" rejoicing that the Library Company was "the first [institution] that encouraged Knowledge and Learning in the Province of Pennsylvania." [24] The directors of the Library did not at all consider such a gift a strange thing with which to burden a library. Indeed, they asserted it to be directly in support of their objective which was "the Improvement of Knowledge." [25] Members of the proprietary family continued to donate scientific apparatus, contributing a pair of sixteen-inch globes, a telescope, and an electric machine before 1749.[26]

This breadth of function demonstrated by the Library Company occurred at many other places in the colonies. In Massachusetts, the dominant intellectual institution was Harvard College, so it was about Harvard that scientific apparatus and a heterogeneous collection of curiosities as well as a library clustered.[27] At Newport, a learned society dating from 1730, "The Society for the Promotion of Knowledge and Virtue by a Free Conversation," had a library engrafted upon

22. Samuel Miller, *A Brief Retrospect of the Eighteenth Century* (New York, 1803), II, 351-52.
23. Austin K. Gray, *Benjamin Franklin's Library* (New York, [1936]), 15.
24. Minutes of the Directors of the Library Company of Philadelphia, I, 74.
25. *Ibid.*, 79.
26. *Ibid.*, 131.
27. See especially I. Bernard Cohen, *Some Early Tools of American Science* (Cambridge, 1950), 26-44.

it when Abraham Redwood, inspired by a trip to Philadelphia, donated £500 for the purpose of buying books. The Redwood Library followed.[28] At Charleston, the example of the Philadelphians was followed in 1748 with the formation of the Charleston Library Society which came to embrace apparatus and a museum as well as a collection of books.[29] Few intellectual agencies of this period were very specialized.

Philadelphia's much copied subscription library was only one evidence of its continually expanding intellectual vitality. Behind this increasing ability to sustain cultural activities lay a profitable trade and a population that was beginning to grow so fast that it would soon exceed Boston's.[30] Its people came of many nations and many faiths and they retained varied contacts with the Old World. As wealth, learning, and leisure became more common, the city began to provide more encouragement for writers, teachers, musicians, painters, and those who were interested in science. Throughout the eighteenth century, it pioneered in more and more ways. For science, one of the most important foundations of these years was the college which opened its doors in 1751, the first colonial college not under the control of any one religious sect. The following year, the Pennsylvania Hospital began to receive patients as the first general hospital in the colonies. Both of these institutions acquired libraries, museums, and collections of apparatus of their own.

John Bartram's suggestion of a scientific academy came just as Philadelphia was entering this period of flowering. Very significantly, the natural history circle in America was just beginning at this time too. Because of both of these developments, Bartram was not defeated by Collinson's lack of encouragement. He found a receptive audience in America.

Benjamin Franklin was approaching a turning point in his career when John Bartram began to interest him in his ideas. Bartram said that Franklin was the only printer in the city who had ever made a success

28. Adolph B. Benson (ed.), *Peter Kalm's Travels in North America* (New York, 1937), II, 638; William E. Foster, "Some Rhode Island Contributions to the Intellectual Life of the Last Century," American Antiquarian Society, *Proceedings*, 8 (1892), 109-10.

29. *South-Carolina Gazette*, April 23, 1750; *The Rules and By-Laws of the Charlestown Library Society* (3d ed., Charlestown, 1770), 4-6.

30. Carl and Jessica Bridenbaugh, *Rebels and Gentlemen* (New York, 1942), 230.

of that trade, but however that may be, Franklin's success was unquestionable.[31] In 1743, he began to plan for ultimate retirement by inviting David Hall to come over from Edinburgh with the idea of making him a partner if the arrangement worked out. He did not make his famous withdrawal from the business until 1748 but that was no sudden decision. By that time, his energies were turning in many directions at once: to the organization of fire companies, to interest in the Masons, to education, and more significantly to science. Franklin's interest in learning was an essential part of him; he once said that he could not remember a time when he was unable to read. His Junto of 1727 had been only a mutual improvement club but one of the questions asked of new members had been, "Do you love truth for truth's sake?"[32] The Library Company had been even more clearly an effort to further learning. Franklin's promotion of the subscription scheme of 1742 in favor of Bartram's explorations had shown the scientific direction in which his thoughts were steadily turning.

Franklin always displayed the warmest affection toward John Bartram but in 1742 the relationship between the two men was almost the reverse of what it became after the Philadelphia printer had attained world renown. In 1742, John Bartram was a recognized member of the natural history circle who corresponded not only with Peter Collinson but also with Gronovius, Dillenius, Catesby, and Sir Hans Sloane, president of the Royal Society, as well as with several American naturalists. Franklin knew at firsthand the value Collinson placed upon Bartram's work. To Franklin, Bartram represented attainment in an area which was increasingly claiming his own attention. When his friend talked of his European correspondence and of such Americans as Colden, Clayton, and Mitchell, whose work was applauded by the Europeans and who already constituted the beginning of an American scientific community, Franklin listened carefully. Together the two men tried to plan a society which could be built upon the American natural history circle.

Franklin knew little of this group except through Bartram, but he did know Philadelphia and it was clear that any scientific society must have a nucleus at some one point where the efforts of the corresponding

31. Bartram to Collinson, May 30, 1756, Darlington, *Memorials*, 207.
32. "Rules for the Junto," Albert Henry Smyth (ed.), *The Writings of Benjamin Franklin* (New York, 1905-7), II, 90.

naturalists throughout the colonies could be centered. There were several men in the city who were ready to support such an effort, the most important being Thomas Bond, who was personally acquainted with some of the leading European naturalists as a result of his medical studies abroad. The others were men with whom Franklin had been associated in the Junto and in the management of the Library Company. There were Thomas Godfrey, who had won recognition from the Royal Society for his quadrant; Phineas Bond, who had studied in Europe like his brother; Thomas Hopkinson, an Oxford man; Samuel Rhoads, a successful builder; William Parsons, a surveyor; and William Coleman, a merchant.[33]

Another Philadelphian belonged in this group, too. Franklin and Bartram were ready to place the name of James Logan at the head of their list for Logan was undoubtedly the most distinguished scientist in the area. Yet, despite his own scientific accomplishments, despite his wide European acquaintance, and despite his means and past examples of patronage, he refused to countenance the new learned society. A Quaker aristocrat of a touchy temperament, James Logan had already discouraged Bartram and Colden from depending upon him.[34] His defection was a serious loss but in Bartram's stalwart words, the projectors "resolved that his not favoring the design would not hinder our attempt and if he would not go along with us we would Jog along without him." [35]

In 1743 Franklin devoted much thought to the promotion of knowledge. It was in that year that he drew up a proposal, which he did not publish, for the formation of an academy in Philadelphia. He did publish, however, his *Proposal for Promoting Useful Knowledge among the British Plantations in America* urging the establishment of an American learned society. Although it was issued from Franklin's press and printed in broadside form over Franklin's signature, Bartram called it "our Proposals." There are many indications that it was in fact a joint product.[36] It was clearly written with the natural history circle in mind. It appeared to be a full implementation of the sugges-

33. Franklin to Colden, April 5, 1744, Smyth (ed.), *Writings of Franklin*, II, 276-77.
34. Alice M. Keys, *Cadwallader Colden* (New York, 1906), 6.
35. Bartram to Colden, April 29, 1744, Gratz Collection, Historical Society of Pennsylvania.
36. Bartram to Collinson, April 29, 1744, *ibid.*

tion Colden had made to Douglass some fifteen years earlier. The announced purpose was identical: "That one Society be formed of *virtuosi* or ingenious men residing in the several Colonies, to be called *The American Philosophical Society* who are to maintain a constant correspondence." [37]

The enumeration of subjects to which the society would devote its attention revealed the basis of the whole plan. First mention was made of botany and medicine: "All new discovered plants, herbs, trees, roots, &c, their virtues, uses, &c.; methods of propagating them, and making such as are useful, but particular to some plantations, more general, improvements of vegetable juices, as cyders, wines, &c.; new methods of curing or preventing diseases." [38] These were exactly the things that interested the natural history circle in general and John Bartram in particular. The proposal also mentioned geology, geography, agriculture, mathematics, chemistry, and the crafts but omitted entirely any mention of physics or astronomy and showed little understanding in the suggestions relating to the physical sciences. It was intended to cover all knowledge, "all philosophical experiments that let light into the nature of things, tend to increase the power of man over matter, and multiply the conveniences or pleasures of life," but it was formulated from the viewpoint of a naturalist.

The Philadelphians who associated themselves with the enterprise were expected to serve the essential role of receiving scientific papers from distant members, considering them, and redirecting them to other members. Each resident member was given a particular post, Bartram being the botanist and Thomas Bond the physician while Franklin was simply the secretary. These three were the effective leaders of the group.[39] They counted upon support from the "many in every province in circumstances that set them at ease, and afford[ed them] leisure to cultivate the finer arts and improve the common stock of knowledge." [40] Members were to bear their own expenses without any plan to seek public or private aid except that Franklin agreed to frank all of their correspondence. Each member would pay for a quarterly abstract of society papers. The basis of support would be individual means and leisure just as it was in the natural history circle.

37. Smyth (ed.), *Writings of Franklin*, II, 228-32.
38. *Ibid.*, 229-30.
39. Bartram to Colden, Oct. 4, 1745, *Colden Papers*, III, 160.
40. Smyth (ed.), *Writings of Franklin*, II, 228.

At first, the society grew very well. During the summer of 1743, Franklin, meeting Cadwallader Colden for the first time, won his support.[41] Both Colden and John Mitchell visited Philadelphia in 1744, after their election to membership. John Clayton was at least informed of the society.[42] In New York and New Jersey, a surprising number of men proved eager to accept membership. James Alexander and Chief Justice Robert Hunter Morris, who a little later was elected to fellowship in the Royal Society, were among this group, although most of the men were political figures of wealth and position without notable scientific interests.[43] Franklin reported that there were "a Number of others in Virginia, Maryland, Carolina, and the New England Colonies, who we expect to join us, as soon as they are acquainted that the Society has begun to form itself."[44] Despite this promise, Bartram refused to mention it to any of his European correspondents "for fear it should turn out poorly."[45]

When the Europeans did find out about it from Colden and from Clayton, they all became anxious to know how it was progressing.[46] Colden was noncomittal in his report, declaring, "I cannot tell what expectations to give you of that undertaking," but Collinson immediately became enthusiastic.[47] His tepid view of Bartram's first suggestion was forgotten when it appeared that a society had actually been formed. "I can't enough commend the Authors and promoters of a Society for Improvement of Natural knowledge," he wrote, "Because it will be a Means of uniteing Ingenious Men of all Societies together and a Mutual Harmony be got . . . I expect Something New from your New World, our Old World as it were Exhausted."[48] To Peter Collinson, the advantages to be anticipated from such a society lay within

41. Franklin to Colden, Nov. 4, 1743, *Colden Papers*, III, 34; see also Carl Van Doren, "The Beginnings of the American Philosophical Society," American Philosophical Society, *Proceedings*, 87 (1944), 280.

42. Bartram to Colden, Nov. 2, 1744, *Colden Papers*, III, 79; [Thomas Penn to ?], n.d., Library Company of Philadelphia.

43. Alexander to Colden, Nov. 12, 1744, *Colden Papers*, III, 82; Franklin to Colden, April 5, 1744, Smyth (ed.), *Writings of Franklin*, II, 277.

44. *Ibid.*

45. Bartram to Colden, Oct. 4, 1745, Gratz Collection.

46. Gronovius to Bartram, June 2, 1746, Darlington, *Memorials*, 357; Collinson to Colden, Aug. 23, 1744, *Colden Papers*, III, 69.

47. Colden to Collinson, June 1744, *Colden Papers*, III, 61.

48. Collinson to Colden, Aug. 23, 1744, *Colden Papers*, III, 69.

the realm of natural history. He repeated what he had often said before, "I certainly think [they] cannot Labour Long when Such wonders are all around them Ready brought forth to their hands and to Which Wee are great strangers Butt because you See them Every Day they are thought Common and not worth Notice." [49]

Unexpected support was given the struggling group by one of those travelling lecturers who were so much a part of the colonial scene, a Dr. Adam Spencer of Edinburgh. Spencer was not a quack, like many of the itinerants; he came with recommendations from Dr. Richard Mead and other prominent London physicians as "a most judicious and experienced Physician and man midwife." [50] He was working his way down the coast from Boston to the West Indies when he reached Philadelphia. There, Franklin became so interested in his demonstrations that he agreed to act as his agent. [51] To Dr. Thomas Cadwalader he lent important assistance in the preparation of his *Essay on the West-India Dry-Gripes*, a pamphlet which inaugurated medical publication in Philadelphia with a commendable treatment of several cases of lead poisoning caused by distilling rum through lead pipes. Spencer's lectures on physiology and the diseases of the eye proved popular but he found that the greatest response followed his lectures on electricity. After attending one of the meetings of the Philosophical Society he extended his encouragement to the Philadelphia group. He even offered to promote the society as well as he could by taking copies of Franklin's *Proposal* with him to the West Indies. [52] Nearly a year later he was still thinking about the society when he talked with Dr. John Mitchell on his way through Virginia. [53]

Encouragement was offered from many quarters, but it was not sufficient to sustain the society. After his visit to Philadelphia in the fall of 1744, Colden wrote Franklin in concern lest the whole project be abandoned. [54] The following summer, Franklin was still not ready to admit that it had collapsed though he did declare, "The Members of our Society here are very idle Gentlemen; they will take no

49. Collinson to Colden, April 26, 1745, *Colden Papers*, III, 114.
50. [Thomas Cadwalader], *Essay on the West-India Dry-Gripes* (Philadelphia, 1745), v.
51. *Pennsylvania Gazette*, April 26, 1744.
52. Bartram to Colden, April 29, 1744, Gratz Collection.
53. Mitchell to Colden, Sept. 10, 1745, *Colden Papers*, VIII, 321.
54. Colden to Franklin, Dec. 1744, *Colden Papers*, III, 93.

Pains." [55] Thomas Bond, however, said that Franklin was at fault. Out of this bickering came a resolution by Bartram, Bond, and Franklin to revive the society and make it effective. John Bartram felt this might easily be done, "if we could but exchange the time that is spent in the Club, Chess and Coffee House for the Curious amusements of natural observations." [56] The real trouble was that there had never been an integrated scientific community in Philadelphia and the society failed to create one. It was because of inadequate local support that the noble aspirations for an intercolonial learned society were dissipated.

Benjamin Franklin was the only one who did not give up. Upon Cadwallader Colden's suggestion, he decided to publish an American Philosophical Miscellany. This would be a scientific journal issued monthly or quarterly and it was hoped that it might "in time produce a Society as proposed by giving men of Learning or Genius some knowledge of one another." [57] By taking upon himself the whole job of compiling and by permitting anonymous papers to appear in it, Franklin would not jeopardize the reputation of anyone. After the Pennsylvania Assembly adjourned in 1746, he was able to get together enough papers to make up five or six numbers of the miscellany. Many of them had been submitted to the society while it was attempting to carry out its announced aims; others Franklin obtained independently. Most of them have since been lost although copies of some and hints about others remain—enough to reveal the general character of the collection. [58]

The most important known papers came from or through members of the natural history circle. John Mitchell submitted to the society an essay on yellow fever which was later given to Benjamin Rush by Franklin and became the basis of his famous "cure" in the yellow fever epidemic of 1793. When Franklin asked Mitchell for permission to publish this account in his miscellany, Mitchell refused because he said it was not complete enough. [59] He may have been more ready to see his

55. Franklin to Colden, Aug. 15, 1745, *Colden Papers*, III, 143.
56. Bartram to Colden, Oct. 4, 1745, *Colden Papers*, III, 160.
57. Colden to Franklin, Dec. 1744, *Colden Papers*, III, 93-94; Franklin to Colden, Aug. 15, 1745, Nov. 28, 1745, *ibid.*, 143, 182.
58. Franklin to Colden, Oct. 16, 1746, *Colden Papers*, III, 275. As late as 1768, William Franklin believed his father still held the papers: William Franklin to Cadwalader Evans, Jan. 28, 1768, Franklin Papers, XLVII, 43, American Philosophical Society.
59. Mitchell to Franklin, Sept. 12, 1745, *Colden Papers*, III, 152-53.

essay on the pines of Virginia published, however, and another which he had ready for the society in September 1745.[60] Cadwallader Colden wrote one medical paper himself which he sent to Franklin and he forwarded another written by Dr. Evan Jones on the effects and treatment of rattlesnake bites.[61] Joseph Breintnall also wrote a letter on a rattlesnake bite which he may have offered to Franklin before he sent it to the Royal Society for publication in the *Philosophical Transactions* of 1746. Dr. Isaac DuBois wrote a paper on an epidemic fever about this time which found its way to Colden's hands. It may well have been intended for the society or for the miscellany.[62] A description of a newly invented wooden cannon was sent to Franklin from an undisclosed source.[63] Franklin himself worked out a theory to account for the shorter time consumed in crossing the Atlantic from America to Europe than from Europe to America, a theory which Colden offered to support if he wanted to publish it in his miscellany.[64] Fortunately this was not published, for his solution, involving the rotation of the earth, was incorrect. He kept working on the problem until he did find the correct explanation—the gulf stream.

The proposed American Philosophical Miscellany never appeared. The enthusiasm that had led to the organization of the Philosophical Society had expended itself, leaving insufficient energy to establish this sort of journal. The dream of an American learned society could not be fulfilled until other factors had been added to the environment. When Peter Kalm visited Philadelphia in 1748, he was told that the impact of the War of Austrian Succession on the colonies had killed the American Philosophical Society.[65] That judgment was not fair. It is true that the declaration of war was published in Pennsylvania in July 1744 but the province raised no troops until 1746, and Franklin did not become involved in the preparations until the middle of 1747. By that time both the society and the miscellany were quite dead.

The American Philosophical Society had failed even though it had been based in part upon an already existing intercolonial scientific

60. Bartram to Colden, [1745], *Colden Papers*, III, 180.
61. Evan Jones to Colden, July 17, 1744, *Colden Papers*, III, 65; Franklin to Colden, Aug. 15, 1745, *ibid.*, 139-43.
62. Isaac DuBois to Colden, [1745?], Colden Mss.
63. Franklin to Colden, Oct. 16, 1746, *Colden Papers*, III, 276.
64. Colden note on back of letter from Franklin, 1746, *Colden Papers*, III, 187.
65. Benson (ed.), *Kalm's Travels*, I, 31.

group—the natural history circle. It had failed because its local base had been too feeble and limited. It had not made use of the nonscientific but "curious" elements of the city's population. Merchants and gentlemen, in particular, had not been brought into the society as they were brought into another project of scientific interest which was launched in 1753 and 1754. In those years, two expeditions were sent from Philadelphia by the North West Company in search of the northwest passage. Franklin reported that the subscription for the first voyage in the sixty-ton schooner *Argo* amounted to £1300.[66] Not only were Philadelphia merchants involved, but some even from Maryland, New York, and Boston.[67] They were ready to contribute generously because of the great possibilities of profit that lay in finding the passage, but at the same time they realized the scientific importance of the exploration. Although both expeditions failed in their objective, they did bring many leading Philadelphians into direct contact with an effort to advance knowledge. Some of the Eskimo utensils brought back by Captain Charles Swaine were presented by the North West Company to the Library Company of Philadelphia to add to its collection of curiosities.[68] The American Philosophical Society had not succeeded similarly in enlisting the support of the mercantile community—a group which in England formed one of the most important sources of support of the Royal Society.

As a matter of fact, before Franklin became interested in the northwest passage and even before the war claimed his attention he had already turned away from the abortive society to a new enthusiasm. His developing interest in science was first attracted and then absorbed in the study of electricity which was then becoming something of a popular fad in Europe. Peter Collinson recognized it as another wave of interest in a scientific wonder, reporting, the "Phenomena of the Polypus Entertained the Curious for a year or two past but Now the

66. Franklin to Colden, Feb. 28, 1753, *Colden Papers*, IV, 373; Franklin to Jared Eliot, April 12, 1753, Smyth (ed.), *Writings of Franklin*, III, 123.

67. *Pennsylvania Gazette*, Nov. 15, 1753, Nov. 29, 1753, Oct. 24, 1754; Edwin S. Balch, "Arctic Expeditions Sent from the American Colonies," *Pennsylvania Magazine of History and Biography*, 31 (1907), 420; Howard N. Eavenson, *Map Makers and Indian Traders* (Pittsburgh, 1949), 46-51.

68. *The Charter, Laws, and Catalogue of Books of the Library Company of Philadelphia* (Philadelphia, 1770), 4.

Vertuosi of Europe are taken up In Electrical Experiments." [69] The people of Europe and America began to read articles on the subject in their magazines and to see advertisements of lecturers who drew sparks from subjects' eyes and gave electrical shocks to their patrons. Half parlor magic and half academic study, it offered a rare opportunity for scientific achievement to those who held truth as their goal, even though they might not have had the formal mathematical training required for astronomical studies or the years of botanical observation needed to permit the recognition of new genera.

It was probably the itinerant lecturer, Adam Spencer, who gave Franklin his first knowledge of electricity when the Philadelphian visited Boston in the summer of 1743. The following spring, Spencer provided more information while he was lecturing in Philadelphia.[70] In 1746, Franklin saw William Claggett in Newport and talked to that colonial lecturer about electricity.[71] If he read the *American Magazine* in 1745 or 1746, he ran across accounts of French and German electrical experiments, and he could have found electrical papers in the *Philosophical Transactions* even earlier.[72] Electricity was in the air and no alert man could have avoided knowing about it. Adam Spencer was, at best, only one source of information.

Peter Collinson was much more. Sometime before 1747, he sent to the Library Company of Philadelphia a glass tube for conducting electrical experiments and an account of German experiments in electricity with directions for repeating them.[73] These things arrived at an opportune moment—just after Franklin had seen the philosophical society collapse and the philosophical miscellany prove unrealistic. His one scientific publication of these years, *An Account of the New-Invented Pennsylvanian Fireplace*, described an important invention and was well

69. Collinson to Colden, March 30, 1745, *Colden Papers*, III, 109-10.

70. Franklin, "Autobiography," Smyth (ed.), *Writings of Franklin*, I, 417; I. Bernard Cohen (ed.), *Benjamin Franklin's Experiments* (Cambridge, 1941), 49-50.

71. Carl and Jessica Bridenbaugh, *Rebels and Gentlemen* (New York, 1942), 324.

72. *American Magazine and Historical Chronicle*, 2 (1745), 530; 3 (1746), 461-64; *Philosophical Transactions*, 41 (1739-40), 98-125, 209-10, 634-40, 661-67; 42 (1742-43), 14-18, 140-43; 43 (1744-45), 239-49, 290-92; 307-15, 419-21, 481-501.

73. Franklin to Collinson, March 28, 1747, Cohen (ed.), *Franklin's Experiments*, 169; see also N. H. de V. Heathcote, "Franklin's Introduction to Electricity," *Isis*, 46 (1955), 29-35.

received, but it was not enough to satisfy his growing appetite for scientific work.[74] Confronted with Collinson's donations, Franklin threw himself into the study of electricity to a point where it "totally engrossed" his attention as it had never been engaged before.[75] The "Philadelphia Experiments" followed.

Several Philadelphians shared Franklin's enthusiasm and explored with him the possibilities of electricity. His three most important associates were Thomas Hopkinson, who had been designated president of the American Philosophical Society; Philip Syng, a silversmith and Junto crony; and Ebenezer Kinnersley. Kinnersley, an unemployed Baptist clergyman, became the most original and important of the men behind Franklin. He worked out significant experiments himself describing them in letters which were included in later editions of Franklin's *Experiments and Observations* and in the *Philosophical Transactions*. He lectured successfully on electricity in several American towns.[76] Franklin, in his own letters, made careful reference to the aid he received from these associates, but as one reviewer aptly put it, "These acknowledgements . . . are neither numerous or important enough to produce any considerable diminution of the Author's fame as a philosopher." [77]

Franklin's work also demonstrated his continuing relationship with the natural history circle. One of his published letters was addressed to John Mitchell, now in England, another to Cadwallader Colden, and a third to John Lining.[78] Mitchell did only a little work on electricity, but Lining came to devote to this study all the time he could spare from his indigo planting.[79] Colden showed interest for a time, and his son, David, did some worthwhile experimental work.[80] Some years

74. Smyth (ed.), *Writings of Franklin*, II, 246-76.
75. Franklin to Collinson, March 28, 1747, Cohen (ed.), *Franklin's Experiments*, 170.
76. *Philosophical Transactions*, 53 (1763), 84-97; 63 (1773), 38-39; Cohen (ed.), *Franklin's Experiments*, 58-60, 348-58, 409-21; *American Magazine and Monthly Chronicle*, 1 (1757-58), 164-66.
77. Quoted by Cohen (ed.), *Franklin's Experiments*, 125.
78. Franklin to Mitchell, April 29, 1749, *ibid.*, 201-11; Franklin to Colden, 1751, April 23, 1752, *ibid.*, 245-46, 323-27; Franklin to Lining, March 18, 1755, April 14, 1757, *ibid.*, 331-38, 339-45.
79. Garden to Colden, Jan. 14, 1755, *Colden Papers*, V, 3.
80. *American Magazine and Monthly Chronicle*, 1 (1757-58), 164-66; David Colden to Silvanus Americanus, March 9, 1759, Colden Mss.; David Colden to John Canton, June 30, 1759, Oct. 2, 1760, *ibid.*

later, Alexander Garden experimented with the electrical eel in a study which nicely combined electricity and natural history. John Bartram, however, would not be deflected from his botany. It was not that he was unresponsive but he told Colden, "I take this to be the most Surpriseing Phenomena that we have met with and is wholy incomprehensible to thy friend John Bartram." [81]

First and last, it was Peter Collinson, at the heart of the natural history circle, who was Franklin's most important support. He was even more essential in communicating Franklin's experiments to the world than he had been in initiating them. Collinson did not read Franklin's first letter containing experiments to the Royal Society, but to a small circle of friends including William Watson, one of the leading experimenters in England. Watson quoted extensively from it in a letter of his own which he did read to the Royal Society with the result that portions of Franklin's very first letter found their way into the *Philosophical Transactions*.[82] Collinson directed later letters to Edward Cave who put two of them in his *Gentleman's Magazine* and then, in 1751, published a collection of Franklin's letters with a preface by John Fothergill under the title, *Experiments and Observations on Electricity*.[83] It was Collinson who introduced Franklin to the world.

Franklin's book was the most important scientific contribution made by an American in the colonial period. In it, his ability to construct hypotheses that were often brilliant and to check them with adequate experiments was made clear for the first time. His most important accomplishments were in the realm of pure theory where he was able to show how his single-fluid concept fitted the observed conditions more satisfactorily than the two-fluid theory then widely accepted under the stimulus of Charles Dufay's writings.[84] Franklin also suggested the terminology that became so widely useful, calling one charge negative and the other positive—the presumption being that a deficiency of electric fluid caused the negative state. According to Franklin, the flow of current was from positive to negative and although modern knowledge has revealed that the electrons flow in the opposite direction, the

81. *Philosophical Transactions*, 65 (1775), 102-10; Bartram to Colden, March 6, 1746/47, *Colden Papers*, III, 363.
82. *Philosophical Transactions*, 45 (1748), 98-100.
83. *Gentleman's Magazine*, 20 (1750), 34-35, 208; Cohen (ed.), *Franklin's Experiments*, 84-90.
84. Cohen (ed.), *Franklin's Experiments*, 41-43.

Franklinian terminology and direction of flow are still used by electricians. At many points, notably when confronted with an example of induction, Franklin showed the ability to grasp the essential nature of the phenomenon and to fit it into his conceptual scheme.[85]

Obscuring his more basic conclusions were the dramatic suggestions Franklin made concerning lightning. First, he suggested that lightning was an electrical phenomenon—not a new idea but now based upon a careful comparison of their properties. Then he devised two experiments to test this hypothesis. The guard house experiment, in which a man stood inside an insulated house on top of a steeple surmounted by a lightning rod from which he could draw off electricity, was performed first in France, but Franklin performed his more famous kite experiment before anyone else.[86] The hypothesis was proven and the way cleared for the great utilitarian accomplishment of his study—the lightning rod. Franklin was led to the suggestion of the lightning rod by his study of the electrical effect of points. It is not clear whether he actually used one before he made the suggestion in his almanac, but at any rate, the idea was quickly taken up.[87] Lightning rods soon sprouted in London and Paris as well as in Philadelphia until they became Franklin's greatest advertisement.

The response that met Franklin's work was overwhelming. His book passed through five English and three French editions, and was translated into German and Italian as well.[88] In England, the Royal Society awarded him its Copley Medal and admitted him to fellowship without requiring him to request the election, sign the register, or pay the usual fees—a rare honor. In France, the King directed his own "Thanks and Compliments in an express Manner to *Mr. Franklin of Pennsylvania*" while the Académie des Sciences elected him one of its eight associés étrangers.[89] In America, Harvard, Yale, and William and Mary granted him honorary degrees. Franklin became a world celebrity.

His work in electricity was significant but it was not enough to account for the reputation he won—a reputation which seemed to transcend that of any of his contemporaries, European or American,

85. *Ibid.*, 98.
86. *Pennsylvania Gazette*, Oct. 19, 1752.
87. Carl Van Doren, *Benjamin Franklin* (New York, 1938), 160-62.
88. Cohen (ed.), *Franklin's Experiments*, 141-48.
89. Franklin to Jared Eliot, April 12, 1753, Smyth (ed.), *Writings of Franklin*, III, 124.

literary or political. His experiments were only the foundation of the great esteem he met. He had made his discoveries when electricity was a world-wide fad and while general interest was at a peak. His invention of the lightning rod was a spectacular justification of the faith of the eighteenth century that all knowledge was useful. The very fact that he was an American added to the charm of his achievement in an age which celebrated the virtues of the simple society and distrusted scholastic learning. His work in electricity, however, gave the world only its first view of the capacities of a genius. Franklin was then able to pass from the little world of Philadelphia to the great world of London and Paris, there to demonstrate the same breadth of interest, keenness of perception, and warmth of humanity that had distinguished his earlier performances. Indeed, Franklin improved with age until he became the very epitome of the learned philosopher.

Franklin's achievements were accepted as the achievements of an American and in him, for the first time, an American was admitted as a peer of the leading scientists of the western world. This was recognition of an altogether different character from that which had been accorded the natural history workers in America. They, too, had been honored, encouraged, and regarded with continuing esteem. Yet, much of their accomplishment had resulted from their great advantage of position. They had been field workers doing work that no one in Europe could do. Franklin had enjoyed no similar advantage from being in America. Presumably, he should have been able to carry through his experiments and make his speculations even more successfully in Europe where he would have had the advantages of the great libraries and the conversation of the leading thinkers. Instead, he had flowered in America—not as a species of wild life, but as a product of the cultural milieu of Philadelphia and an associate of the natural history circle. Franklin's work was not simply field work which could be fitted into some European's scheme of organization. He developed the theory himself from facts he had found himself and in concert with other Americans. This was a prodigy; an American scientist of the first magnitude! It is likely that some part of the great recognition Franklin received was a result of the understanding of that important fact.

Chapter Five

Teachers and Preachers

I n colonial America, the group of men who contributed to the advancement of natural history was not equally capable of contributing to that other great division of scientific knowledge, natural philosophy. Natural philosophy embraced what today might be called the physical sciences: chemistry, astronomy, and physics—especially mechanics. There was general agreement in the eighteenth century when the scientific publicist, Benjamin Martin, proclaimed: "*Natural Philosophy* is ... a Mystery that has been hid from Ages, and from Generations; but is now made manifest, to all Nations, by the divine Writings of the immortal Sir Isaac Newton." [1] Natural philosophy had little meaning apart from the Newtonian system which brought to one focus much of the most significant work of the whole scientific revolution. Because Newton and his predecessors had made such constant use of mathematics, it was necessary for all who studied natural philosophy to have a mathematical background. The naturalists worshipped at the shrine of Newton, too, and sought to use the key he provided to unlock the secrets of the three kingdoms of nature. Most of them, however, did not have the facility in mathematics or the knowledge of the almost scholastic physics necessary to advance natural philosophy. This was obvious in the case of the self-educated John Bartram, but it was strikingly true also of Cadwallader Colden who was better prepared and who made efforts to contribute to the advance of natural philoso-

1. B[enjamin] Martin, *A Panegyrick on the Newtonian Philosophy* (2d ed., London, 1754), 3.

phy. The custodians of natural philosophy in colonial America came generally from another group altogether.

It was among the teachers in the colleges that the best mathematicians, astronomers, and natural philosophers were found, for they devoted their lives to the study and teaching of those subjects. Their primary concern was the dissemination of knowledge, although some of them did seek to make creative contributions as well. The teachers offered rational, ordered courses in mathematics and natural philosophy but steered shy of similar courses in natural history.[2] It was not that they were disinterested in natural history. To them, it appeared of secondary importance.

On occasion, the academic group responded to the encouragement of the natural history circle and to the obvious advantages that Americans enjoyed in these pursuits. Francis Alison, Presbyterian divine and vice-provost of the College of Philadelphia, lamented, "Natural History is not regularly Studied in our Colleges." He urged that at Harvard and Yale, "the Spirit of making discoveries in Natural history . . . that now so much prevails in Europe" be excited.[3] John Winthrop, second Hollis Professor of Mathematics and Natural Philosophy at Harvard, sent botanical, zoological, and mineralogical specimens to the Royal Society before his appointment to the Harvard chair. After 1738, he might write a friend about the varieties of mankind, but he restricted his publications to astronomy and the physical sciences.[4] The best of the teachers retained a strong interest in natural history but none of them became naturalists.

No one would have admitted that there was any kind of separation of the naturalists from those who sought to advance natural philosophy. The members of the natural history circle stood ready to contribute to the advance of knowledge wherever their efforts promised to be helpful. Franklin quickly learned how much these men were able to help him and he displayed the same readiness to aid the mathematically-trained college teachers. Neither Franklin nor Collinson understood the processes of calculating a comet's orbit but their superior

2. Theodore Hornberger, *Scientific Thought in the American Colleges, 1638-1800* (Austin, 1945), 69.
3. Francis Alison to Ezra Stiles, Dec. 10, 1762, Stiles Papers, Yale University.
4. C. R. Weld, *History of the Royal Society* (London, 1848), I, 471; John Winthrop to Stiles, July 15, 1759, Isabel M. Calder (ed.), *Letters and Papers of Ezra Stiles* (New Haven, 1933), 5-8.

connections with European centers of learning could be a great advantage to those who did.

An effort to stimulate the academic community into action was made in 1753 in connection with the transit of Mercury of that year. This attempt was an accidental result of inadequate communications between France and her colonies in North America. The Marquis de la Galissonière was forced to forward instructions for observing the transit at Quebec through New York because that was the quickest route.[5] When La Galissonière's covering letters and a memorial on observation of the transit prepared by Joseph-Nicolas Delisle of the Académie des Sciences came into the hands of James Alexander, they opened important hopes in his mind for he was himself a competent astronomer.[6] He obtained permission to publish the French papers and sent translations to both Cadwallader Colden and Benjamin Franklin. From his reading of the *Abridgement of the Philosophical Transactions*, he understood the advantages anticipated by scientists from the transit Venus was to make across the sun's face in 1761. It would yield a better value for the distance from the earth to the sun. The transit of Mercury had no such use but Alexander viewed it as an excellent opportunity to train and prepare observers in several parts of America so that they would be capable of making competent observations of Venus in 1761. To Franklin, he wrote, "It Would be a great honour To our young Colledges in America if they forthwith prepared themselves with a proper apparatus for that Observation and made it. . . . I thought no person so proper as your self to think of the ways and means of persuading these Colledges to prepare themselves for takeing that Observation." [7]

Franklin had already been informed of the transit of Mercury before Alexander's letter arrived, and in his almanac for 1753 he had referred also to the significance of the coming transit of Venus.[8] His

5. [Marquis de la Galissonière to ?], Oct. 10, 1752, *The Letters and Papers of Cadwallader Colden*, IV, New-York Historical Society, *Collections for 1920*, 53 (1921), 346.

6. Alexander to Franklin, Jan. 29, 1753, *Colden Papers*, IV, 367-68; I. Bernard Cohen, "Benjamin Franklin and the Transit of Mercury in 1753," American Philosophical Society, *Proceedings*, 94 (1950), 222-32.

7. Alexander to Franklin, Jan. 29, 1753, *Colden Papers*, IV, 367-68.

8. Franklin to James Bowdoin, Feb. 28, 1753, Albert Henry Smyth (ed.), *The Writings of Benjamin Franklin* (New York, 1905-7), III, 122; [Franklin], *Poor Richard Improved...for the Year of Our Lord 1753* (Philadelphia, [1752]), 9.

first knowledge, however, was based upon the instructions distributed by the Royal Society for observing the transit of Mercury in America, but these were distinctly inferior to the French papers that Alexander forwarded. Specifically, the Royal Society's instructions erred in asserting that the beginning as well as the end of the transit would be visible.[9] Franklin had already thought of observing the transit with the three-foot reflecting telescope owned by the Academy of Philadelphia, but it was Alexander who suggested to him a wider effort.[10]

What he did was to print off copies of Alexander's translations and send them to his friends and correspondents in different parts of America with the hope of stimulating observations. He also printed an explanatory letter of Alexander in which the appeal was directed particularly to the colleges: "As there are now sundry Nurseries of Learning springing up in Pennsylvania, Jersey, New-York, Connecticut, and Boston, all Ways should be thought of, to induce each of those to provide a proper Apparatus for making such Observations, long before the Year 1761, that they may be expert at taking Observations of that Kind before the Transit happens." [11] He urged that each college prepare itself for the observations by ascertaining its own latitude and longitude and referred them to the *Abridgement of the Philosophical Transactions* for details about the apparatus required. If any college did not possess the abridgements, it "should, in the first Place get them." [12]

The response was disappointing. Peter Collinson, who had interested himself in the affair after he heard of Franklin's, Colden's, and Alexander's efforts, reported that the only American observations to reach London came from Antigua.[13] The Antigua report was inspired, at least, by Franklin's publication on the transit, six copies of which he had sent to his nephew, Benjamin Mecom, then on the island.[14]

9. [Franklin], *Poor Richard ... 1753*, 30-32; Alexander to Colden, April 19, 1753, *Colden Papers*, IV, 385.

10. Franklin to Bowdoin, Feb. 28, 1753, Smyth (ed.), *Writings of Franklin*, III, 122.

11. *Letters Relating to the Transit of Mercury over the Sun, which is to Happen May 6, 1753* [Philadelphia, 1753], [1].

12. *Ibid.*, [2].

13. Collinson to Colden, March 6, 1754, *Colden Papers*, IV, 433; Collinson to Franklin, Aug. 12, 1753, Franklin Papers, LXIX, 65, American Philosophical Society.

14. *Philosophical Transactions*, 48 (1753-54), 318-19; William Shewington to Colden, June 20, 1753, *Colden Papers*, IV, 393.

Alexander and a group of men in New York were "pretty well prepared" to observe the event, but they were able to do nothing because of cloudy skies.[15] Perhaps that is what happened elsewhere, for the apparatus and interest were certainly adequate in Philadelphia and in Cambridge where John Winthrop was the best qualified man in America to conduct such observations. Winthrop had observed the transits of Mercury which had occurred in 1740 and 1743, and both of his reports had been published in the *Philosophical Transactions*.[16] He did not observe the transit of 1753.

Although no observations were reported from continental America, the transit was significant in awakening the interest of Alexander, Franklin, Colden, and Collinson in an area which presented the Americans with some of the advantages they possessed so bountifully in the study of natural history. In most phases of natural philosophy and mathematics the Americans labored under difficulties because their intellectual environment was so inferior to that of Europe. In some descriptive astronomy, however, they enjoyed an advantage of position very similar to the advantage enjoyed in natural history. Where parallax was a factor in observing specific events within the solar system, observations by the Americans might have a value no European observations could duplicate. The colleges did not take advantage of this opportunity in 1753 but they remained the best qualified agencies for work of that sort.

The character of the colonial colleges was an important determinant of the nature of the science they could support. In all but one case they were religious foundations, yet not one was exclusively a divinity school. Rector Thomas Clap's assertion that at Yale it was the principal design "to educate Persons for the Work of the Ministry" might have been said of most of the others.[17] At Harvard, the highest function remained the training of graduate students in theology; even at the nominally non-denominational College of Philadelphia, a clerical administration directed the school.[18] Where there was a conflict about the religious nature of the college, as at King's College and the College

15. Alexander to Colden, May 10, 1753, *Colden Papers*, IV, 388.
16. *Philosophical Transactions*, 42 (1742-43), 572-78; 59 (1769), 505-6.
17. Thomas Clap, *The Annals or the History of Yale College* (New Haven, 1766), 84.
18. Samuel E. Morison, *Harvard College in the Seventeenth Century* (Cambridge, 1936), I, 272.

of Rhode Island, the dispute revolved about the effort of one denomination to gain control.[19] In even the largest of the college libraries, theological titles predominated.[20] Everywhere, however, the religious mission became less marked in the second half of the eighteenth century. Even the radical "new light" College of New Jersey moved steadily in the direction of secularization and broader service to the community.[21] The clerical presidents of each of the colleges and the several clerical teachers were devoted to a comprehensive education that willingly embraced the sciences. Provost William Smith argued strenuously in favor of the proposition that "the interests of Christianity will be advanced, by promoting the interests of Science." [22] From the ranks of the clergy—particularly those associated with the colleges —many of the men best able to contribute to the physical sciences were drawn.

From the beginning, the colleges had been designed for "training up men that will be useful in the other learned professions—ornaments of the State as well as the Church." [23] Each college was a community institution, serving a recognizable region and dependent upon that region for its support. The earliest colleges, Harvard, Yale, and William and Mary, were able to rely upon substantial provincial support. The newer ones had to count more upon voluntary giving, two-thirds of this income coming from the locality of the college.[24] The communities had much to do with forming the character of the colleges; those at Philadelphia, New York, and Providence demonstrated a broader tolerance which mirrored the cosmopolitanism of the communities that supported them. As the communities supported the colleges, they ex-

19. Herbert and Carol Schneider (eds.), *Samuel Johnson . . . His Career and Writings* (New York, 1929), IV, 117-214; Walter C. Bronson, *The History of Brown University* (Providence, 1914), 9-33.

20. Louis Shores, *Origins of the American College Library, 1638-1800* (New York, [1935]), 80-82, 86.

21. Francis L. Broderick, "Pulpit, Physics, and Politics: The Curriculum of the College of New Jersey, 1746-1794," *William and Mary Quarterly*, 3d ser., 6 (1949), 57.

22. William Smith, *Discourses on Public Occasions in America* (2d ed., London, 1762), 119.

23. Quoted by Broderick, "Pulpit, Physics, and Politics," *William and Mary Quarterly*, 3d ser., 6 (1949), 56-57.

24. Beverly McAnear, "The Raising of Funds by the Colonial Colleges," *Mississippi Valley Historical Review*, 38 (1951-52), 592, 611; Shores, *Origins of the American College Library*, 4.

pected them to educate the youth of the region for leadership in business, government, and teaching as well as in theology.

The character of the colleges depended very heavily also upon selective European influences. Although Harvard was in the first instance patterned after one of the colleges of Cambridge University, it was the influence of the Scottish universities and the English dissenting academies that became paramount there as well as at the other American colleges in the eighteenth century. The strong dissenting character of the American population and the fact that the English universities were barred to dissenters laid the basis for this condition. It was further strengthened by the patronage extended to America by prominent English dissenters and by the disposition of Scottish teachers to emigrate to the colonies in company with the Scottish physicians. Even in the two Anglican colleges in the colonies, the best teaching of mathematics and natural philosophy was done by Scottish professors while the Anglican priest who presided over the one nonsectarian college had himself been educated in Scotland.

The Scottish and dissenting influence had happy results for the promotion of science in America. The English universities had made progress in incorporating the new science into their curriculum, but they still offered a predominantly literary education. Most of the scientific advance of the eighteenth century was made outside their walls. The English dissenting academies, on the other hand, offered university grade work that was often superior to that of the English universities, especially in science. Both the English dissenting academies and the Scottish universities paid more attention to the advance of science and housed more creative scientists on their staffs. They were more favorable to the use of English and the modern languages. They were more concerned about the relationship of learning to life.[25]

The character of both domestic and European support was made dramatically clear for Harvard when Harvard Hall burned in 1764 and with it very nearly the whole of the library and the scientific equipment that had been accumulated through the years. The General Court quickly demonstrated the relationship between college and province by providing funds to reconstruct the building. Private individuals and groups of men responded to appeals for aid in replacing the

25. See Irene Parker, *Dissenting Academies in England* (Cambridge, 1914), 76, 77, 108-9, 112, 132.

books and apparatus in a remarkable way. Such wealthy Bostonians as James Bowdoin and Colonel John Hancock gave generously, as might have been expected. So did less affluent men and men of other provinces. New Hampshire contributed some 700 books in recognition of the meaning of Harvard to the people of that region. Benjamin Franklin gave books, instruments, and cash in addition to his personal help in purchasing the new equipment. From England came other gifts, the most important being the £200 from Thomas Hollis, enthusiastic English dissenter, and the 1,200 books he presented. The loss was repaired with surprising speed because so much interest was shown not only by the provincial legislature and the city of Boston but by nearby provinces and by the dissenting circles in England.[26]

Harvard's ability to train its own teachers of mathematics and natural philosophy was clear indication of a higher level of intellectual development than the other colleges had attained. The Harvard teachers were more prolific and probably contributed more to the advance of science in this period than all the other teachers combined. This tradition was begun by Thomas Robie, Harvard tutor from 1712 to 1723, who developed a correspondence particularly in astronomy and mathematics "with Mr. [William] Derham and other learned persons in those Studies abroad."[27] Robie sent astronomical observations as well as miscellaneous accounts of combustion products, inoculation, and spiders' venom to the Royal Society; he published a pamphlet describing an aurora borealis; and in the newspapers he printed articles on the annular eclipse of 1722.[28] His student, Isaac Greenwood, went to London in 1723 to improve his education under Jean T. Desaguliers and other exponents of the Newtonian system. While in England, he was introduced to many dissenters including Thomas Hollis, who had

26. Francis A. Foster, "Burning of Harvard Hall, 1764 and its Consequences," Colonial Society of Massachusetts, *Publications*, 14 (1911-13), 8, 9; William C. Lane, "New Hampshire's Part in Restoring the Library and Apparatus of Harvard College," *ibid.*, 25 (1922-24), 24, 27; Franklin B. Dexter (ed.), *Extracts from the Itineraries and other Miscellanies of Ezra Stiles* (New Haven, 1916), 206.

27. Quoted by Frederick G. Kilgour, "Rise of Scientific Activity in Colonial New England," *Yale Journal of Biology and Medicine*, 22 (1949), 135.

28. Frederick G. Kilgour, "Thomas Robie," *Isis*, 30 (1939), 473-90; Thomas Robie, *A Letter to a Certain Gentleman Desiring a Particular Account May be Given of a Wonderful Meteor* (Boston, 1719); *Philosophical Transactions*, 31 (1720-21), 121-24; 33 (1724-25), 67-70.

already founded the Hollis professorship of divinity at Harvard and who now decided to complete plans for founding a professorship of mathematics and natural philosophy with Greenwood specifically in mind as the first occupant of the chair.[29] After being established in that dignity, Greenwood published a small pamphlet describing his *Experimental Course of Mechanical Philosophy* and he published the first arithmetic text written by an American. He, too, sent observations and speculations on astronomy and the physical sciences to the Royal Society, some of which were published in the *Philosophical Transactions.*[30] Greenwood made a good beginning and, had his intemperance not caused his discharge in 1738, he might have contributed much more.

Colonial Harvard's finest flower in the field of science was Greenwood's successor, John Winthrop. Winthrop was similarly born and bred in Massachusetts and received all his education there without any visits to Europe. At twenty-three, after an examination in the sciences but pointedly without any examination of "his Principles of Religion," Winthrop was given his old teacher's chair without hesitation.[31] He became a great teacher and a creative scientist—one of the few to appear in natural philosophy in the colonial period. As his first recorded scientific writing was on sunspots, so astronomy continued to be the field in which he most delighted although he wrote on earthquakes, weather, and mathematics as well. He published occasional lectures and pamphlets and eleven of his papers were printed in the *Philosophical Transactions.* His writings were distinguished for their clarity and sweep as well as for their occasional flashes of brilliant insight. Alone among the teachers, Winthrop was elected to the Royal Society. He was twice offered and twice he refused the presidency of Harvard and he was suggested for the presidency of the College of Philadelphia.[32]

29. I. Bernard Cohen, *Some Early Tools of American Science* (Cambridge, 1950), 31-32.

30. *Philosophical Transactions,* 37 (1731-32), 55-69; 35 (1727-28), 390-402; 36 (1729-30), 184-91.

31. Hollis Book, VI, 43, Harvard University Archives.

32. Schneider (eds.), *Samuel Johnson,* I, 155; Josiah Quincy, *History of Harvard University* (Cambridge, 1840), II, 149-50, 161; *Philosophical Transactions,* 42 (1742-43), 572-78; 50 (1757), 1-18; 52 (1762), 6-16; 54 (1764), 185-88, 277-83; 57 (1767), 132-54; 59 (1769), 351-58, 505-6; 60 (1770), 358-62; 61 (1771), 51-52; 64 (1774), 153-57; Frederick G. Kilgour, "Professor John Winthrop's Notes on Sun Spot Observations," *Isis,* 29 (1938), 359.

Ezra Stiles reflected his incredible reputation when he declared, "In Math. and nat. Phil. I believe he had not his equal in Europe." [33]

Yale had been quite undistinguished in its attention to science before the Reverend Thomas Clap became rector in 1739. Clap, who had been a student at Harvard in Thomas Robie's day, brought with him great interest in science and a devotion to the Newtonian system. He brought in tutors with similar interests in science, the most notable being Ezra Stiles. With such Englishmen as William Whiston and Peter Collinson, Clap developed a correspondence in the cause of science. [34] His own particular concern was with the study of meteors upon which he published a pamphlet and several newspaper articles and sent communications to the Royal Society. To his fellows Clap appeared a scientist of real merit, although he was not of the same stature as Winthrop. [35]

Neither the College of New Jersey nor the College of Rhode Island made permanent provisions for the teaching of science until quite late. Scientific apparatus was provided at New Jersey and occasional lectures offered but it was not until 1768 that a chair of mathematics and natural philosophy was established. Even then, it was not filled until 1771 when William Churchill Houston, a Scot, was placed in it. After the Reverend John Witherspoon assumed the presidency of the institution in 1768, the Scottish Enlightenment with the high value it placed on science was transplanted to New Jersey. [36] At Rhode Island, President James Manning showed his interest in science by making the first tutorial appointment in 1767 in mathematics and natural philosophy. The post went to David Howell, a New Jersey graduate, who was raised to the rank of professor two years later. [37]

Earlier progress was made at King's College in New York which drew its first president, the Reverend Samuel Johnson, from Yale.

33. Franklin B. Dexter (ed.), *The Literary Diary of Ezra Stiles* (New Haven, 1901), II, 334.

34. Clap to William Whiston, July 1, 1752, Stiles Papers; Clap to Stiles, June 28, 1768, *ibid*.

35. Naphtali Daggett, *A Sermon Occasioned by the Death of the Reverend Thomas Clap* (New Haven, [1767]), 32; Thomas Clap, *Conjectures upon the Nature and Motion of Meteors, which are above the Atmosphere* (Norwich, 1781).

36. Broderick, "Pulpit, Physics, and Politics," *William and Mary Quarterly*, 3d ser., 6 (1949), 52; Hornberger, *Scientific Thought*, 56.

37. Bronson, *History of Brown*, 38.

During his tenure as tutor at Yale, Johnson had developed strong scientific interests which he never lost despite his later submission to the idealism of Bishop Berkeley.[38] In 1758, one of John Winthrop's students, Daniel Treadwell, was appointed professor of mathematics and natural philosophy. Upon his death, he was succeeded in 1761 by a Glasgow-educated Scot, Robert Harpur, who was given the post just three days after he landed in New York. Although competent, Harpur did little more than Houston, Howell, or Treadwell to advance science and published only a few astronomical observations in the newspapers.[39]

Strong foundations for the encouragement of science were laid at the College of Philadelphia by Provost William Smith who inaugurated a novel program calling for nearly 40 per cent of the students' classroom time to be devoted to scientific subjects.[40] In establishing this pattern, Smith's own inclinations, formed during his student days at the University of Aberdeen, were strengthened by the influence of Benjamin Franklin, who had much to do with bringing him to Philadelphia, and by Peter Collinson, who helped him during his visit to England in 1753. It was Smith who taught natural philosophy but there were several other men at Philadelphia who were almost as much interested in science. The Glasgow training of Vice-Provost Francis Alison had awakened in him interests which led him to seek the devotion of still more student time to science. Ebenezer Kinnersley pursued important research in electricity while serving as master of English and oratory. Theophilus Grew, professor of mathematics until his death, was a professional school teacher who made the calculations for almanacs published at Philadelphia, New York, Annapolis, and Williamsburg and who also wrote a text on *The Use of the Globe* for students in the academy and in the college.[41] He was a frequent con-

38. Theodore Hornberger, "Samuel Johnson of Yale and King's College," *New England Quarterly*, 8 (1935), 378-97, stresses his abandonment of Newtonian science, but he demonstrated continuing interest in science in his correspondence with Colden. See Schneider (eds.), *Samuel Johnson*, II, 287-305.

39. William M. and Mabel S. C. Smallwood, *Natural History and the American Mind* (New York, 1941), 288-89; David E. Smith and Jekuthiel Ginsburg, *History of Mathematics in America before 1900* (Chicago, 1934), 45; Schneider (eds.), *Samuel Johnson*, IV, 55, 246, 247, 248.

40. Hornberger, *Scientific Thought*, 29.

41. Theophilus Grew, *The Description and Use of the Globes, Celestial and Terrestrial* (Germantown, 1753).

tributor of mathematical puzzles and answers to the newspapers. His extensive reputation in the Middle colonies led to his employment in making surveys and determining boundaries. Grew was succeeded by one of his own students, Hugh Williamson, who later attained some recognition in science. When the Reverend John Ewing took over the provost's teaching during his absence he showed that he too had interest and capacity in science. An atmosphere strongly favoring science and mathematics pervaded the College of Philadelphia even though it had no single man of the caliber of John Winthrop.

In fact, the only school with a teacher who demonstrated some of Winthrop's capacity was William and Mary. Dr. William Small, who became professor of mathematics there in 1758 and professor of philosophy a little later, was a remarkable man. He was a Glasgow graduate who had studied medicine, although he did not go to Virginia to practice it. At Williamsburg, he introduced the modern lecture system in the college and brought to the community a sparkling and erudite personality. His courses were "rational and elevated"; they were presented with "eloquence and logic." To his greatest pupil, Thomas Jefferson, he offered "enlightened and affectionate guidance," introducing him to the company of Virginia's accomplished governor, Francis Fauquier, and to George Wythe who guided Jefferson in his study of law. These four men came to meet often at dinners which, in Jefferson's golden memory, produced "more good sense, more rational and philosophical conversations than in all my life besides." [42] Small contributed nothing to the advance of knowledge while in America, but after his return to England, he furnished proof of the reality of his capacity.

William Small left Virginia in 1764, taking with him a letter of recommendation from Benjamin Franklin to the manufacturer, Matthew Boulton, in Birmingham.[43] There, he settled, opening a medical practice and joining a remarkable circle of men. Small, indeed, did much to create the circle. With Erasmus Darwin, physician, poet, and scientist, and Matthew Boulton, he formed the Lunar Society, an intellectual group that met at dinner in much the same manner as the

42. Thomas Jefferson to Louis H. Girardin, Jan. 15, 1815, Edith Philips, *Louis Hue Girardin and Nicholas Gouin Dufief and their Relations with Thomas Jefferson* (Baltimore, 1926), 10.

43. T. Whitmore Peck and K. Douglas Wilkinson, *William Withering of Birmingham* (London, 1950), 125.

Williamsburg circle of which he had also been a part. The eight or ten members of the Lunar Society met on the Monday closest to the full moon so that they might have light on their way home. They discussed almost any topic except religion or politics, but according to Joseph Priestley who was a frequent visitor, they "were united by a common love of *Science*." Even in this group, in which none "failed to distinguish themselves in science or literature," Small was remarkable for his "benevolence and profound sagacity." [44] Boulton considered a business partnership with him; and James Watt, whom Small had known while a student at Glasgow, regarded him in the highest terms. Small often acted as a kind of consulting engineer in that mushrooming industrial community which included Josiah Wedgwood as well as Watt and Boulton. On his own, he took out several patents for improvements in the mechanism of clocks. In the midst of this activity, he continued to aid the growth of science in America; in 1768, he acted as agent in the purchase of an extensive scientific apparatus for William and Mary.[45]

The American colleges never came to form a well-integrated community, although there was continuous intercourse between them, particularly in the movement of personnel. Thomas Clap proposed at one time that the colleges agree together to adopt uniform standards of admission. Francis Alison agreed with the need for raising standards but it proved impossible to form a general association.[46] About 1760, Thomas Hollis, who had done so much to strengthen science at Harvard, urged the formation of a philosophical society that would have bound together those in the academic community who were interested in the promotion of science. Despite Hollis's influence with the Harvard group, the plan could not then be executed.[47] Dr. Thomas Moffatt of Newport concluded that all of the American colleges should have been united into a single, more substantial educational founda-

44. Henry C. Bolton (ed.), *Scientific Correspondence of Joseph Priestley* (New York, 1892), 195, 218.

45. Galen W. Ewing, *Early Teaching of Science at the College of William and Mary in Virginia*, College of William and Mary, *Bulletin*, vol. 32, no. 4 (April, 1938), 7-9; Hornberger, *Scientific Thought*, 62.

46. Alison to Stiles, May 27, 1759, Dexter (ed.), *Itineraries of Stiles*, 423.

47. T. Brand Hollis to Joseph Willard, Aug. 15, 1783, Sept. 3, 1783, "Joseph Willard Letters, 1781-1822," Massachusetts Historical Society, *Proceedings*, 43 (1909-10), 612-13.

tion. Each one, singly, he felt, was "too narrow and poor at bottom to produce the liberal fruits of art and knowledge." [48]

Moffatt's analysis was correct—at least as far as the fruits of knowledge were concerned. The function of teaching mathematics and natural philosophy was reasonably well performed in the colleges but they added very little to the store of knowledge. College theses which were defended at graduation, course notes, and occasional letters give a good picture of the general shape of the science taught in the colleges. In mathematics, it became usual to teach some arithmetic, algebra, geometry, and trigonometry—occasionally, conic sections and fluxions were taught. The sciences embraced under the term natural philosophy varied, but in 1764 John Winthrop taught pneumatics, hydrostatics, mechanics, statics, optics, astronomy, geography, natural history, navigation, and surveying. Even under the best circumstances, time did not permit a thorough treatment of any topic. One of John Winthrop's students recalled, "He touched on a few matters rapidly: the subjects of course very familiar to him—but to the novitiates, 'it was all Greek'." [49] The students were given an acquaintance with the sciences but no mastery of them. Nevertheless, by the 1760's, the majority of college theses were being chosen in mathematics or the sciences. [50]

Many of the original writings published by the academic group were ancillary to their teaching. This was especially true of mathematics for in that study Americans felt nothing but disadvantage from their geographical removal from Europe. The bulk of the titles in mathematics issued by the colonial press were reprints of English textbooks. Isaac Greenwood's arithmetic text did not succeed although it was a competent piece of work. [51] In New York, another arithmetic text appeared in Dutch. Other than that, there was nothing to be found but slight modifications of some of the English texts. The essays and problems printed in the magazines and newspapers often showed dexterity

48. Thomas Moffatt to Franklin, May 12, 1764, Franklin Papers, I, 88, American Philosophical Society.

49. Smith and Ginsburg, *History of Mathematics,* 55; copy of undated "Reminiscences of Col. Timothy Pickering," Harvard College Papers, I, 96, Harvard University Archives.

50. Winthrop Tilley, The Literature of Natural and Physical Science in the American Colonies from the Beginnings to 1765 (Doctoral Dissertation, Brown University, 1933), 152, 159, 167.

51. [Isaac Greenwood], *Arithmetick Vulgar and Decimal* (Boston, 1729).

with figures but they made no contribution. Even the *American Magazine* which William Smith and a group of "literary gentlemen" founded in 1757 with the specific design of promoting original writing, did not produce anything of value in mathematics.

Few of the sciences were as sterile as mathematics, for most of them had descriptive phases which the Americans could illuminate even though they were unable to improve upon basic theory. In addition, a certain amount of experimental work was done, from Greenwood's attempt to measure the density and analyze the air in wells to Ezra Stiles' experiments on lowering temperature.[52] None of this work was so important as the more numerous descriptive accounts published in the *Philosophical Transactions* and elsewhere. Observations which could not be duplicated at some other time and place gave the Americans their best opportunity. Astronomical events were particularly eligible for observation and description. Eclipses of sun and moon, transits of the planets across the sun, eclipses of Jupiter's moons, observations of meteors and comets, and accounts of the aurora borealis were reported by the college teachers. Meteorological conditions, magnetic variation, and earthquakes were similarly observed and described by the Americans.

Earthquakes were phenomena which called forth all kinds of speculation because of the traditional belief that they were supernatural in origin. Some eleven American accounts of earthquakes were published in the *Philosophical Transactions* before 1765. These accounts were factual reports, many of them written by the naturalists—by Paul Dudley, Joseph Breintnall, and Cadwallader Colden.[53] In America, however, other accounts of a theological character were published by the clergy—particularly in Boston. The earthquake of 1727 produced twenty-seven such publications, nearly all of them sermons.[54] The quakes were not pictured as miracles but very clearly as "works of God" even though they might be understood to proceed from natural causes.[55] By 1755, when the great earthquake which destroyed Lisbon

52. Stiles, Meteorological Observations, I, Stiles Papers; *Philosophical Transactions*, 36 (1729-30), 184.

53. *Philosophical Transactions*, 29 (1714-16), 62-71; 36 (1729-30), 124-27; 38 (1733-34), 119-21; 39 (1735-36), 63-73; 41 (1742), 359-60; 42 (1742-43), 33-42; 49 (1755-56), 439-42, 443-44, 544; 50 (1757-58), 1-18.

54. Tilley, The Literature of Science, 116.

55. Thomas Prince, *Earthquakes, the Works of God* (Boston, 1727).

was felt at many points in America, the clerical essays had become less prominent. Descriptions of the effects in Boston, New York, and Philadelphia were printed in the *Philosophical Transactions*.[56] One of them was penned by John Winthrop, but it was in America that he published his most important account of this earthquake.

A dramatic setting for Winthrop's performance was provided by the Reverend Thomas Prince when in 1755 he decided to reprint his 1727 sermon entitled *Earthquakes the Works of God and Tokens of his Just Displeasure*. In the new edition, he added an appendix in which he made the suggestion that the secondary cause of earthquakes might be electrical in nature. The Bostonians, he implied, might even have brought it all upon themselves by erecting so many lightning rods. Winthrop replied to this suggestion by publishing a lecture he had given to the college students upon the subject of earthquakes. In addition to much of the same material he had sent to the Royal Society, this pamphlet also contained an appendix in which Winthrop specifically contradicted Prince.[57] This was followed by a further interchange between the two men in the newspapers, at the end of which it was quite clear, as Jared Eliot declared, that the professor had "laid Mr. Prince flat on [his] back." [58] It was not that Prince's idea was wholly untenable to men of science in his day, for Stephen Hales in England supported a similar concept.[59] Nevertheless, Winthrop did give it the proper tag when he declared, "Philosophy, like everything else, has had its fashions; and the reigning mode of late has been to explain everything by Electricity." [60] Earthquakes were not to be explained thus. Winthrop presented a more satisfactory explanation at the same time that he ranged himself not so far from Prince in theology by proclaiming that *"natural* effects" were to be referred to the agency of God by all true "philosophy." [61]

Flattening the Reverend Mr. Prince gave pleasure to John Winthrop but it was his insight regarding the character of earthquakes that had lasting significance. He interpreted the shock as a "kind of *undulatory*

---

56. *Philosophical Transactions*, 49 (1755-56), 439-42, 443, 444; 50 (1757-58), 1-18.
57. John Winthrop, *A Lecture on Earthquakes* (Boston, 1755).
58. Eliot to Stiles, March 24, 1756, Dexter (ed.), *Itineraries of Stiles*, 480.
59. See Cohen, *Some Early Tools*, 127-28.
60. *American Magazine and Monthly Chronicle*, 1 (1757-58), 111.
61. Winthrop, *Lecture on Earthquakes*, 8.

motion" which might include both horizontal and vertical components just like a wave of water.[62] In fact, he referred to a *"wave of earth."* Winthrop's recognition of the wave character of earthquakes long preceded its statement by the Reverend John Mitchell who is still often given credit for originating the concept.[63]

Winthrop also interested himself in meteorological observations. From 1742, he carefully recorded temperature, barometric pressure, wind, and state of weather—three times a day.[64] Neither Winthrop nor other members of the academic group sought to tie such data to any larger theory as the physicians did who were trying to find in weather a cause of disease. They confined themselves to recording the bare facts which sometimes were published by the general magazines.[65] On one occasion, the Reverend Aaron Smith presented the opportunity for controversy when he asserted in a printed sermon that only prayer should be relied upon to provide the proper amount of moisture for crops.[66] He even interpreted rain during a wheat harvest as a sign of divine anger. No one replied.

Among the many inducements toward astronomical observation, the startling appearance and ancient fear of the comet were two of the most compelling. Interest of another sort attached to the comet of 1758 because its return had been predicted by Edmund Halley. The accuracy of the prediction delighted the academic community and led Winthrop in Cambridge, Stiles in Newport, Clap in New Haven, and Grew in Philadelphia to trace its course in their telescopes.[67] The comet also led to an anachronistic work entitled *Blazing Stars Messengers of God's Wrath* of which the teachers took no notice.[68] Once again, the clearest voice the academic community could find was that

62. *Ibid.*, 6.

63. *Philosophical Transactions*, 51 (1759-60), 566-634; A. Wolf, *A History of Science, Technology, and Philosophy in the Eighteenth Century* (New York, 1939), 398.

64. Winthrop, Meteorological Journal, 3 vols., Harvard University Archives.

65. *New American Magazine*, 1 (1758-59), 56.

66. Aaron Smith, *Some Temporal Advantages in Keeping Covenant with God* (Boston, 1749).

67. Winthrop, Ms. Interleaved Diary, April 3, 1759, Harvard Library; Alison to Stiles, May 27, 1759, Dexter (ed.), *Itineraries of Stiles*, 422; Louis W. Mc-Keehan, *Yale Science, The First Hundred Years, 1701-1801* (New York, 1947), 32.

68. Lawrence C. Wroth, *An American Bookshelf* (Philadelphia, 1934), 76.

of John Winthrop. He published an essay consisting of two lectures on comets which he had first delivered to the college.[69] This pamphlet was a piece of careful exposition which began a train of thought and calculation leading to a more important scientific paper. In response to his election to the Royal Society in 1765, Winthrop submitted a paper entitled "Cogitata de Cometis" in which he calculated the mass and density of several recorded comets.[70]

Less awe inspiring than comets, meteors similarly led to numerous observations by the Americans. Isaac Greenwood called for a program of regular observations of meteors at sea, and John Winthrop described several meteors, but it was Thomas Clap who studied them most seriously.[71] His interest began with the fiery meteor of 1742. In ensuing years, he collected as much information as he could, corresponding to that end with Ezra Stiles, John Winthrop, and Chauncey Whittlesey. When he wrote out his theory of meteors in 1756, he circulated it among his American friends. In 1765, he sent it to England where it was read to the Royal Society through the efforts of Peter Collinson, who served Clap with the same diligence he had shown to so many other American observers of nature.[72] Collinson reported that the society was interested and wanted Clap to make further observations. Before his theory was finally published, posthumously by Ezra Stiles in 1781, it had gained attention on both sides of the Atlantic. Clap knew that some meteors fell to earth but he did not find any evidence that the high, so-called fiery meteors did. He developed the theory that these meteors were "terrestrial comets" revolving about the earth at a height of not more than five hundred miles.[73] This general idea had been presented in the *Gentleman's Magazine* in 1755, and in 1759 Sir John Pringle suggested several of the concepts Clap had by then elaborated himself.[74] At the time, the theory was not unreasonable to many men of education.

Even so, Clap's most important contributions were unquestionably made as a teacher and expounder of science. To this end, he designed and built a very simple orrery or planetarium to help his students to

69. Winthrop, *Two Lectures on Comets* (Boston, 1759).
70. *Philosophical Transactions*, 57 (1767), 132-54.
71. *Ibid.*, 35 (1728), 401; 52 (1761-62), 6-16; 54 (1764), 185-88.
72. McKeehan, *Yale Science*, 23; Clap to Stiles, June 26, 1768, Stiles Papers.
73. Clap, *Conjectures upon Meteors*, 5, 6.
74. McKeehan, *Yale Science*, 35, 39.

understand the solar system. Instead of the elaborate mechanism and gear trains common in European orreries, Clap's represented the planets by beads which were strung on wires bent in the form of their orbits.[75] Harvard had had an orrery before Yale upon which Isaac Greenwood gave a course of lectures, but that was of English manufacture.[76]

Astronomical observations did not often engage the active participation of the naturalists, but they, as well as members of the academic community, did observe displays of the aurora borealis. John Bartram and Joseph Breintnall both sent accounts to the Royal Society as did Isaac Greenwood.[77] One of the few efforts to solve the cause of the aurora was published anonymously in William Smith's *American Magazine*.[78] It demonstrated acquaintance with European thinking upon the subject agreeing that the phenomenon was electrical in nature but holding that motion and dryness in the air were important too. Manuscript notes of John Winthrop give some indication that he connected an aurora borealis with the sunspots he had been observing. If he did, he might have seen the idea in a review which had appeared in the *Philosophical Transactions* a little earlier.[79]

The transit of Venus of 1761 presented the academic community with a rare opportunity but it was not so much an opportunity as James Alexander and Benjamin Franklin had anticipated in 1753. At that time, they had hoped that the colleges would prepare themselves to observe this important event. In 1761, it appeared that one of their basic assumptions had been wrong for no part of the transit would be visible in any of the thirteen continental colonies. Several European nations intended to send out expeditions to observe at more favorable spots and the question immediately arose whether the Americans might not plan something of the sort. If not, they would be unable to take any part in observing the much touted event.

75. C. W. [Chauncey Whittelsey?], in *American Magazine and Historical Chronicle*, 1 (1743-44), 202-3.
76. Isaac Greenwood, *Prospectus of Explanatory Lectures on the Orrery* (Boston, 1734).
77. *Philosophical Transactions*, 41 (1739-41), 359-60; 52 (1761-62), 474; 37 (1731-32), 55-69.
78. *American Magazine and Monthly Chronicle*, 1 (1757-58), 25-28.
79. Kilgour, "Winthrop's Notes on Sun Spot Observations," *Isis*, 29 (1938), 359-61.

It was Edmund Halley who in 1716 had clearly described a procedure for using observations of the transit of Venus to determine the solar parallax and hence the distance from the earth to the sun.[80] A reliable value for this distance was very much in demand by eighteenth-century astronomers because all of the distances in the solar system were known only relative to one another. Once the actual distance from the earth to the sun were known, then the distances of all the planets would immediately be known as well. The trouble was that Venus passed across the face of the sun only very rarely. The last time it had done so was in 1639 when no useful observations had been made. It would transit in 1761, once more in 1769, and then not again for 105 years. It was necessary, therefore, to make the most of the opportunities presented in 1761 and 1769, for this method could never thereafter be attempted by any man then living.

What was required was that different observations be made of the transit at points widely separated on the surface of the earth. The latitude and longitude of the observation points had to be known very precisely in order to establish the distance between them. The duration of the transit then had to be measured by a clock whose rate of "going" had been exactly determined. This would require accurate observations of the time of contact of Venus with the sun upon either ingress or egress and precise knowledge of the geographical location of the observer.

The academic community in America was well acquainted with the presumed importance of observing the transit, but in most of the provinces there was no disposition to dispatch an expedition to a point where it could be seen. Francis Alison reported that there was only one telescope in Philadelphia that could have been used for the purpose and that was temporarily unusable while its speculum was being re-silvered in London.[81] Ezra Stiles made careful observations of the face of the sun as soon as it rose just in case the predictions might be faulty, but Venus was not to be found.[82] Elsewhere, no reported activity took place except in Massachusetts.

It was John Winthrop who pressed for an expedition to Newfoundland where the end of the transit would be visible. When he brought

80. *Philosophical Transactions*, 29 (1714-16), 454-64.
81. Alison to Stiles, July 10, 1761, Stiles Papers.
82. Dexter (ed.), *Itineraries of Stiles*, 106.

the matter to the attention of his friend James Bowdoin, a member of the provincial council, Bowdoin presented it to Governor Francis Bernard as a highly desirable object.[83] Bernard, in turn, presented the assembly with Winthrop's letter offering to serve and with his own request that the legislature finance the expedition. The assembly responded handsomely by assigning the "Province-Sloop" to carry Winthrop and his party to Newfoundland and back.[84] Just as eager to "serve the Cause of Science, and do Credit to the Province," the Harvard College Corporation voted to permit Winthrop to take any of the college instruments he needed, provided they were insured against loss or damage.[85]

With the aid of two of his college students, Winthrop set up his apparatus on a hill in St. John's, Newfoundland. His first step was to establish the longitude of the spot by measuring the distance of a star from the moon—not a particularly accurate method but the only one available to him at that time. Using a pendulum clock and a refracting telescope, he obtained the time of the contacts of Venus with the limb of the sun as the transit came to an end. James Short, the English astronomer to whom Winthrop sent his observations, declared that the work had been done "with great care, and as much exactness as the low situation of the sun at that time would permit." [86] Short's final estimate of the mean horizontal parallax was 8.68″ or 94,030,000 miles from the earth to the sun. He found that Winthrop's observations compared with the observations made at the Cape of Good Hope yielded a value of 8.25″.[87] All of his figures yielded a greater distance from the earth to the sun than is now accepted. Moreover, there was a very wide variation in results when single pairs of the observations were compared. Winthrop's were the only American figures made available to the world of science, but in 1769, everyone would have another chance.

The teachers, like the naturalists and the physicians, were a part of

83. Winthrop to James Bowdoin, Jan. 18, 1769, *Bowdoin and Temple Papers*, I, Massachusetts Historical Society, *Collections*, 6th ser., 9 (1897), 116. Hereafter cited as *Bowdoin and Temple Papers*.

84. Winthrop, *Relation of a Voyage from Boston to Newfoundland, for the Observation of the Transit of Venus, June 6, 1761* (Boston, 1761), 7-8.

85. *Ibid.*, 22; Harvard Corporation Records, II, 142, Harvard University Archives.

86. *Philosophical Transactions*, 52 (1761-62), 625.

87. *Ibid.*, 621; 54 (1764), 283; Winthrop, *Two Lectures on the Parallax and Distance of the Sun* (Boston, 1769), 41.

the Western world of science. Those with Scottish educations and those who had studied at Harvard were distinctly superior in their accomplishments. The immigrants and those with European educations were most likely to maintain ties with European scientific circles. Yet even the best of the teachers contributed very little to the advance of the physical sciences except for observations which were descriptive in nature. They were more concerned with the dissemination of science and less with its advancement than were the naturalists. The principal advantage the naturalists enjoyed was that they discovered greater need for descriptive activity and more support for it than the teachers found for their astronomy and meteorology. The teachers did not form so coherent a community as the naturalists although their relationships with one another and with the Europeans were growing.

PART TWO

‡‡‡‡‡‡‡‡‡‡‡‡‡‡‡‡‡‡‡‡‡‡‡‡‡‡‡‡‡‡‡

Along the Road to Revolution

1763-1775

Chapter Six

AGITATION AND ORGANIZATION

T HE PEACE OF PARIS of 1763 introduced a train of problems that stimulated an increasing awareness among Americans of their differences with England and their own existence as a people. The imperial reorganization attempted at this time caused a reaction, political on the surface but far more deep-running than the overt events would have indicated. It led to increasing demands that an incipient American nationality be fulfilled in many spheres of life. In particular, it led to a cultural nationalism which had implications of the most important sort in science. Some of the cultural developments of the feverish years after 1763 were directly stimulated by English political action, but basically they derived from the stage of maturity which had then been attained. The two factors cannot be untangled. The fullness of American development was precisely the reason that English policies aroused the vigorous response they did.

An American sense of destiny—even an American nationalism—had long been developing. As early as 1748, Peter Kalm had been given to understand that "the English colonies in North America, in the space of thirty or fifty years, would be able to form a state by themselves entirely independent of Old England." [1] This was a pleasant contemplation; an evidence of deep faith in the American destiny rather than an indication of planned conspiracy, as Joseph Galloway would have

1. Adolph B. Benson (ed.), *Peter Kalm's Travels in North America* (New York, 1937), I, 139-40.

had it.[2] The faith in the American future was more happily, but just as fervently, expressed by Benjamin Franklin who predicted that in only a century, "the greatest number of Englishmen will be on this side of the water." [3] Emotional views of American destiny became particularly evident during the French and Indian War when, even as they celebrated English victories, the Americans reminded themselves, "*A new world has arisen, and will exceed the old!*" [4]

Although the American destiny advanced a step closer to fulfillment with the removal of the French from the continent in 1763, the new English imperial policies showed little comprehension of either American dreams or actualities. As each new step in the program unfolded, the Americans complained and protested—and very often formed an organization to combat the menacing policy. Organization was already a mode of response characteristic of the Americans.[5] The years after 1763 saw the formation of a profusion of American organizations: some of them in direct reaction to English policy; others, spurred perhaps by the awakened sense of destiny, sought a variety of ends that became increasingly desirable as the nation matured.

The most immediate and some of the most effective organizing was done by the business community as it either experienced or anticipated injurious English policies. In April 1763, the Boston merchants formed a Society for Encouraging Trade and Commerce with the specific purpose of opposing Parliament's renewal of the Molasses Act of 1733 which was due to expire in 1764.[6] The Philadelphia merchants appointed a committee early in 1764 to work for the same end.[7] Both of these groups failed in their lobbying efforts. The Sugar Act that was passed in 1764 was more serious than the old Molasses Act. In the case

2. Galloway asserted that a conspiracy had developed by 1754. [Joseph Galloway], *A Candid Examination of the Mutual Claims of Great-Britain and the Colonies* (New York, 1775), 2.

3. Franklin, "Observations Concerning the Increase of Mankind, Peopling of Countries, etc.," Albert Henry Smyth (ed.), *The Writings of Benjamin Franklin* (New York, 1905-7), III, 71.

4. *New American Magazine*, 1 (1758-59), 113.

5. Arthur M. Schlesinger, "Biography of a Nation of Joiners," *American Historical Review*, 50 (1944), 1-5.

6. Charles M. Andrews, "Boston Merchants and the Non-Importation Movement," Colonial Society of Massachusetts, *Publications*, 19 (1916-17), 161.

7. Arthur M. Schlesinger, *The Colonial Merchants and the American Revolution* (New York, 1917), 61.

of the Boston merchants, however, the society continued as Boston's first board of trade. It held annual meetings and appointed a committee which held monthly meetings.

The Sugar Act was popularly coupled with the trade stringency that developed at about the same time although in reality it was a normal post-war depression rather than a "Sugar Act depression." Whatever its cause, it alarmed the merchants and called forth a still different response in New York. There, late in 1764, a type of organization, new to America, was formed: "The Society for the Promotion of Arts, Agriculture, and OEconomy, in the Province of New York, in North America." It sought to correct the troubles that were so disturbing to the merchants: "The present declining State of our Trade, the vast Luxury introduced during the late War, our Want of sufficient staples for Returns, the extream Scarcity of Cash, the great inconveniences resulting from the Prohibition of Issuing our paper Currency in the usual Form, and the numerous Restrictions with which our Commerce is lately encumbered." [8] The society sought thoroughgoing remedies that from the beginning transcended questions of profit and loss.

The New York Society of Arts was directly patterned upon a London model, "The Society Established at London for the Encouragement of Arts, Manufactures, and Commerce"—better known as the Society of Arts or the Premium Society. Its primary function was to award premiums for the production of certain favored commodities and for the discovery of new techniques and inventions.[9] Since much of its effort was directed toward the colonies, its activities were widely known in America. Each year it offered premiums for the production of designated items—all of which were expected to supplement the economy of England rather than compete with it. Some of these products were at the same time eligible for bounties offered by the government. In one year, the society offered premiums ranging from £10 to £100 for the production of logwood, olive trees, potash, safflower, and wine.[10] In less than ten years, it paid out over £1,100 to residents of Georgia, Connecticut, and Pennsylvania for their efforts in the

8. *New-York Gazette; or the Weekly Post-Boy*, Nov. 29, 1764.

9. William Shipley to Franklin, Sept. 13, 1755, Franklin Papers, I, 38, American Philosophical Society; Sir Henry T. Wood, *A History of the Royal Society of Arts* (London, 1913), 6-9.

10. *Premiums by the Society, Established at London for the Encouragement of Arts, Manufactures, and Commerce* (London, 1758), 25-28.

production of silk. It was asserted, indeed, that the society was responsible for the establishment of potash and pearl ash manufacture and for the cultivation of grapes in America.[11]

The London Society of Arts did not function as a learned society in its early days, but it did provide another link between American men of science and intellectual circles in Great Britain. Franklin and the naturalists were the most useful American members.[12] On one occasion, the secretary of the society addressed several of these men in an effort to discover grasses more suitable for England. In doing so, he revealed aims that were not unlike those of a learned society. Peter Templeman prefaced his circular letter with the remark, "The surest Method of improving Science is by a generous intercourse of the Learned in different Countries, and a free communication of Knowledge." [13] Useful knowledge of this sort was not only gathered but also exported. The society gave one English inventor £300 in recognition of a new type of sawmill he had devised—and then gave him another £60 for a model of it which was sent to America where it was utilized to advantage.[14] In 1763, Jared Eliot was voted a gold medal in recognition of his success in producing malleable iron from black sea sand.[15]

Alexander Garden had failed in his enthusiastic attempt to stimulate the foundation of American societies subsidiary to the London Society of Arts.[16] It was a different thing when the New York merchants decided to form a society of arts in 1764. Garden had immediately jumped from commercially useful knowledge to the hope that "Natural History and Philosophy must likewise be enriched." [17] The New Yorkers used the same model but they applied it to a different purpose.

11. Robert Dossie (ed.), *Memoirs of Agriculture* (London, 1768), I, 24-26.
12. Shipley to Franklin, Sept. 1, 1756, Franklin Papers, I, 44, American Philosophical Society; Garden to Colden, Oct. 27, 1755, April 20, 1756, *The Letters and Papers of Cadwallader Colden*, V, New-York Historical Society, *Collections for 1921*, 54 (1923), 33, 70.
13. Templeman to Bartram, Sept. 16, 1760, Bartram Papers, IV, 110, Historical Society of Pennsylvania; Templeman to Colden, [Sept. 16, 1760], *Colden Papers*, V, 342.
14. Wood, *Society of Arts*, 92; *Pennsylvania Chronicle*, Aug. 2, 1767, reported methods of cutting lumber received from the Society of Arts.
15. Wood, *Society of Arts*, 88.
16. Garden to Whitworth, April 27, 1757, James Edward Smith (ed.), *A Selection of the Correspondence of Linnaeus and Other Naturalists* (London, 1821), I, 385-86.
17. *Ibid.*, 386; *South-Carolina Gazette*, April 1, 1757.

The New York Society of Arts was formed in response to an economic crisis that was temporary and a political crisis that was not. It would cooperate with the London society in its effort to encourage the production of goods that supplemented the English economy. It would open a correspondence between the two societies, it would publicize the offers of the London group, and it would attempt to facilitate the collection of premiums by Americans who had earned them. It would encourage agriculture and fishing in the province and try to check luxury and extravagance. Alongside this unexceptionable cooperation which could only have been pleasing to the King's ministers, the New York society decided to inaugurate a program that was more questionable. In addition to the premiums it would offer on hemp, mules, and sturgeon, it planned premiums on linen yarn, linen cloth, shoes, gloves, and stockings.[18] The manufacture of linen cloth, it soon became clear, was the primary object of the society.

As the imperial crisis deepened, the success of the society grew, reaching a peak after the passage of the Stamp Act. Whatever the meaning of one of its addresses to "every Lover of his Country," the activities of the society were of interest to neighboring provinces. Years later, a similar organization was established in Connecticut.[19] Still, the New York society remained a local affair which sought to stimulate subsidiary groups within the province but was most effective within the City of New York.[20] There its efforts led to the employment of three hundred people in the manufacture of linen and the establishment of a successful semi-monthly home market. The royal governor apologized to the Board of Trade, pointing out that no broad cloth was manufactured and no more than fourteen looms were in use at the "manufactory," but he did admit that something had been accomplished by "those, who were desirous of distinguishing themselves as American Patriots." [21]

18. *New-York Gazette; or the Weekly Post-Boy*, Nov. 29, 1764, Dec. 20, 1764, Dec. 27, 1764.

19. *Ibid.*, Nov. 29, 1764; *Connecticut Courant*, March 11, 1765, June 4, 1770; *Maryland Gazette*, March 21, 1765.

20. Michael Kraus, *Intercolonial Aspects of American Culture on the Eve of the Revolution* (New York, 1928), 172; *New-York Gazette; or the Weekly Post-Boy*, Dec. 27, 1764.

21. *New-York Gazette; or the Weekly Post-Boy*, May 30, 1765, Oct. 24, 1765, Dec. 5, 1765; Schlesinger, *Colonial Merchants*, 77; Gov. Moore to the Board of Trade, Jan. 12, 1767, E. B. O'Callaghan (ed.), *Documents Relative to the Colonial History of the State of New York* (Albany, 1856-87), VII, 888.

With the repeal of the Stamp Act, the society declined, to be re-vived, just once again, at the time of the Townshend Acts. Unlike the model after which it was patterned, the New York society did not ad-vance the cause of science or intellectual life. The closest it came to this was its recommendation of the best techniques in certain economic pursuits. It was effective in stimulating a little manufacturing in the city and the influence of that success continued after the society had ceased to exist. More important, it showed how well a voluntary group could mobilize money and support when strongly enough motivated.

At the same time that the commercial community was forming suc-cessful organizations, the medical community showed a disposition to organize—but from different causes. The medical activities were on the one hand logical developments of the stage of maturity then reached by the country. On the other hand, the medical community was in-fluenced by the emotional temper of the times. The most specific influ-ence followed from the military experiences American physicians and surgeons had shared during the French and Indian War, which brought them in contact with British military medicine.[22] The eyes of many were opened, especially of those who had had no academic training. They were exposed to a much better trained and organized profession in which certain standards of performance were insisted upon. All the Americans came to recognize more clearly their need of better educa-tion and of regulations which would bar the incompetent from prac-tice. War experiences coupled with post-war patriotism and enthusiasm for organizing led to surprising activity.

Imperative to many of the physicians was the need to bring some order into the unregulated practice of the day. Nearly all of the colo-nies had enacted legislation pertaining to medicine but the bulk of it related to the control of contagious disease through isolation or inocu-lation. In 1760 New York enacted the first law requiring that practi-tioners of medicine and surgery be examined and licensed.[23] Elsewhere, success was not met so early but regulation became a primary objective of the many medical societies attempted shortly after the Peace of

22. Stephen Wickes, *History of Medicine in New Jersey* (Newark, 1879), 43.
23. N. S. Davis, *Contributions to the History of Medical Education and Medical Institutions* (Washington, 1877), 10-11; Sidney I. Pomerantz, *New York, an American City, 1783-1803* (New York, 1938), 399.

Paris. The first of these groups was organized in 1763 in New London County, Connecticut, but it did not endure.[24]

In Boston a more promising attempt to organize the medical fraternity began in March of 1765. The first meeting of this group was held in secret because opposition was anticipated to any effort to "suppress empirics." Cotton Tufts, a leading practitioner of the province, was called upon to draw up a plan of organization. Although keenly conscious of the fact that the profession was not on the most "respectable footing," he went beyond the mere question of regulation to project a "society for promoting medical knowledge." [25] It would meet quarterly to discuss discoveries in medicine, surgery, anatomy, chemistry, and botany. Rules of conduct for the profession might come later. The membership was restricted to an invited group of physicians who, however, did not demonstrate sufficient interest to prevent the collapse of the society after its third meeting.

In Philadelphia, the establishment of a medical society was the work of one of the first men to return to the city with a medical degree from the University of Edinburgh. Early in 1766, John Morgan invited several physicians to join in forming an organization which became known as the Philadelphia Medical Society.[26] Morgan was an exceedingly ambitious young man who had graduated from the College of Philadelphia, served as a surgeon in the French and Indian War, studied medicine under Dr. John Redman in the city, and completed his medical education abroad. He had obtained his medical degree at Edinburgh and in London he had become acquainted with Dr. John Fothergill and other great friends of America. With the help of such men, he had met leading medical figures not only in England but in France and Italy as well. He returned to America laden with honors, the most conspicuous of which was his fellowship in the Royal Society. He returned, also, with great plans including the formation of a medical society in Philadelphia that would seek powers to license and regulate the medical profession similar to those granted to the Royal Colleges

24. Gurdon W. Russell, *Early Medicine and Early Medical Men in Connecticut* (reprinted from Connecticut Medical Society, *Proceedings*, 1892), 149.
25. Quoted by Walter L. Burrage, *A History of the Massachusetts Medical Society* (n.p., 1923), 3, 6.
26. American Society Minutes, Nov. 4, 1768, 133, American Philosophical Society.

of Physicians at London and Edinburgh.[27] Morgan was able to get the support of Governor John Penn and Chief Justice William Allen of the Province of Pennsylvania. Allen, indeed, caught some of Morgan's enthusiasm for he saw the project as one step toward making Philadelphia "the Seat of the Sciences, and, in the phisical way, the Edinburgh of America." [28] The attempt to get a charter was opposed on the other hand by William Smith in America and by John Fothergill in England, with the result that the society never gained power to regulate the profession.[29] As a private club, meeting to discuss medical questions, it succeeded.

In 1766, another medical society was established upon such stable foundations that it continued to be an effective body for several years. It went under the title of the Medical Corporation of Litchfield County, Connecticut. Its purpose was to promote medical science as well as to regulate the profession, but its success was most conspicuous in the matter of regulation.[30] On a local basis, it conducted examinations and admitted physicians to practice and to membership in the corporation.[31] It also listened to medical papers. Its efforts to obtain legislation establishing effective and uniform licensing requirements did not succeed, although widespread support was mobilized.[32]

The Medical Society of New Jersey, established in July of 1766, was the most comprehensive of the societies of these years. Embracing the physicians of old East New Jersey, it was divided into four "inferior" or local societies centered in different towns of the colony.[33] It was the "low state of Medicine in New Jersey" which led to this organization and it was the intention of its founders, at the very outset, to seek "Legislative Interposition" to regulate the practice of medi-

27. This story is best told in a forthcoming biography of John Morgan by Whitfield J. Bell, Jr., a portion of which Mr. Bell showed to the author.
28. William Allen to Thomas Penn, Nov. 13, 1766, Penn Papers, Official Correspondence, X, 65, Historical Society of Pennsylvania.
29. Smith to Penn, Nov. 14, 1767, *Pennsylvania Magazine of History and Biography*, 31 (1907), 454; Penn to Morgan, Feb. 18, 1768, Penn Letter Book, IX, 226, Historical Society of Pennsylvania; Thomas Penn to John Penn, Feb. 20, 1768, *ibid.*, 227.
30. Russell, *Early Medicine*, 125.
31. *Connecticut Courant*, Sept. 7, 1767.
32. *Ibid.*, May 8, 1769, Nov. 27, 1769, Dec. 25, 1769.
33. *The Rise, Minutes, and Proceedings of the New Jersey Medical Society* (Newark, 1875), 4-7, 13.

cine.[34] Many years passed before such action could be accomplished. In the interim the society attempted to introduce several regulations and agreements among its own members. It approved a code of ethics, it agreed that certain prerequisites and requirements be demanded of all medical apprentices, and it adopted a schedule of fees and rates which members agreed to charge. The fixing of fees quickly aroused community antagonism and the schedule proved impossible to enforce.[35] The society listened to medical papers and debated remedies —including one originated by Jared Eliot—but its greatest energy was spent in the effort to improve medical practice.[36] The society was sustained in its early years by the strong sense of need among physicians for regulation which would protect them, as well as their patients, against the mountebank and the unqualified.

Medical schools, like medical societies, were first successfully established in the period of imperial reorganization and American agitation that followed the French and Indian War. The usual practice of serving a medical apprenticeship or studying medicine in Europe had not been altered by occasional lecture courses such as that offered by Dr. William Hunter in Newport in 1754, 1755, and 1756.[37] In fact, since even the men who studied abroad, served first as apprentices in America, the pattern of medical apprenticeship was really unchallenged. Then, in 1762, William Shippen, Jr. began to deliver lectures on anatomy in Philadelphia with the hope that his exertions would be followed by the creation of a medical school.[38] In 1763, Dr. James Jay suggested that a medical professorship be established at King's College.[39] In 1764, a Boston newspaper spoke of the appointment of a "Professor of Physic and Anatomy" at Harvard when the college revenues should become adequate.[40] Interest in a medical school appeared at many points, but it was not in America that the foundations of success were laid.

34. *Ibid.*, 3; *New-York Gazette; or the Weekly Post-Boy*, Feb. 26, 1767.
35. *Minutes of the New Jersey Medical Society*, 9-13, 15, 18, 27.
36. *Ibid.*, 55-61, 16.
37. James Thacher, *American Medical Biography* (Boston, 1828), I, 305.
38. Caspar Wistar, *Eulogium on Dr. William Shippen* (Philadelphia, 1818), 26.
39. *Ear y Minutes of the Trustees of Columbia University* [New York, 1932], not paged.
40. Cited by Henry R. Viets, *A Brief History of Medicine in Massachusetts* (Boston, 1930), 76.

Increasing numbers of Americans went abroad in search of adequate medical training during the last years of the war itself. More and more they came to prefer the University of Edinburgh for their basic courses. Earlier students had favored Leyden and Paris but they had never gone to those centers in so steady a stream as their successors went to Edinburgh after 1760. From 1761 to the Revolution, Americans were granted degrees at Edinburgh in every year save one. In 1765, five Americans received their medical doctorates—a year in which only five native Scots were granted the same degree.[41] Still more Americans studied at Edinburgh for a year or more and then left without receiving their degrees.[42] After Edinburgh, whether the degree had been obtained or not, the usual course was to walk the hospitals in London and to attend related lectures there.

Among the Americans studying at Edinburgh and London in these years, a small group talked seriously about establishing a medical school in Philadelphia. The first of these men to return to America was William Shippen, Jr., who was encouraged in his plan to begin anatomical lectures by Dr. John Fothergill. The London doctor made a gift to the Pennsylvania Hospital of a set of anatomical drawings in color which would be very useful for a lecturer in anatomy. He also wrote a letter of recommendation in which he added the significant note that Shippen would "soon be followed by an able assistant, Dr. Morgan, both of whom, I apprehend, will not only be useful to the province in their employments, but if suitably countenanced by the legislatures, will be able to erect a school for Physic amongst you."[43] Samuel Bard, who, like the others, studied at Edinburgh, could only lament that the whole project was not to be planted near his own home in New York so that he might take part.[44]

John Morgan did not reach Philadelphia until three years after Ship-

41. John D. Comrie, *History of Scottish Medicine* (London, 1932), I, 340; Samuel Lewis, "List of the American Graduates in Medicine in the University of Edinburgh from 1705 to 1866," *New England Historical and Genealogical Register*, 42 (1888), 159-65.

42. Whitfield J. Bell, Jr., "Some American Students of ... Dr. William Cullen," American Philosophical Society, *Proceedings*, 94 (1950), 279-81.

43. Fothergill to James Pemberton, April 7, 1762, Etting Collection, Pemberton Papers, 47, Historical Society of Pennsylvania.

44. Samuel Bard to John Bard, Dec. 29, 1762, Bard Papers, New York Academy of Medicine.

pen had begun his lecturing. When he arrived, he brought with him a letter from Thomas Penn, proprietor of Pennsylvania, recommending that Morgan be empowered to organize a medical school.[45] He brought also the ambition to assume primacy in the movement and enough honors to sustain him in that effort. When the trustees of the college accepted Morgan's proposal, he launched his program in an epoch-making two-day *Discourse upon the Institution of Medical Schools in America*. Offhandedly, he referred to Shippen as a capable teacher of anatomy who might be asked to serve on the faculty of the school.[46] That easily were Shippen's claims to leadership disposed of. He accepted the post when it was offered, but he neither forgot nor forgave. Morgan, ambitious, calculating, and superficially accomplished, never again felt himself free of the hatred of the apparently affable, secure, and well-placed Shippen.

At the time, Morgan was occupied with a noble vision which did not leave him opportunity to speculate upon personality problems. In his discourse, he capably presented a complex of ideas with which he had become acquainted in Europe. His most essential doctrine was that, since medicine belonged to the sciences, the practitioner must be trained as a scientist. "Observation and physical experiments," he declared, "should blend their light to dissipate obscurity from medicine."[47] Pathology must be based upon physiology; upon "a philosophic knowledge of the Human body, or a science of all the conditions arising from the structure of its parts."[48] He demanded that the student acquire a liberal education before entering medical school, a requirement that proved beyond fulfillment until much later. In practice, he tried to divorce medicine from surgery and especially from pharmacy. This was another idea he had picked up in Europe which was less applicable in America than abroad where it was proving hard enough to introduce.[49] The course Morgan charted could not be held to, but the goal he set was admirable. After a long passage of time,

45. Thomas Penn to Board of Trustees, Feb. 15, 1765, Francis R. Packard, *History of Medicine in the United States* (New York, 1932), I, 346-47.
46. John Morgan, *Discourse upon the Institution of Medical Schools in America* (Philadelphia, 1765), 34-35.
47. *Ibid.*, 22.
48. *Ibid.*, 12.
49. *Ibid.*, 5; Courtlandt Canby, "The Commonplace Book of Doctor George Gilmer," *Virginia Magazine of History and Biography*, 56 (1948), 392.

other captains on other ships won the port that Morgan had sought, and they made some use of his chart.

His school succeeded, even though his higher objectives were not all realized. John Morgan became professor of the theory and practice of medicine; Shippen, professor of anatomy. Adam Kuhn, after studying under Linnaeus himself and winning an Edinburgh degree, took the post of professor of botany and materia medica. Benjamin Rush went to Edinburgh in 1765 with some assurance that when he returned he would be made professor of chemistry—he was. Never officially attached to the medical school faculty, Thomas Bond offered clinical lectures at the Pennsylvania Hospital which were an integral part of the student's course of study. William Smith, provost of the College of Philadelphia, gave a course in natural and experimental philosophy. The program was complete and the school flourished. Inaugurated in 1766, the school graduated its first class in 1768 by which time it had nearly forty enrolled students.[50] In John Morgan's eyes, this impressive accomplishment showed, "(I mention it for the reputation of my country) what a Spirit for cultivating Science prevails in this Western World."[51]

Morgan, indeed, was much influenced by the tide of emotional feeling that swept the colonies in the very years when he was laying the first foundations of the medical school. As the differences between England and America became more apparent, he remarked in a prize essay attempting to bridge the gap, "I consider myself at once as a *Briton* and an *American*."[52] Morgan and his colleagues became more conscious of the American destiny, but they were able to harmonize the conflicts in the situation better than others. In Charleston, opposition to the Stamp Act led Alexander Garden to declaim, "The die is thrown for the sovreignty of America!"[53] William Smith reflected more quietly, "When I review the history of the world, and look on the progress of Knowledge, Freedom, Arts, and Sciences, I cannot

50. *Pennsylvania Gazette*, June 30, 1768.
51. Morgan to Sir Alexander Dick, March 28, 1768 [copy], American Philosophical Society.
52. John Morgan, "Dissertation on the Reciprocal Advantages of a Perpetual Union Between Great-Britain and her American Colonies," *Four Dissertations* (Philadelphia, 1766), 2.
53. Garden to Ellis, Dec. 16, 1765, Smith (ed.), *Correspondence of Linnaeus*, I, 543-44.

but be strongly persuaded that Heaven has yet glorious purposes to serve thro' *America*." [54] Such achievements as the medical school were an indication of these purposes. Still studying in Britain, Benjamin Rush caught the mood of the moment: "Methinks I see the place of my nativity becoming the *Edinburgh of America*. The student now no longer tears himself from every tender engagement and braves the danger of the sea in pursuit of knowledge in a foreign country." [55]

In New York the establishment of a medical school presented more difficulties than it had in Philadelphia. New York, John Bard was quick to point out, lacked a hospital while Philadelphia could operate its medical school in conjunction with its justly celebrated Pennsylvania Hospital. [56] Additional obstacles appeared to Samuel Bard as he talked with his friends in Edinburgh. New York lacked "a good library of medical books" and it was too close to Philadelphia especially with the head start that the Philadelphians had gained. [57] Moreover, the medical school would have to be founded upon the college and that "alone would be sufficient to make the Presbeterian partie our Enimys." [58] Personal ambition arose to cloud the issue just as quickly as it had in Philadelphia. When Dr. James Jay suggested introducing European professors into the proposed medical school, he was immediately charged with "building a ladder by which he may climb to the top of the profession." [59]

It was by action of the trustees of King's College on August 14, 1767, that the medical school was created in response to letters from six of the town's leading physicians. Each of them was given a professorship, providing a full program of study at the very outset. Dr. Samuel Clossy became professor of anatomy; Dr. Peter Middleton, professor of physiology and pathology; Dr. John Jones, professor of surgery; Dr. James Smith, professor of chemistry and materia medica; Dr. Samuel Bard, professor of the theory and practice of medicine;

54. William Smith, "An Eulogium, On the Delivery of Mr. Sargent's Prize-Medal at the Public Commencement in the College of Philadelphia, May, 1766," *Four Dissertations* (Philadelphia, 1766), 11.
55. Benjamin Rush to Morgan, Nov. 16, 1766, L. H. Butterfield (ed.), *Letters of Benjamin Rush* (Princeton, 1951), I, 29.
56. John Bard to Samuel Bard, April 9, 1763, Bard Papers.
57. Samuel Bard to John Bard, Sept. 4, [1763?], Dec. 29, 1763, Bard Papers.
58. Samuel Bard to John Bard, Dec. 29, 1763, Bard Papers.
59. Samuel Bard to John Bard, Sept. 4, [1763?], Bard Papers.

and Dr. John V. B. Tennent, professor of midwifery.[60] This was a capable group though one with a less uniform medical training than the Edinburgh graduates at Philadelphia. Middleton was a Scot and Clossy an Irishman who had delivered a course of lectures on anatomy under the auspices of King's College in 1763. The others had all studied medicine at one or more of the European centers.[61] The school awarded its first bachelor's degree in 1769 and its first doctorate in 1770. It did not become so large and influential as the Philadelphia school but it continued effective instruction during the balance of the colonial period.

Although the medical school came first, Samuel Bard inaugurated a formal campaign for the establishment of a hospital at the graduation of the first class of medical students. A hospital to him was not merely a humane institution but an aid to the promotion of knowledge. "Every Country," he declared, "has its particular Diseases; the Varieties of Climate, Exposure, Soil, Situation, Trades, Arts, Manufactures, and even the Character of a People, all pave the Way to new Complaints, and vary the Appearance of those with which we are already acquainted." [62] Only in public hospitals, he felt, could these things be properly studied and cures be found. Turning back again to the medical school, he pointed out that hospital training was an essential element in the education of physicians.

Bard's plea was followed by a rapid and impressive development of support in many quarters. The royal governor of New York, Sir Henry Moore, immediately opened a subscription for building the hospital. The physicians formed "a Society for promoting the Knowledge, and extending the Usefulness of their Profession" which labored diligently to promote a hospital.[63] Bard and Peter Middleton both published pamphlets which included strong statements of the necessity for such an institution. With John Jones, they united in preparing a petition favoring the incorporation of "The Society of the Hospital in the City of New-York in America." [64] It was presented to the council by

60. *Early Minutes of the Trustees of Columbia*, Aug. 14, 1767, not paged.
61. Packard, *History of Medicine*, I, 395.
62. *New-York Gazette; and the Weekly Mercury*, May 22, 1769; Samuel Bard, *A Discourse upon the Duties of Physicians* (New York, 1769), 16.
63. Bard, *Discourse*, i.
64. *Ibid.*; Peter Middleton, *A Medical Discourse* (New York, 1769), 60n; *Charter for Establishing an Hospital in the City of New-York* (New York, 1771), 4.

Cadwallader Colden and soon enacted into law. In a short time, the subscription amounted to £800 sterling. To this the corporation of the city added £3,000 and the provincial legislature granted an annual allowance of £800 for twenty years. With the encouragement of Dr. John Fothergill and Sir William Duncan, contributions were received from Great Britain too.[65] Enduring foundations were laid in this period even though recurring troubles, including fire, prevented the hospital from becoming operative until after independence had been attained.

Virginia's accomplished governor, Francis Fauquier, was almost single-handedly responsible for the erection in that colony of another sort of hospital. In 1766 he first called the attention of the council and the House of Burgesses to the plight of that "poor unhappy set of People who are deprived of their Senses and wander about the Country, terrifying the Rest of their Fellow Creatures." He reported to the legislature, "Every Civilized country has an Hospital for these People, where they are confined, maintained and attended by able Physicians, to endeavour to restore them their lost Reason."[66] America could boast no such hospital although the Pennsylvania Hospital did receive mental patients and gave them reasonably good care. The Virginians seemed favorably disposed to act upon their governor's recommendation, but despite his repeated requests, no positive action was taken until after his death in March of 1768. A law was finally passed providing for the establishment of a hospital to care for the insane, and trustees were appointed to supervise its establishment—among them, Fauquier's old friend, George Wythe.[67] By 1773, the hospital was ready to receive patients.

It was possible to establish medical institutions because their obvious utility made a wide appeal; for more basic scientific organizations, very little support could be mobilized. One man who gave much thought to the establishment of a general philosophical society was the Rever-

65. David Hosack, "Sketch of the Origin and Progress of the Medical Schools of New-York and Philadelphia," *American Medical and Philosophical Register,* 2 (1812), 230.

66. *Journals of the House of Burgesses of Virginia,* ed., J. P. Kennedy and H. R. McIlwaine (Richmond, 1905-15), 1766-69, 12.

67. *The Statutes at Large; Being a Collection of all the Laws of Virginia,* ed., W. W. Hening (Richmond, 1820), VIII, 378; James A. Siske, A History of Eastern State Hospital under the Galt Family, 1769-1862 (Master's Thesis, College of William and Mary, 1950).

end Ezra Stiles. At Newport, Rhode Island, Stiles dwelt in a very pleasant, intellectual environment, presiding over the Redwood Library
and enjoying the catholicity of the community. He maintained a considerable correspondence with his fellow clergymen throughout New
England, particularly with those of his friends who were close to Yale
College. Many of his correspondents and acquaintances shared his
broad interests in science as well as his conspicuous patriotism. The
setting as well as the man contributed assurance that any society Stiles
planned would have both scientific and patriotic objectives.

Just a few months after the passage of the Stamp Act by an enthusiastic Parliament, Stiles drew up a plan for an "American Academy of
Sciences." Conditioned by such comments as that of his good friend,
the Reverend Chauncey Whittlesey of New Haven, who reported
that the English feared American independence and were ready to
punish the colonists for disloyalty before they had manifested any,
Stiles's temper flared.[68] He poured his feelings into the projected academy, declaring that it was "designed . . . for the Honor of American
Literature, contemned by Europeans. Therefore let the Associates be
all Americans and if born in America of the 2d Gen[eration], it shall
be indifferent whether of English, Scotch, Irish, French or German
Blood all these distinctions being lost in American Birth." [69] He provided that two-thirds of the associates must always be Presbyterians
or Congregationalists "to defeat episcopal Intrigue by which this Institution would be surreptitiously caught into an anti-american Interest." The council of the academy must be particularly careful to keep
the Anglicans in a minority, "Because they are incessantly intriguing
themselves into the Monopoly or Supreme Controll of every Institution literary or political, to the Exclusion of Americans." If the in-
triguers should stoop so low as to obtain a royal charter for a society
using the same name but under the control of "Europeans," the academy must then augment its title to read "the American *Anti-European*
Academy of Sciences." [70] Stiles, perhaps, displayed some confusion of
patriotism with sectarianism but his experience with the Anglicans

68. Chauncey Whittlesey to Stiles, April 16, 1765, Franklin B. Dexter (ed.),
*Extracts from the Itineraries and Other Miscellanies of Ezra Stiles, D.D., LL.D.,
1755-1794* (New Haven, 1916), 587.
69. Draft of a Constitution for an Academy of Sciences, Aug. 15, 1765,
Stiles Papers, Yale University.
70. Italics by the author. Stiles later crossed out the words "Anti-European."

during the Revolution indicates that his ideas were not entirely without foundation.

Except for this blatant patriotism, Stiles's society conformed to the general outline of the great European academies. It would "collect all the curious Things in Science especially in America, and maintain a Correspond[ence] over all the World." Scientific papers would be received and annual memoirs would be issued. Fossils, books, instruments, and apparatus of use in scientific work would be collected.[71]

One of Stiles's principal pleasures in constructing his paper academy was the naming of members and officers. For the presidency of the institution he had no trouble in naming Professor John Winthrop of Harvard. On the council he placed such prominent men as Benjamin Franklin, John Bartram, and Benjamin Gale, and beside them, local men with some interests if little general reputation. He was probably not acquainted with the naturalists and physicians in the colonies south of Pennsylvania, but one conspicuously missing name was neither unknown nor is it probable that it was forgotten. Cadwallader Colden was certainly eminent enough to be included but he had a fault that to Stiles was inexcusable. "He had a superlative Contempt for American Learning." [72]

Ezra Stiles continued to play with his academy, drawing up new plans in 1766 and 1767, none of which were ever translated into reality. It was a game he enjoyed, and after the Revolution had been completed, he was able to use a later plan as the basis of a real academy. His paper academy was more significant as an indication of the emotion generated by the imperial crisis and of the manner in which one man connected his patriotism with the promotion of science.

The one successful attempt to establish something in the nature of a scientific academy began in a much more humble way. The American Society held at Philadelphia for Promoting and Propagating Useful Knowledge ran a curious course of evolution. It grew out of a secret, self-improvement club patterned after the famous Junto of Benjamin Franklin. Founded in 1750, this "young Junto" consisted of twelve young men, two of them sons of members of Franklin's "old Junto." It engaged in discussion and used the same rules as the "old Junto." It

71. Draft of a Constitution, Aug. 15, 1765, Stiles Papers.
72. Franklin B. Dexter (ed.), *The Literary Diary of Ezra Stiles* (New Haven, 1901), II, 78.

gave little promise of gaining any wide influence in the community or of developing into an important organization for the encouragement of science. After many vicissitudes, its meetings declined until in 1762 they ceased altogether.[73]

The revival of the "young Junto" and its conversion into a different type of organization was a result of the influence of the Stamp Act crisis upon one of the members of the club, Charles Thomson. Thomson was an immigrant from Ireland who had taught both at the academy and at the Friends' school and then had become a merchant. He did not attain his greatest prominence until the Stamp Act brought the imperial contest to the boiling point. For a time, most of Thomson's energies were absorbed in the Stamp Act protest. He served on the committee which persuaded Philadelphia's stamp agent, John Hughes, to resign. Despite his affection for Benjamin Franklin, he refused to consider his advice to accept the act. When Franklin urged the Americans to "light candles" and "make as good a night of it" as possible, Thomson replied with high feeling, "Be assured the Americans will light lamps of a different sort from those you contemplate." [74] He worked unceasingly to force the repeal of the hated act, not venturing to relax until his goal came into sight. Only then did he feel free to launch a project he had been considering for some time—the revival of the "young Junto." Even before final confirmation of repeal had been received, Thomson had turned happily to his club. The first meeting of the "young Junto" since 1762 was held on April 25, 1766—five weeks after the Stamp Act had been repealed, but only two weeks after rumors of repeal began to circulate in Philadelphia.[75] The meeting was called in an atmosphere of anticipation.

The nine members of the revived society were an interesting group of ambitious young Philadelphians. Only Thomson, Isaac Paschall, and Edmund Physick had been members in 1762. The new members included two of John Bartram's sons, Isaac and Moses, who were building a successful drug business; Joseph Paschall, like his brother a merchant; Owen Biddle, a clockmaker; James Pearson, a hatter; and

73. Brooke Hindle, The Rise of the American Philosophical Society, 1766-1787 (Doctoral Dissertation, University of Pennsylvania, 1949), 31-42.

74. Lewis R. Harley, *The Life of Charles Thomson* (Philadelphia, [c. 1900]), 35-38, 61-65.

75. *Pennsylvania Gazette*, April 10, 1766, April 17, 1766, April 24, 1766, May 22, 1766; American Society Minutes, April 25, 1766, 1.

Isaac Zane, Jr., who became a prosperous iron master. The most striking characteristic of this group was their Quaker coloration; most of them were Friends—Thomson who was not, had taught at the Friends' School and married a Quaker. Edmund Physick, the only other non-Quaker, sent his son to the Friends' school. The Quaker mantle, however, was lightly worn by these young men; two of them associated themselves with the Free Quaker movement during the Revolution and a third was disowned for other reasons. They were intellectually eager in a way that was not markedly true of such pillars of the society as Israel Pemberton, John Reynell, and Abel James.[76]

The club went through a gradual and fascinating metamorphosis. In May, rules were adopted which called for discussions relating to "Philosophical and other usefull subjects," and enjoined the secretary to record "all new discovery's or improvements in Arts and Sciences made by or communicated to the Company."[77] At first many topics in history or economics were discussed but gradually questions of science and technics came to predominate. In September, it was decided to expand the membership by inviting friends to join—the friends being chosen so as to preserve the Quaker, merchant cast of the society. Next it was decided to add "foreign" or corresponding members, and "John Morgan, M.D., F.R.S., Professor of the Theory and Practice of Physick" who was elected to regular membership, proved capable of implementing this plan by proposing several of his European acquaintances as members.[78] In December, a name was finally chosen which would better dignify the expanded outlook of the society. It was unanimously agreed to call it "The American Society for promoting and propagating usefull knowledge, held in Philadelphia." Then, after this noble beginning, the society languished; its meetings declined until by the summer of 1767 they had become uniformly unproductive. The few men who came out to meetings could think of nothing better to do than fine absent members.

The decline might have continued unimpeded except that the Townshend Revenue Act was passed by Parliament on June 29, 1767 with the provision that it go into effect on November 20, 1767. At the

76. Brooke Hindle, "The Quaker Background and Science in Colonial Philadelphia," *Isis*, 46 (1955), 243-50.
77. American Society Minutes, May 23, 1766, 2.
78. *Ibid.*, Nov. 28, 1766, 28; Dec. 13, 1766, 29; Dec. 5, 1766, 28-29.

same time, the New York Assembly was ordered suspended and a new, more efficient system of customs collection established. Tension at once replaced the comparative calm that had marked the relations of the colonies with the mother country ever since the repeal of the Stamp Act. Reaction was not uniform, effective measures being taken by the merchants of New England and New York long before Philadelphia and the South could be brought into line. Some of the more ardent Philadelphians felt thwarted in their failure to promote action, and Charles Thomson was conspicuous among them.

On September 18, 1767, he read to the now declining American Society a sketch of a new vision he had had of its potentialities; a vision born of his reflections upon the deterioration of relations between the colonies and England. As the secretary recorded his remarks:

He conceived the North American Colonies, on examining the State of their Trade and their present Situation with regard to the Mother Country to be already Embarassed and declining, and therefore thought it necessary to examine into our own natural Resources.—On this Occasion he consider'd the Several Supports of Mankind at large, Agriculture, Manufactures and Commerce—The first he concluded to be the chief, and compared with the other two, the most favorable to the Increase of Mankind and the Preservation of their Lives and Morals.—This Society being instituted for promoting Knowledge really useful, he therefore proposed Agriculture for the Object of their Attention and Study: and the first Regard being due to our nearest Connexions, to begin with the Province where we live and to enquire into 1st The Climate, Soil and natural Productions of it. 2dly The present Mode of Planting and Farming 3dly The Improvements that may be made in them and 4thly What articles can be raised to add to the Staples of our Trade.[79]

In answer to the problem of the hour, he had formulated a specific program which was in conformity with the high aspirations of the American Society to become a learned society seeking useful ends.

Enthusiasm in the society was rekindled. The members set to work upon the development of Thomson's program, even going beyond questions of agricultural improvement to consider how manufactures might be encouraged. Franklin's old friend, Dr. Cadwalader Evans, interpreted these efforts rather well when he wrote about the plan that the "young Junto, ever since last Sepr., had been fabricating ... from

79. *Ibid.,* Jan. 1, 1768, 64.

that of the Royal Society, and the Society of Arts, Commerce, &c." [80]
The society that was planned did begin to look like a fusion of the
Royal Society with the London Society of Arts, while at the same
time it contained a liberal element of American patriotism.

On New Year's Day of 1768, Charles Thomson read a new paper
which he presented as "the Sense of the Company." [81] His beginning
glowed with anticipation of the American destiny as he exulted in the
extent and resources of "the Country which we inhabit." One of the
few sections suppressed by the society before publication was his para-
graph calling for contributions to basic science in the pattern of the
Royal Society, "even in the higher Branches of Mathematics and Astronomy." Paralleling this deletion was the addition of another para-
graph extolling useful knowledge:

Knowledge is of little use when confined to mere Speculation; But when
speculative Truths are reduced to Practice, when Theories, grounded upon
experiments, are applied to common Purposes of life, and when, by these
Agriculture is improved, Trade enlarged, and the Arts of Living made
more easy and comfortable, and of Course, the Increase and Happiness of
Mankind promoted, Knowledge then becomes really useful.[82]

The resultant emphasis was placed squarely upon the application of
science to economic improvement—not alone of the colonies, but at
the same time, for "the emolument of the Mother Country." [83]

Other men in New York, Newport, Boston, Litchfield, and Williamsburg were dreaming dreams of American attainment in new
realms, but none of them showed more clearly the relationship of
patriotism with the desire to achieve in science than did Charles Thomson. In his peroration, he declared:

the Spirit of Enquiry is awake, and nothing seems wanting but a public
Society as the Am[erica]n is now proposed to be, formed on a plan to
encourage and direct Enquiries and experiments, collect and digest discoveries and inventions made, and unite the labours of many to attain one
grand End, namely the Advancement of useful Knowledge and improvement of our Country.

80. Cadwalader Evans to Franklin, Jan. 25, 1768, Franklin Papers, LVIII,
52, American Philosophical Society.
81. American Society Minutes, Jan. 1, 1768, 61.
82. *Pennsylvania Chronicle*, Mar. 7, 1768.
83. Morgan to Dick, March 28, 1768 [copy], American Philosophical Society.

Nor did he refer to Pennsylvania alone:

As Philadelphia is the Centre of the Colonies, as her Inhabitants are remarkable for encouraging laudable and useful undertakings why should we hesitate to enlarge the plan of our Society, call to our Assistance Men of Learning and ingenuity from every Quarter and unite in one generous Noble attempt, not only to promote the Interest of our Country but to raise her to some eminence in the rank of Polite and learned nations.[84]

84. American Society Minutes, Jan. 1, 1768, 64.

Chapter Seven

THE AMERICAN PHILOSOPHICAL SOCIETY

W HILE THE PLANS of the American Society were still being formulated in the last weeks of 1767, another group of Philadelphians came
together for a strikingly similar purpose. Thomas Bond explained to
Franklin that he had "long Meditated a Revival of our American Philosophical Society." [1] What finally led him to make the attempt was
the circumstances surrounding the formation of John Morgan's Philadelphia Medical Society. A significant group of Philadelphia physicians
had been antagonized by Morgan's manner of forming the society
among a group of the younger doctors, and then sending "Tickets to
the old Physicians to join as Members, which, some did and more declined." [2] Even more insulting was Morgan's failure to invite either of
the two Shippens to join his organization. Thomas Bond and his
brother both refused to accept membership on this account. Even Cadwalader Evans, who was bitterly opposed to the political principles of
the Shippens, held himself aloof from Morgan because he felt this ill-
tempered action seemed to "imply something defective or attrocious
in their characters, which rendered them unfit members of the society,
and might be a real injury to them." Around the nucleus of the injured
Shippens and the other disgruntled physicians, Thomas Bond, "then,
strenuously endeavored to revive the old society." [3]

1. Bond to Franklin, June 7, 1769, Franklin Papers, II, 179, American Philosophical Society.
2. Smith to Penn, May 14, 1767, *Pennsylvania Magazine of History and
Biography*, 31 (1907), 454.
3. Cadwalader Evans to William Franklin, Franklin Papers, Jan. 25, 1768,
LVIII, 52, American Philosophical Society.

The society of 1743 had faded away until in 1767 hardly any of the original resident members could be found who were interested in making a second attempt. Seven of them were still alive but only Samuel Rhoads was ready to join with Thomas Bond and his brother. Franklin, in England, was altogether out of touch with things. Bartram, still living at a distance from the city, was well enough cared for by the patronage of the King and old enough not to be interested in joining. The only other old associate who was ready to support a new attempt was the Reverend Francis Alison. Alison had been a corresponding member in 1743; now as vice-provost of the college, he lived in the city.[4] If the society were to succeed at all, it would need many new members.

Thomas Bond sought to avoid party feeling in the society by developing a broadly-based membership. To that end, he invited into the organization the Quaker, Cadwalader Evans, and John Lukens, an Anglican close to the Quaker group, as well as the Presbyterian Shippens who supported the Proprietary party. Next, however, several men of a Proprietary party cast accepted membership: Edward Shippen, member of the governor's council; William Smith, provost of the college; John Ewing, leading Presbyterian clergyman; and George Bryan, rising politician who had recently fought well against the Quaker effort to make Pennsylvania a royal colony. Philip Syng, the Anglican, Junto crony of Samuel Rhoads, was a less partisan addition.[5] Despite Bond's wishes, the society seemed to be assuming a Proprietary-Anglican-Presbyterian complexion. This development was hastened when Cadwalader Evans, good friend of Bond though he was, refused to take part in the society. Although it was "very disagreeable" to differ with him, Evans declared that he "did not like his company," particularly Smith and Ewing who Bond had indicated were to be secretaries. He did not see how it was possible to cooperate in philosophical pursuits with such "tainted conduits" of truth.[6] Evans received a characteristically mild reply from Dr. Bond: "he hoped on consideration I shoud be able to reconcile it to myself, if not they would be sorry

4. American Philosophical Society Minutes, n.d. [I], 1, American Philosophical Society.
5. *Ibid.*, Jan. 19, 1768, [I], 6.
6. Evans to Benjamin Franklin, Jan. 27, 1769, Franklin Papers, II, 201, American Philosophical Society.

for it." [7] With Evans' defection, the hope of a philosophical society that could surmount the distractions of party politics was frustrated for the time being, at least. At the first and second meetings thirty-two more members were elected, the preponderance of them men who favored the Proprietary faction.[8]

In political orientation, the newly revived American Philosophical Society became the converse of the reorganized American Society which had a Quaker-Assembly inclination. The two organizations had not been created in opposition to each other, but they immediately found themselves in competition. Each of them had reached approximately the same early stage of development and aspired to become a general scientific society. Because each was dominated by one of the city's political factions, antagonism showed itself at every turn.

At the very outset, while each society was engaged in strenuous efforts to increase its membership, an attempt was made to unite the two. Members of the American Society were offended by the claim of their rival to have "subsisted ever since" 1743; they were sure that no meeting had been held for at least fifteen or twenty years.[9] Nevertheless, the first proposal for merger came from the American Society where it had been introduced by Edmund Physick, the only Proprietary officer in that den of liberal Quakers. When John Morgan offered to Thomas Bond a resolution favoring union "on an equal fotting and terms equally honorable to both ... but on no other," he met an enthusiastic response.[10] Bond suggested a conference of representatives from each society to attempt to restore "that harmony which should subsist amongst the Lovers of Science." [11] The conference decided that the American Society should submit proposals of union to the Philosophical Society. The Philosophical Society then upset the whole program by electing all the members of the American Society, en masse, into membership in their own organization.[12] This was resented

7. Evans to William Franklin, Jan. 25, 1768, Franklin Papers, LVIII, 52, American Philosophical Society.

8. American Philosophical Society Minutes, Jan. 19, 1768, Jan. 26, 1768, [I], 6, 7.

9. *Pennsylvania Gazette*, Jan. 28, 1768; American Society Minutes, Feb. 8, 1768, 79.

10. Evans to William Franklin, Jan. 25, 1768, Franklin Papers, LVIII, 52, American Philosophical Society; American Society Minutes, Jan. 29, 1768, 75.

11. Bond to Morgan, Jan. 28, 1768, American Society Minutes, 83.

12. American Philosophical Society Minutes, Feb. 2, 1768, [I], 3.

as an assumption of superiority which would have entirely swallowed up the patriotic group.

The action of the American Philosophical Society was a logical result of the characteristics that group had already assumed. Its membership boasted a handful of physicians, a couple of college teachers, and the self-taught astronomer, David Rittenhouse, elected because of their scientific interests. At the same time, all the important political and civic leaders of the province had been elected despite the fact that most of them had no scientific interests worth mentioning. All of the Proprietary party leaders became members including two of the Penns, former Governor James Hamilton, Chief Justice William Allen and three of his sons, and Attorney General Benjamin Chew. At the same time, the membership included such Assembly party leaders as Joseph Galloway, John Dickinson, Joseph Fox, and Israel Pemberton, and the prominent merchants: Thomas Willing, John Reynell, Hugh Roberts, and John Ross.[13] None of these men had more than a passing acquaintance with science but they did have the political influence and private means to make the society succeed. Through such members, the society easily procured permission to hold its meetings at the State House while its rivals continued to pay rent to the Union Library Company. In a like manner, permission was obtained to use the college buildings and apparatus when desired.[14] On the other hand, it was because of the large non-scientific membership and particularly because of their prevailing Proprietary affiliation that Bond's intention to elect Franklin president was defeated. Instead, James Hamilton was elected and Governor John Penn agreed to become the patron of the society.[15] The American Philosophical Society assumed a sparkle that made the American Society look just a little callow and dull beside it.

Even so, the American Society was unready to acknowledge its inferiority. Its membership expansion had not been rapid but among the new additions were the testy Quaker doctor, Cadwalader Evans and the former Quaker, Thomas Mifflin, who had a mercantile background and a political future. John Dickinson, then prominent as author of the "Farmer's Letters," accepted membership in this society which was

13. Brooke Hindle, The Rise of the American Philosophical Society, 1766-1787 (Doctoral Dissertation, University of Pennsylvania, 1949), 76-77.
14. American Philosophical Society Minutes, Jan. 26, 1768, [I], 7.
15. Ibid., Feb. 9, 1769, [II], 11.

more congenial to him than the Philosophical Society to which he had already been elected a member.[16] He, too, had formerly been a Quaker. Another new face at society meetings was that of the cultivated Samuel Powel, Anglican heir to a great Quaker fortune, who with John Morgan provided a leadership which was not likely to bow to anyone.

The American Society rejected the mass election of all its members by its rival in a short, polite note, but the members felt very keenly about the episode. They admitted that the action of the American Philosophical Society, "might be deemed an honour to us as individuals: yet as a society we cannot consider it in that light." [17] Convinced that theirs was the senior society, they contended that their rivals had not met before 1768. Their answer ended negotiations and accelerated a competition between the two bodies which lasted for the better part of a year.

This contest was a remarkably invigorating affair. It spurred all of their activities with a sense of urgency that led them to mobilize all possible sources of support and to publicize their work in the newspapers. Much of their energy went into the membership duel that brought wholesale increases in both resident and corresponding members to both societies. The outlook of the societies was markedly different at the start but as they began to compete with each other, they drew closer and closer in viewpoint. The American Society began as a liberal Quaker patriotic group seeking to contribute to the scientific knowledge and development of its own country. The American Philosophical Society, dominated by Proprietary politicians and professional men, sought to advance science in the pattern of the Royal Society. These distinctions became less sharp as the societies developed but they were never entirely lost.

The contest for members did as much as any other single thing to lessen the differences between the societies. Yet, even by the end of the year when the Philosophical Society boasted ninety-two resident members and thirty-six corresponding members to the American Society's seventy-eight resident and sixty-seven corresponding members, some differences could still be detected. Of the resident members who belonged to only one society, the Philosophical Society attained a bet-

16. American Society Minutes, Jan. 19, 1768, 72.
17. *Ibid.*, Feb. 8, 1768, Feb. 9, 1768, Feb. 12, 1768, 79, 81-82, 82.

ter distribution in terms of occupation, religion, and politics. With the single exception of medicine, the professions were better represented in the Philosophical Society. The advantage of the American Society in numbers of physicians was solely a result of its absorption of Morgan's twelve-member Philadelphia Medical Society.[18] The Philosophical Society had a larger share of government officers and of men who might, with approbation, call themselves "gentlemen." The American Society was less broadly based with especially scanty representation among Proprietary party leaders. Its greatest strength was in its merchant and artisan membership which accounted for at least half of its enrollment exclusive of members common to both societies. Each society established a wide distribution of corresponding members in the colonies and in Europe, but in neither case was the list an honor roll of the greatest scientists of the day. The caliber of corresponding members was higher than that of domestic members in either society, more of the best scientists being associated with the Philosophical Society.[19]

The membership contest and the mechanics of organization did not deflect all attention from the scientific objectives for which each society existed. The American Philosophical Society announced its first meeting by reprinting Franklin's *Proposal* of 1743 as an indication of the society's aims.[20] Charles Thomson's New Year's Day address became the platform of the American Society. The programs that followed were not wholly disinterested attempts to advance knowledge, although some of the individual papers undoubtedly were. Papers and donations of curiosities were received by each society with an eye to their use in the contest with the other. Scientific essays and news of society projects were regularly printed in the newspapers. The American Society made use of the *Pennsylvania Chronicle* while the American Philosophical Society more often used the *Pennsylvania Gazette* although all three Philadelphia papers often carried the same items. In this phase of the contest, the American Society had an advantage because of its backlog of papers which it had never thought of publishing before the contest developed.

The American Society was particularly successful in attracting at-

18. *Ibid.*, Nov. 4, 1768, 133.
19. Hindle, Rise of the American Philosophical Society, 87-89.
20. American Philosophical Society Minutes, Jan. 19, 1768, [I], 6; *Pennsylvania Gazette*, Jan. 28, 1768.

[132]

tention to American natural history, agriculture, and invention. Numerous specimens of plants, leaves, fish, and minerals were received. To care for these items, three curators were elected, one for each of the three kingdoms of nature: vegetable, animal, and "fossil" or mineral.[21] The society interested itself in the extraction of oil from sunflower seed, the distillation of persimmons, and the production of wine. To encourage the production of wine, it even paid a £10 premium in the pattern of the Society of Arts.[22] From some of the artisan members came such inventions as Owen Biddle's file-cutting machine, William Henry's steam-operated "sentinel register" which controlled heat, and Richard Wells' water-driven ship's pump.[23] In another vein altogether, Dr. Lionel Chalmers of Charleston sent the society several medical essays including one upon the relation of climate and disease which he later expanded into a book. With his first essay, Chalmers sent along an explanation that although such topics were not "expressly mentioned" in the published proposals of the society, he was sure the design was not so narrow as to exclude them.[24] Most of the society's other correspondents restricted themselves to natural history or applied science.

Basic science found more encouragement from the American Philosophical Society. Rather oddly, considering the derivation of the 1743 society and the emphasis of Franklin's *Proposal*, very little attention was paid to natural history. In agriculture, it accomplished at least as much as the rival group, hearing papers upon the production of wine, the elimination of garlic from fields, and two upon an insect then attacking wheat crops.[25] It received some ideas upon agricultural machinery, too, but no inventions or gadgets of the type that came to the

21. American Society Minutes, March 18, 1768, March 25, 1768, Aug. 26, 1768, Sept. 23, 1768, 93, 95, 111, 116.

22. *Ibid.*, Nov. 11, 1768, Nov. 25, 1768, Dec. 16, 1768, 134, 140, 149.

23. *Ibid.*, May 13, 1768, May 27, 1768, June 24, 1768, Sept. 16, 1768, Nov. 25, 1768, 103, 104, 106, 113, 140; *Pennsylvania Gazette*, July 7, 1768; American Philosophical Society, *Transactions*, 1 (1771), 286-92, 300-2; *Pennsylvania Chronicle*, Nov. 16, 1767.

24. Lionel Chalmers to the American Society, Nov. 2, 1768, American Society Minutes, 144.

25. American Philosophical Society Minutes, March 22, 1768, April 19, 1768, May 18, 1768, June 21, 1768, July 19, 1768, Aug. 16, 1768, Nov. 15, 1768, [I], 14, 15, 16, 18-19, 23, 26, 23.

American Society. Very little interest in medicine was demonstrated although a paper from Dr. John De Normandie did present a chemical analysis of the mineral waters of Bristol, Pennsylvania.[26] In basic science, astronomy received the most attention—a subject entirely neglected by the American Society. Here, the society benefited particularly from William Smith's discovery and promotion of David Rittenhouse.

David Rittenhouse was a native-born Pennsylvania prodigy who earned his living as a clockmaker. Whether or not he did devise a system of calculus himself, he became thoroughly familiar with Newtonian fluxions, with astronomy, and with mechanics.[27] In 1763, he was employed by the government of Pennsylvania to conduct some surveys and calculations in connection with the determination of the southern boundary of the province. He demonstrated mechanical skill in the development of a metallic thermometer which could fit in the pocket. In 1767, Rittenhouse turned to the construction of his famous orrery or mechanical planetarium that would represent the motion of the planets about the sun with an accuracy hitherto unapproached. Impressed by this fresh instance of Rittenhouse's capacity, William Smith caused the College of Philadelphia to award him a master of arts degree in recognition of his "felicity of natural genius, in mechanics, mathematics, astronomy, and other liberal arts and sciences." [28] The first scientific paper to come before the American Philosophical Society was a description of Rittenhouse's projected orrery.[29]

It was in connection with the plans for the observation of the transit of Venus expected in 1769 that the first real cooperative project in science began to take form. The society was very receptive to the Reverend John Ewing's proposal that provisions be made for observing the transit. Ewing was a Presbyterian clergyman with a good education, a flair for science, and ambition for academic achievement. Both he and David Rittenhouse laid before the society preliminary projections of

26. *Ibid.*, April 19, 1768, [I], 15.
27. Benjamin Rush, *An Eulogium Intended to Perpetuate the Memory of David Rittenhouse* (Philadelphia, 1796), 9, asserted that Rittenhouse had worked out a system of calculus by himself before he had heard of the work of Newton or Leibnitz.
28. Quoted by Edward Ford, *David Rittenhouse* (Philadelphia, 1946), 28.
29. American Philosophical Society Minutes, March 22, 1768, [I], 14.

the path of Venus across the sun.[30] Plans were made for observing the transit at two different points: at Philadelphia and at Rittenhouse's home in Norriton.[31]

The society demonstrated the wisdom of its membership selections when it came to raising money for the observations. Expenditures beyond the capacity of the society were called for in purchasing necessary apparatus that could not be borrowed. In this difficult situation, a cleverly-worded memorial was submitted to the provincial legislature announcing that the observations would entail expenses that could not be "borne by private Persons, and therefore must be defrayed by the Public" if they were to be met at all.[32] It properly emphasized the attention being lavished upon the event by every civilized nation of Europe. Regarding the purpose of the transit observations, however, the memorial was not entirely truthful. It declared the observations, "an Object, on which the Promotion of Astronomy and Navigation, and consequently of Trade and Commerce so much depends." [33] In reality, astronomy would benefit by getting a better value for the sun's parallax but commerce would not. As a by-product, separate observations might obtain the longitude of the observation points more accurately but that would be no benefit to navigation for the longitude of such spots as Philadelphia was already known with much more accuracy than it could ever be obtained at sea. The assembly, however, was very responsive to utilitarian goals—particularly those which would aid commerce. The expediency of this appeal as well as its inaccuracy was certainly known to Smith and Ewing. The assembly was more than generous. It immediately voted £100 to purchase the telescope the society had said it needed and, two months later, Smith was informed that an additional £100 would be forthcoming to cover other expenses.[34]

The American Society was neither able to mobilize monetary sup-

30. *Ibid.*, June 21, 1768, [I], 17; American Philosophical Society, *Transactions*, 1 (1771), 4-7.
31. American Philosophical Society Minutes, May 18, 1768, June 21, 1768, [I], 16, 17-18.
32. *Votes and Proceedings of the House of Representatives of the Province of Pennsylvania, Pennsylvania Archives*, 8th ser., VII (1935), 6289. Hereafter cited as *Votes of Assembly*.
33. *Ibid.*
34. *Ibid.*; William Smith to James Hamilton, Dec. 18, 1768, Archives, American Philosophical Society.

port of this sort nor could it develop a comparable scientific program, but it did score one notable success over its rival. When a president was finally elected in November, 1768, the society's choice was Benjamin Franklin.[35] The failure of the Philosophical Society to place Franklin in its chair made it possible for the American Society to draw to itself some of the prestige of America's most celebrated scientist. Franklin was claimed as a member of each society during their great contest, but he had contributed nothing whatever to either of them. In fact, he had never even officially acknowledged his election to membership in the American Society when he found himself chosen president.[36]

The two societies were still sparring in November, but by that time many of the members in each organization had come to resent the contest. On the very evening of a day that had seen a strong newspaper attack upon it, the Philosophical Society was confronted by a request from the American Society that union be again considered.[37] This time committees from each society were able to clear up the principal differences between them. On December 20, 1768, each society met separately and agreed to the terms of union that had been worked out. They exchanged lists of members and they agreed to hold the first meeting of the united society on the second of January at a meeting place used by neither of the constituent societies.[38] They adopted a cumbersome title, the American Philosophical Society, held at Philadelphia, for Promoting Useful Knowledge. Embracing elements of the names of each of the old societies, it was an evidence of "union on terms of perfect equality."

The contest between the two Philadelphia societies ended just as the imperial contest took a new turn for the Philadelphians. Despite the efforts of John Dickinson, Charles Thomson, and other ardent leaders, the Philadelphia merchants had temporized throughout 1768, refusing to follow the Boston and New York lead in adopting some kind of non-importation agreement. Philadelphia trade was not so depressed

35. American Society Minutes, Nov. 4, 1768, 132.
36. Charles Thomson to Benjamin Franklin, Nov. 6, 1768, *The Papers of Charles Thomson*, New-York Historical Society, *Collections for 1878*, 11 (1879), 19.
37. American Philosophical Society Minutes, Nov. 15, 1768, [I], 27.
38. *Ibid.*, Dec. 20, 1768, [I], 28-29; American Society Minutes, Nov. 18, 1768, Nov. 29, 1768, 135-36, 141.

and the Townshend Acts did not create there the same sense of urgency that had followed the Stamp Act. Although John Dickinson told James Otis, "The Liberties of our Common Country appear to me to be at this moment exposed to the most imminent Danger," and in Massachusetts, men spoke of "open Rebellion," the Philadelphia merchants were hard to move.[39] In September, the assembly sent petitions to King and Parliament, and finally, in November, the merchants took the decisive step. They sent a memorial to the British merchants urging that they work to have Parliament remove the trade restrictions. At the same time, they agreed that if no relief had been obtained by spring, they would adopt non-importation measures. This decision removed the uncertainty from the air but it hardly eased the tension. For the next few months, men could take no positive action; they could only wait through an uneasy truce. Nervous men, eager to be doing something, did put some of their energy into the united American Philosophical Society.

In this charged atmosphere, the united society presented a drama of general interest at its first meeting—devoted to the election of officers. Because the competition between the two constituent societies had reflected the internal political and social cleavages of the province, the election was important to many men. It drew the remarkable attendance of eighty-nine. Discord and contention were masterfully sidestepped in filling most of the offices by the simple expedient of increasing the number of posts to include the most important officers of the constituent groups. The united society had three vice-presidents, four secretaries, three curators, and one treasurer. The whole slate showed evidence of careful balancing. Two of the secretaries had served in the American Society; two in the Philosophical Society. Two of the curators were from the American Society and one from the Philosophical Society, but the treasurer had been a member of the Philosophical Society. Two of the vice-presidents were members of both societies, each of them ranged on opposite sides of political disputes in the province. Thomas Bond was also elected a vice-president and although he had been the force behind the success of one of the societies, he was not a partisan leader. The office of the presidency could not be

39. John Dickinson to James Otis, Dec. 5, 1767, *Warren-Adams Letters*, I, Massachusetts Historical Society, *Collections*, 72 (1917), 3.

disposed of in so satisfactory a manner. On this question, the meeting split down the middle.[40]

Here, the choice lay between the president of the old Philosophical Society, James Hamilton, and the president of the American Society, Benjamin Franklin. Franklin's political opponents were strongest in the old Philosophical Society, but his victory over Hamilton did not represent a victory for the American Society or for the Quaker-Assembly party. Franklin won because of his many friends in each society and because he had attained a scientific reputation that transcended party considerations. The most inveterate of his political opponents were alienated from the society by his election; John Penn, for example, refused to accept the office of patron in the united society.[41] In the other camp, a few like Cadwalader Evans were unable to accept union with their political opponents and they drifted away from the society. To the majority, the election was a good omen.

Benjamin Franklin did come to play a very important role in the society after his election, but for many years it was an indirect role. The contribution he made was vital to the success of the combined society —a contribution that could have been made by no other man. Franklin admitted that he hesitated to recommend European friends for membership until the society's first volume of transactions was published, for fear that it would discredit itself.[42] After that, it was through Franklin that much of the outside world became acquainted with the society. To Europeans, the American Philosophical Society became "Franklin's Society"—an accolade that did not have to be precisely interpreted to be effective.

As far as leadership within the society was concerned, Thomas Bond provided the greatest sustaining force. He rarely failed to attend the meetings, and, more often than not, he presided. Most of the leaders of the old American Society lost their interest in the united society. Samuel Powel attended only one meeting, Thomas Mifflin stopped coming after the first year, and Charles Thomson after the second. John Morgan and Owen Biddle did continue to attend a few meetings each year but neither of them became a leader of the combined society. After

40. American Philosophical Society Minutes, Jan. 2, 1769, [II], 2.
41. *Ibid.*, Jan. 16, 1769, [II], 3.
42. Franklin to Le Roy, March 30, [1773], Franklin Papers, XLVI, 14, American Philosophical Society.

the first year leadership passed to members of the old Philosophical Society. Men associated with the college were particularly prominent, with the provost, William Smith, attending very regularly and being re-elected to office every year before the Revolution. Adam Kuhn served as an officer one year; William Shippen, Jr., two years; and Benjamin Rush, every year after the first. This record gave some justification to the complaints of Cadwalader Evans, who felt that the society was being captured by the clever intrigue of Smith and Ewing.[43] William Smith did think of the society as an organization which complemented the college—a higher institution dedicated to the advancement of knowledge as well as to its diffusion.[44]

There is considerable evidence that it was the momentum of interest developed during the competition between the two separate societies in 1768 and the contest for control in 1769 that kept the society going through the balance of the colonial period. Attendance at meetings declined sharply each year after the peak of enthusiasm in 1768 and 1769. No other colonial meeting brought out even half as many members as had attended the exciting election meeting of January 2, 1769. Succeeding imperial crises were not associated with any real revival. War, the greatest crisis of all, proved not a stimulus but an inhibiting factor which finally brought all the activities of the society to an end.[45] Despite this ebbing of enthusiasm, sufficient strength remained to carry the society to gratifying achievements—most of them recorded in the years before 1772.

The membership rolls had been so padded during the 1768 contest that there was no need in the next few years to increase the number of members. Instead, attention was devoted to qualitative improvement. The society was particularly anxious to become more than a local Philadelphia organization—to become an *American* society. During the balance of the colonial period, only a handful of Philadelphians was added to the rolls while forty-four new members were chosen from other parts of America. Most of them were residents of the Middle and Southern colonies, a few came from the West Indies, and only four from New England. That distribution was a reflection of the com-

43. Evans to Franklin, June 11, 1769, Franklin Papers, II, 180, American Philosophical Society.
44. William Smith, *An Oration* (Philadelphia, 1773), 8.
45. American Philosophical Society Minutes, Feb., 1779, [III], 31.

mercial ties and religious relationships of the Philadelphians rather than an evaluation of the cultural merit of the respective regions. Nevertheless, the most outstanding scientific figures in the country did become members—along with others whose association with science was peripheral. New European members were elected too, most of them from Great Britain, several being the nominees of Benjamin Franklin. The rest were known to members of the society either through their work or by personal acquaintance. Few of these corresponding members contributed much to the work of the society but they did help to spread its fame and build its reputation.[46]

Money to sustain its activities was always a problem, but the old Philosophical Society had developed a capacity for meeting these demands which the united society carried to a still higher point. The greatest single source of revenue continued to be the provincial assembly which increased its contributions in support of the transit of Venus observations until they totalled £450.[47] The assembly contributed, also, £1,000 to match another £1,000 raised by subscription to sustain a society formed under the auspices of the American Philosophical Society for the purpose of encouraging silk culture. The merchants of Philadelphia raised close to £200 to finance a survey of possible canal routes between the city and the Chesapeake.[48] Then, for the observation of the transit of Venus, the proprietor made available a telescope which was to be donated to the college after the society had completed its transit studies.[49] Other individuals donated natural curiosities and books in a rather unsystematic fashion. Dues were collected from the beginning, but if the society had had to depend upon them, it would never have been able to accomplish anything. The ten-shilling admission money required from new members and the ten-shilling annual dues were small enough by the standards of the Royal Society but even those amounts could not be collected. Committees were appointed, members in arrears were excluded from voting, and on one occasion even a bill collector was hired, but the dues could never be made to

46. Hindle, Rise of the American Philosophical Society, 119-20.
47. American Philosophical Society Minutes, March 5, 1773, [II], 145; *Votes of Assembly*, VII, 6359.
48. *Ibid.*, VIII, 6846; American Philosophical Society, *Transactions*, 1 (1771), 293.
49. Minutes of the Trustees of the College of Philadelphia, June 20, 1769, II, 170, University of Pennsylvania.

yield much revenue.[50] The society was maintained by grants from the assembly and by private contributions.

One of the most earnestly sought objectives of every learned society was the publication of a journal because this was the accepted mode of presenting its scientific work to the world of learning. Both the American Society and the Philosophical Society of 1768 had received scientific papers with the intention of issuing a collection of them, but no such journal was published until after the separate societies were united. In fact, the precedence which would be accorded to articles in the publication of the united society was one of the conditions of union.[51] After the union, the issuance of a journal was considered to be only a matter of time. As a result, no further papers were published in the newspapers.

Before the society got to the point of publishing its journal, one of the members began to print an unofficial version of its "Transactions." Lewis Nicola, a curator of the society, was an odd figure. Born in France, he had lived for a time in Ireland, and had opened a bookstore in Philadelphia where he finally decided to settle. It was just a few days after the first meeting of the united society that Nicola announced his intention of publishing *The New American Magazine* which was then conceived in terms which would not make it a competitor of the society's journal. Nicola declared, "The natural history of the American and West-India colonies, falling within the plan of the *American Philosophical* Society held at Philadelphia for promoting useful Knowledge, will more properly be communicated to them, and appear in their publications." [52] At the same time, he urged that items relating to civil history be sent to him. It was in his second number that Nicola began to print portions of what he called the "Transactions of the American Philosophical Society."

Even in the first number which featured two original scientific articles and two extracted from European journals, Nicola's magazine demonstrated more than a little interest in science. Each number after the first, contained a separately paged appendix which was made up of

50. American Philosophical Society Minutes, Feb. 3, 1769, April 6, 1770, Nov. 15, 1770, Feb. 14, 1772, Feb. 19, 1773, March 6, 1772, [II], 5, 98, 115, 135, 142, 135; John Morgan, Journal, 1-7, Historical Society of Pennsylvania.

51. American Philosophical Society Minutes, Dec. 20, 1768, [I], 29.

52. *Pennsylvania Gazette*, Jan. 12, 1769, Jan. 19, 1769; the title he gave it, however, was the *American Magazine or General Repository*.

papers in the files of the American Philosophical Society. It was intended that the appendices later be bound together with a separate title page after enough had been printed to make a sizeable volume.[53] Twenty different papers were published in this way before the magazine itself was discontinued. Since all of these papers had been printed before in the newspapers during the contest between the societies, Nicola had access to them in these files without needing permission from the society to use them. This unofficial octavo edition of the "Transactions of the American Philosophical Society" was completed in September 1769 with the appearance of the last paper which had been published in the newspapers.[54] By that time the society was actively planning an official volume of transactions.

The society appointed a committee in June 1769 to prepare materials for its transactions, but it was not until February 1771 that the finished quarto volume was ready for distribution. It was prefaced by a statement copied from the *Philosophical Transactions* of the Royal Society in which responsibility for facts or reasoning contained in any of the papers was disavowed. It was pointed out that "neither the Society, nor the Committee of the Press, do ever give their opinion as a body, upon any paper they may publish, or upon any subject, of Art or Nature that comes before them." [55] The public was not convinced; the society continued to be regarded as a court of review in scientific matters. The keynote of the volume was stated in a modified version of Charles Thomson's address of New Year's Day, 1768. The actual papers were divided into four sections, by far the most important consisting of descriptions and calculations in connection with the observations of the transit of Venus. The sections on agriculture, medicine, and inventions were less related and more typical of the usual learned society journal.

An attempt was made to distribute the volume widely and judiciously. Free copies were presented to most of the individuals and groups which had made important contributions to the society. Every member of the assembly of Pennsylvania, the governor, and the proprietors as well as all merchants who had contributed to the canal sur-

53. *American Magazine or General Repository*, 1 (1769), 132.
54. A bound copy of these papers, without any title page, is held by the Historical Society of Pennsylvania.
55. American Philosophical Society, *Transactions*, 1 (1771), iv.

vey were so honored. The society planned to send copies to the library of every college in America, to the libraries of seven British universities, and to eleven foreign philosophical societies. Some distinguished foreigners were sent copies and Franklin, in England, was sent a few more to distribute as he saw fit. In America, every member who had paid his dues received a copy and others could buy copies at fifteen shillings apiece. With this distribution, the edition was soon exhausted.[56]

The volume was the first indication to many that the society was more than a name. In America, it sold well and became a source of pride to the members. Pierre Eugene du Simitière reported to a friend that it was "a handsome quarto of the Same Size of the Philosophical Transactions of England, with several Copperplates." [57] In Newport, Ezra Stiles felt a flush of pride as he read the collection, recording in his diary that it had been issued by a society "of which I am a Member." [58]

The European reception of the volume surpassed the hopes of the most sanguine promoters. Across the Atlantic it served as the organization's first general announcement of the society. That so new a society was able to issue a journal within so short a period after its foundation was, by contrast with the many European academies that had never published papers, remarkable in itself. The chorus of praise swelled on every hand. Franklin reported that his friends in England demonstrated satisfaction with the performance.[59] Another report from England was less restrained, declaring that the *Transactions* were "much sought after by the Literati in London, many of whom have declared that they will do you great Honour." [60] From Sweden, Dr. Charles M. Wrangel wrote that the transit observations had given "infinite satisfaction to our astronomers; as will the rest of your Transactions to the

56. American Philosophical Society Minutes, June 16, 1769, Feb. 22, 1771, [II], 43, 122; Presentation of the Transactions, Feb. 1 to May 12, 1772, Archives, American Philosophical Society.
57. Pierre E. du Simitière to Evarts Bancker, March 27, 1771, Miscellaneous Mss., American Philosophical Society.
58. Franklin B. Dexter (ed.), *Literary Diary of Ezra Stiles* (New York, 1901), I, 357.
59. Franklin to Bond, Feb. 5, 1772, Albert Henry Smyth (ed.), *The Writings of Benjamin Franklin* (New York, 1905-7), V, 387.
60. *Pennsylvania Chronicle*, Sept. 30, 1771.

Literary World." [61] From Mannheim, Christian Meyer recorded pleasure at finding that astronomy was cultivated "even in Philadelphia." [62]

Generous reviews appeared in the English literary journals, the *Critical Review* showing enthusiasm over the "first literary production with which we ever were presented by any society beyond the Atlantic Ocean." The volume afforded "convincing proof of the encreasing greatness and prosperity of our American colonies," and served "as a presage of their future improvement in wealth and grandeur; and what adds to our satisfaction is, that a prospect is now opened of importing from the New World fresh discoveries in science, as well as the articles of commerce." [63] The *Gentleman's Magazine*, which had wider currency in the colonies, described the publication as "no inconsiderable earnest of the great progress the arts and sciences will one day make in this New World." [64]

The favorable comments of continental journals were not conditioned by the numerous ties that assured the colonials of many friends in the home islands. In France, the *Journal des Sçavans* reviewed the volume and the *Observationes et Mémoires sur La Physique* was complimentary to the "travail de ces celebres Insulaires" at Philadelphia, "la métropole du Nouveau monde." [65] Venetian and Florentine journals were kind enough in their reviews but the *Giornale de' letterati di Pisa* outdid itself in eulogy, "This nascent Academy appears under the auspices and presidency of the immortal Benjamin Franklin with all the luster of the most ancient and considerable philosophical societies to none of whom need it repute itself inferior, both for the wise laws governing its recent establishment, and for the surprising number of members who compose it, and more especially because of the learned and interesting memoirs with which it has enriched this its first volume." [66] From Berlin, Jean Bernoulli's *Reçueil pour les As-*

61. Carl and Jessica Bridenbaugh, *Rebels and Gentlemen* (New York, 1942), 350.

62. Christian Meyer to the American Philosophical Society, n.d., Misc. Mss., American Philosophical Society.

63. *Critical Review*, 34 (1771), 241.

64. *Gentleman's Magazine*, 41 (1771), 417; *Pennsylvania Chronicle*, Jan. 6, 1772.

65. *Journal des Sçavans*, 1773, 87; *Observationes et Mémoires sur la Physique*, 1 (1773), 79-80.

66. Antonio Pace, "The American Philosophical Society and Italy," American Philosophical Society, *Proceedings*, 90 (1946), 390.

THE AMERICAN PHILOSOPHICAL SOCIETY

tronomes after praising the accuracy with which the transit of Venus had been observed and the detail with which it was reported, concluded "Voila beaucoup D'eloges qu'on donne à nos Astronomes Americains, mais ils sont bien merités."[67] Even in Russia, notice was taken of the collection by the publication of the Russian Academy of Sciences, *Akademicheskie izvestiya*.[68]

It was largely by this volume that the society was known to the European world in the years before the Revolution. No other volume of transactions was published until after the war although the importance of issuing the journal at regular and frequent intervals was recognized by the leaders. The society did publish annual orations in 1773, 1774, and 1775 which were well received, especially the one by "the celebrated inventor of the AMERICAN ORRERY," David Rittenhouse, but even that was no substitute for a regularly issued journal.[69] A French enthusiast, Barbeu Dubourg, who was translating Franklin's works into French, proposed to translate the first volume of *Transactions* and issue a French edition but he never carried the project through.[70] Some of the articles did gain currency in other languages either through comment or direct translation. The single volume of *Transactions* gained an enviable reputation for the society, primarily as a result of the excellent section it devoted to observations of the transit of Venus. Although no comparable attainment was registered in the balance of the colonial period, the character of the American Philosophical Society had been adequately established to satisfy most of its European well-wishers.

67. James Madison reported this comment with much pleasure to William Smith, Sept. 25, 1776, Ms. Communications to American Philosophical Society, Mathematics, II, 11.
68. Eufrosina Dvoichenko-Markoff, "Benjamin Franklin, the American Philosophical Society, and the Russian Academy of Science," American Philosophical Society, *Proceedings*, 91 (1947), 252.
69. *Pennsylvania Journal*, March 1, 1775.
70. Franklin to Barbeu Dubourg, Dec. 16, 1772, Franklin Papers, XLV, 61, American Philosophical Society; Barbeu Dubourg to Benjamin Rush, Feb. 3, 1773, Rush Papers, XLIII, 12, Library Company of Philadelphia.

Chapter Eight

THE TRANSIT OF VENUS

Not only for the American Philosophical Society but for the reputation and organization of scientific endeavor in America generally, the transit of Venus of 1769 was a most fortunate opportunity. The attention of the entire learned world was riveted upon the event so that any sort of activity related to it was bound to be noticed. This time, too, the Americans were more fortunate than they had been on the occasion of the transit of 1761 when observations had not been possible within the geographical limits of the settled colonies. Some portion of the 1769 transit would be visible—barring bad weather—in all of the colonies. The Americans were at last presented with an opportunity for observation in the physical sciences which promised as attentive a reception as their descriptive work in natural history had long commanded.

The pains taken by the nations of Europe to observe the transit of 1761 had served only to whet their appetites for the 1769 transit. The drama of expeditions sent by England to the Cape of Good Hope and to St. Helena, by France to the East Indies, and by Russia four thousand miles overland to Tobolsk in Siberia had seeped out of the calculations presented in the *Philosophical Transactions* and the *Mémoires* of the Académie des Sciences and into the newspapers and the general magazines. Although a great many had been worked out for the sun's parallax, astronomers were not satisfied, in part because of the unreliability of many of the contacts which had been made at very low altitudes. Everyone was conscious that 105 years would have to pass after 1769 before another transit would occur. Yet, the ardor with which the transit of 1769 was anticipated was not altogether a result of intellectual curiosity—it also involved the question of national prestige. As Thomas Hornsby, Savilian Professor of Astronomy at Oxford, put it: "We may be assured the several Powers of Europe will again contend which of them shall be most instrumental in contributing to the solution of this grand problem." [1]

It was not clear in the early stages of planning for the observations that the Americans were awake to the importance of the transit, and there was certainly no indication that they were moved by a patriotism which demanded scientific prestige. Indeed, when Nevil Maskelyne, astronomer royal of England, sought to encourage an American expedition to observe the transit near Lake Superior, Benjamin Franklin was distinctly pessimistic. He could think of no province in America "likely to have a spirit for such an undertaking, unless it be the Massachusetts, or that have a person and instruments suitable." [2] After all, Massachusetts was the only colony that had made any effort to observe the transit of 1761. Harvard not only boasted the sole American astronomer with a European reputation and a good collection of instruments, but Franklin was even at that moment engaged in buying another telescope and an equal altitude instrument for his friend, John Winthrop.

Maskelyne's interest in western observations followed from predictions that only the beginning of the transit would be visible in the settled sections of the East. The sun would set before Venus left its face. The closest spot where the full duration could be observed was Lake Superior so it was to Lake Superior that the astronomer royal wanted some of the American governments to send observers. It was possible to calculate the sun's parallax without both ingress and egress contacts but it was distinctly preferable to use observations of the full transit.

The project aroused an enthusiastic response in John Winthrop as soon as he heard from Franklin of Maskelyne's wishes. This time, however, despite certain favorable circumstances, an expedition would certainly be much more difficult. For one thing, Winthrop's health had deteriorated to such an extent that he did not feel justified in at-

1. *Philosophical Transactions*, 55 (1765), 344.
2. Franklin to Winthrop, July 2, 1768, Albert Henry Smyth (ed.), *The Writings of Benjamin Franklin* (New York, 1905-7), V, 137.

tempting the trip himself.[3] More difficult to surmount were the afflic-
tions besetting the government of Massachusetts as a result of the
Circular Letter its General Court had issued in 1768. The letter had
been sent to the other colonies in an effort to rally general opposition
to the Townshend Acts on the grounds that taxation by a parliament
in which the Americans were not and could not be represented was
unconstitutional. This defiance had led to the dissolution of the General Court of Massachusetts. The assembly which had made possible
Winthrop's expedition in 1761 did not exist in 1768 when the question
of financing an expedition to Lake Superior was presented. Having
pondered the difficulties, Winthrop decided to turn the problem over
to his friend James Bowdoin who in 1761 had been so helpful. Bowdoin was a member of the council and in a much better position to
gage the proper course of action through the turbulent political sea.

Winthrop's letter to Bowdoin enthusiastically urged the importance
of "the grand problem of determining the sun's parallax and distance"
and the particular desirability of obtaining "observations of the whole
duration of the transit." He stressed also the "useful discoveries" which
might be made in exploring the country, obtaining better values for
latitude and longitude of important points, and learning more of magnetic variation. Winthrop suggested a recent Harvard tutor, Thomas
Danforth, as a man capable, healthy, and willing to undertake the expedition. He refrained from pointing out "any particular method for
promoting such an expedition" but suggested that the governor might
interest himself in the matter or that General Thomas Gage, who commanded the British troops recently sent to Boston, might be ready to
help.[4]

Bowdoin decided that Gage might "make the expence of it a con-
tingency within his own department." He presented the proposed
expedition to the General as an opportunity to promote science,
strengthening his case by referring to Maskelyne and Franklin and by
enclosing Winthrop's initial letter.[5] Gage in turn proved friendly and
anxious to help—but not to pay for it. He sketched out the best route,

3. Winthrop to Bowdoin, Feb. 27, 1769, *Bowdoin and Temple Papers*, I,
Massachusetts Historical Society, *Collections*, 6th ser., 9 (1897), 127.
4. Winthrop to Bowdoin, Jan. 18, 1769, *Bowdoin and Temple Papers*,
I, 116-18.
5. Bowdoin to Winthrop, Jan. 23, 1769, *Bowdoin and Temple Papers*,
I, 120; Bowdoin to Gage, Jan. 23, 1769, *ibid.*, 119.

suggested that the party plan to pass through Montreal in early May so that they could accompany the traders leaving that town, and promised "all the assistance in my power to afford them." [6] He was ready to provide the necessary letters and passes and to send out instructions to post commanders as soon as definite plans were formulated.

When Bowdoin presented the project to Governor Francis Bernard and the council, it met such prompt approbation that he was led to expect a speedy provision for financing the expedition.[7] The day was even fixed upon which Danforth and three assistants were to leave for Albany on the first leg of their trip. Upon further consideration, however, the governor and council "found themselves unauthorized to engage in it." [8] This proved to be the final obstacle that could not be surmounted. Had the assembly been in session, there would have been no question of its competence to grant the people's money for an expedition of this sort in concert with the governor and council. Alone, the governor and council did not possess this power. Gage countered a second attempt to get him to pay for the expedition with the suggestion that a public subscription be raised but, at that moment, Danforth fell ill and the entire plan was abandoned.[9] It was a direct victim of the imperial contest which had deprived Massachusetts of her assembly at a moment when its support was essential. The enthusiasts had to content themselves with observations that might be made in Cambridge—by John Winthrop.

Franklin had not considered it worth while to approach any province other than Massachusetts about the desired western expedition, but unknown to him, the Philadelphians also turned their attention to this project. Indeed, when Bowdoin had made his first approach to Gage on the subject, the General told him, "Some gentlemen from Philadelphia made application to me some months ago, concerning the like intentions of sending some astronomers from that province to Lake Superior to observe the transit of Venus." [10] Gage's suggestion that the two provinces join forces in their efforts was readily accepted

6. Gage to Bowdoin, Jan. 30, 1769, *Bowdoin and Temple Papers*, I, 120.
7. Bowdoin to Gage, Feb. 26, 1769, *ibid.*, 129.
8. Bowdoin to Gage, March 1, 1769, *ibid.*, 130.
9. Bowdoin to Gage, March 27, 1769, *ibid.*, 130.
10. Gage to Bowdoin, Jan. 30, 1769, *ibid.*, 121.

THE PURSUIT OF SCIENCE

by Bowdoin and Winthrop, although Winthrop did feel that upon arriving at their destination it would be wise for the two parties to separate to a considerable distance in order to minimize the danger of unfavorable local weather conditions.[11] Despite this willingness to consider a joint expedition, there was no real enthusiasm for it. None of the Massachusetts men was sufficiently interested to open a correspondence with the Philadelphians or to make any attempt to promote intercolonial cooperation of this sort. The two communities worked out their problems in isolation.

The isolation of the Philadelphians led them to make plans which demonstrated an unfortunate lack of knowledge of English projects as well. It was during the period of contest between the American Philosophical Society and the American Society that an obscure member of the Philosophical Society, James Dickinson, first presented a plan for an expedition to James Bay, an arm of Hudson's Bay, to observe the transit. On September 20, 1768, Dickinson offered to make this trip himself, proposing that the Philosophical Society appeal to three different sources for financial support: to the Pennsylvania Assembly, to the King, and to public-spirited Americans.[12] He and the other members of the American Philosophical Society were apparently ignorant of plans which the Royal Society had already completed for sending William Wales and Joseph Dymond to Hudson's Bay on an identical mission. The society had arranged with the Hudson's Bay Company for their passage and maintenance and had agreed to pay Wales and Dymond £200 upon their return from America.[13] To the Royal Society the importance of this expedition lay in the observations of the entire transit that it would yield. Curiously, Dickinson made no mention of this advantage in his memorial, although it was the only real justification for such an expedition. Dickinson merely referred in general terms to the desirability of observations from as many widely separated points as possible. At the same time, he alluded to opportunities that would be presented for exploring the country.

There is every reason to believe that the assembly would have been willing to support an expedition of the sort for which Dickinson vol-

11. Bowdoin to Gage, Feb. 26, 1769, *ibid.*, 129.
12. American Philosophical Society Minutes, Sept. 20, 1768, [I], 24-25.
13. Royal Society Council Minutes, Jan. 21, 1768, V, 267-70 [Photographic Copy], American Philosophical Society.

unteered—if they had ever been asked to do so. They were not. The generosity with which they responded to every demand made upon them in connection with the transit was clear indication of their wish to aid the advance of science. Yet, the petition the society submitted to the assembly requesting aid in support of a western expedition did not present the facts in a way that would have led the average assemblyman to recognize the importance of Hudson's Bay observations. The legislature was informed that "the Solution of this Problem depends upon the Multiplicity of Observations, in different Places, compared together;—that such as may be made at *Pittsburgh,* Fort *Chartres,* or at any Place on the *Mississippi,* or *Hudson's Bay,* may be compared with those that may be made in *Philadelphia, New York, Boston,* and other Parts of the World." The petition gave no indication that observations made at Hudson's Bay would yield more complete data than could be obtained at Philadelphia or even at Pittsburgh.[14]

This strange oversight was not a result of lack of knowledge on the part of the leadership of the society. There were very important advantages to concentrating all the resources the society could muster upon a number of eastern observations even though only the ingress contacts of the transit could then be obtained. Numerous separate observations might be made near Philadelphia at a much more modest cost. This would permit both William Smith and John Ewing, neither of whom evinced any interest in making the arduous trip to James Bay, to direct separate observations, Smith at Norriton and Ewing at Philadelphia. Each man was clearly interested in enhancing his own reputation as well as that of Pennsylvania and America in general. These ends were better served by giving up the western expedition in favor of numerous eastern observations.

When the society had presented its first petition to the assembly requesting financial help in October, 1768, the legislature had responded quickly by appropriating sums not to exceed £100 for the purchase of a telescope. They made no response to the vague suggestion that assembly support of observations or expeditions would be desirable.[15] In February, the united society received intelligence that

14. *Votes and Proceedings of the House of Representatives of the Province of Pennsylvania, Pennsylvania Archives,* 8th ser., VII (1935), 6288.
15. *Ibid.,* 6288-89.

the telescope was "speedily expected."[16] This awakened interest in the transit once again and caused a memorial of thanks to be prepared which turned out to be more a request for further aid than thanks for past contributions. In this second petition, all pretense of a western expedition was dropped. Now, the assembly was specifically asked for permission to erect an observatory upon State House Square and for the funds required for its construction. Responding with the same alacrity it had demonstrated with respect to the first concrete request made of it, the assembly granted £100 in addition to the permission required for the construction of the observatory in the city.[17]

In its approach to the assembly, the society consistently demonstrated an appreciation of the general expectation that science ought to be useful. Even in London the Royal Society had felt it necessary in its memorial requesting financial aid from the King to include the carefully written statement that the transit observations would "contribute greatly to the Improvement of Astronomy, on which Navigation so much depends."[18] The Philosophical Society's first memorial similarly strayed from the whole truth when it asserted that observations of the transit would afford "the best Method of determining the Dimensions of the solar System, together with the Longitudes of the various Places where Observations upon this rare Phenomenon are made."[19] This distortion, aimed at utility-minded legislators, succeeded as well in Philadelphia as in London.

Within the society William Smith and John Ewing maintained their leadership of the project in the fourteen-man committee that was established.[20] Smith led the party that was to observe at the site of David Rittenhouse's home in Norriton where Rittenhouse had begun construction of an observatory in November, 1768. Ewing's group planned to set up their instruments in the observatory which a contractor was building for the society in State House Square.[21] A third

16. American Philosophical Society Minutes, Feb. 3, 1769, [II], 11.
17. *Ibid.*, Feb. 7, 1769, Feb. 19, 1769, [II], 13-15, 17; *Votes of Assembly*, VII, 6356-57, 6359; *Pennsylvania Gazette*, March 30, 1769; *Pennsylvania Chronicle*, March 27, 1769.
18. Royal Society Council Minutes, Feb. 15, 1768, V, 292.
19. *Votes of Assembly*, VII, 6288.
20. American Philosophical Society, *Transactions*, 1 (1771), 8, 42.
21. American Philosophical Society Minutes, May 19, 1769, [II], 26.

observation point was established at Cape Henlopen under the direction of Owen Biddle of the old American Society. Biddle, an enthusiastic supporter of American development, had much interest in science even though he never considered himself a real scientist.

The equipment collected by these amateur scientists was quite surprising, considering Franklin's doubt that a single telescope adequate for making the observations could be found in the entire province. Four telescopes were assembled at Philadelphia. Ewing planned to use the new one which the assembly had provided for the occasion. Joseph Shippen had obtained permission from the proprietors to use a small reflecting telescope belonging to them. Thomas Prior had a telescope of his own and a large refracting telescope was borrowed from Miss Polly Norris for the use of Dr. Hugh Williamson. Prior supplied a good timepiece while an equal altitude instrument was borrowed from the proprietors.[22] At Norriton, Rittenhouse had constructed a timepiece, an equal altitude instrument, and a 144-power refracting telescope. Smith would use the excellent reflecting telescope sent by Thomas Penn for this occasion. A third telescope was put together for the use of John Lukens, surveyor-general of Pennsylvania, from lenses which Franklin had bought for Harvard College but which arrived at Philadelphia too late to be forwarded to Cambridge in time for the transit. A quadrant belonging to the proprietors of East New Jersey was borrowed from William Alexander, surveyor-general of that province and son of Cadwallader Colden's old friend, James Alexander.[23] For the observations at Cape Henlopen, Owen Biddle was equipped with the Library Company's reflecting telescope which Mason and Dixon said was as good as any they had ever used. The directors of the Library Company participated in the enterprise by instructing Biddle to equip their telescope with an endless screw for elevation and charge the bill to them. Biddle's companion, Joel Bailey, was less well provided with a four-and-one-half-foot refracting telescope.[24]

Elsewhere, preparations for the transit were accelerated as the time

22. American Philosophical Society, *Transactions*, 1 (1771), 44, 45, 48, 50.
23. *Ibid.*, 8-12.
24. *Ibid.*, 95; Minutes of the Directors of the Library Company of Philadelphia, II, 14.

THE PURSUIT OF SCIENCE

drew near. John Winthrop expressed "mortification" that he was unable to satisfy Maskelyne's wish for observations at Lake Superior.[25] To observe the transit at Cambridge was no problem; it required no frenzied search for instruments and financial aid. The pendulum clock and eight-foot reflecting telescope belonging to the college which had been used in 1761 were still available. Another reflecting telescope could be borrowed in Boston and Winthrop counted on using the astronomical quadrant belonging to Joseph Harrison, collector of the port. In addition, the apparatus purchased by Franklin in England was daily expected.[26] In comparison with the extensive activities of the Philadelphians, Winthrop had little to do in the way of preparation, but he did prepare the minds of his students for the event in the finest piece of descriptive writing on the transit produced in America. He explained the process of the transit and its use in prose that was characteristically clear, concise, and coherent. His intense interest and sure knowledge were evident in every line. Delivered initially in the form of lectures, it was soon printed—by desire of his students— as *Two Lectures on the Parallax and Distance of the Sun, as Deducible from the Transit of Venus.*[27]

In Providence, Rhode Island, extensive preparations were undertaken upon the initiative of Joseph Brown, a wealthy merchant with a gentleman's taste for science. His interest in the transit was first aroused when he read Winthrop's account of his expedition of 1761. This led him to send to England for a telescope fitted out like the one Winthrop had described. Afterwards, he came across a copy of the memorial presented by the American Philosophical Society to the Pennsylvania Assembly in which the need for rather more elaborate apparatus was suggested. In a quandary, Brown turned for advice to Benjamin West, Providence bookseller, almanac maker, and correspondent of John Winthrop. The order was changed following West's suggestions and before he was finished Brown put nearly £100 into the enterprise. A three-foot reflecting telescope with adjustable

25. John Winthrop, *Two Lectures on the Parallax and Distance of the Sun, as Deducible from the Transit of Venus* (Boston, 1769), 44.
26. Winthrop to Bowdoin, Feb. 27, 1769, *Bowdoin and Temple Papers*, I, 127-28; Franklin to Winthrop, March 4, 1769, Franklin Papers, XLV, 34, American Philosophical Society.
27. Published in Boston, 1769.

cross hairs, micrometer, and other refinements was imported. A sextant belonging to the provincial government and two good clocks were also provided. Although the project was supported by Joseph's brother, Moses Brown, by Stephen Hopkins, a leading political figure in the province, and by other men, it was the driving energy of Joseph Brown and his willingness to expend his fortune freely that made observations possible. West provided the scientific knowledge requisite for competent observations.[28]

A similar pattern of support prevailed in Newport, Rhode Island, where only less extensive preparations were made for observing the transit. Here the patron was Abraham Redwood, and the scientific advisor was the Reverend Ezra Stiles. In this case, however, the real impetus came from Stiles who made up with enthusiasm for any lack in scientific knowledge. Redwood bore the cost of a new astronomical sextant of five-foot radius with telescopic sights. In addition, two clocks and two telescopes, one refracting and one reflecting, were collected.[29]

Less extensive plans to observe the transit were made at many points throughout the colonies. Several gentlemen conducted observations on their own estates using the apparatus they happened to have on hand; some of them invited men of academic learning to help. It was for this reason that Tristram Dalton, of Newburyport, Massachusetts, asked the Reverend Samuel Williams, a Harvard College tutor, to help with the observations he planned at his home.[30] William Alexander, self-styled Earl of Stirling, used his own telescope at his estate at Baskenridge, New Jersey. John Page of Rosewell, Virginia, used per-

28. Benjamin West, *An Account of the Observation of Venus upon the Sun, The Third Day of June, 1769, at Providence, in New-England* (Providence, 1769); *Pennsylvania Chronicle*, June 26, 1769; American Philosophical Society, *Transactions*, 1 (1771), 97; Charles W. Parsons, "Early Votaries of Natural Science in Rhode Island," Rhode Island Historical Society, *Collections*, 7 (1885), 252-53.

29. *Pennsylvania Chronicle*, June 19, 1769; West, *An Account*, 12n; William E. Foster, "Some Rhode Island Contributions to the Intellectual Life of the Last Century," American Antiquarian Society, *Proceedings*, 8 (1892), 127; Franklin B. Dexter (ed.), *Excerpts from the Itineraries and Other Miscellanies of Ezra Stiles, D.D., LL.D., 1755-1794* (New Haven, 1916), 106; Stiles, Transit of Venus Notes, 1769, Stiles Papers, Yale University.

30. American Philosophical Society, *Transactions*, 2 (1786), 246-49.

spective glasses to see the event.[31] Three separate observations were made in New York City, two of them by people connected with King's College. Robert Harpur, former professor of mathematics and natural philosophy, was the best informed of these but he had to make his observations without any assistance at all. Newspapers throughout the continent were full of the event describing its course, its meaning, and the manner in which individuals could observe it. Everywhere, people collected smoked glasses and anything they could find in the way of magnifying equipment: spy glasses, perspective glasses, and tiny telescopes. Most of them did not bother with time pieces or with the problem of the sun's parallax, but they were anxious to see this rare event about which there was so much concern.[32]

All along the coast, the transit day dawned bright and clear. Irrepressible worries over the weather were dispelled. Where the most elaborate preparations had been made—at Philadelphia and Norriton —crowds gathered to watch the observers with deep respect.[33] How many Americans saw the transit with their own eyes and how many of them recorded the times of contacts will never be known. At least twenty-two supposedly independent sets of contacts were ultimately reported in the newspapers and in the journals of learned societies.

They differed widely in value not only because of the difference in quality of instruments used but also because of the wide range of care with which the essential operations were performed. It was necessary to set the time piece accurately and to know the rate at which it gained or lost time. Precise determination of the longitude of the observation station was imperative. When John Leeds, surveyor-general of Maryland, set his pocket watch by the sun and guessed at the distance which separated him from Annapolis whose longitude was

31. *Ibid.*, 125, 152-53, 13; *Virginia Gazette* (Purdie and Dixon), June 29, 1769; James Madison to William Smith, Sept. 25, 1776, Ms. Communications to American Philosophical Society, Mathematics and Astronomy, I, 11; John Page to James Madison, *ibid.*, 13; William Alexander, Notes on the Transit of Venus, in James Alexander, Journal as Boundary Commissioner, New Jersey, Alexander Papers, New-York Historical Society.

32. *New-York Gazette; or the Weekly Post-Boy*, June 12, 1769; *New-York Gazette; and the Weekly Mercury*, June 26, 1769; *Pennsylvania Chronicle*, June 26, 1769.

33. See William Barton, *Memoirs of the Life of David Rittenhouse* (Philadelphia, 1813), 164.

known, his results could not be expected to be as valuable as those obtained at Norriton. There, much time was spent in determining the "going" of the clock and in working out the longitude not only by astronomical means but by using a chain to run off the mileage from Philadelphia.[34] Even the determination of the moment of contact was difficult. As Venus approached the sun, a dark tongue seemed to leap out to connect the sun and the planet before Venus actually touched the sun's disk. Observers were at a loss to know just what point should be declared the moment of contact. The internal ingress contact was scarcely easier to determine exactly. American observers also had an opportunity to record the time when Venus was midway in its path across the face of the sun, provided their telescope was equipped with a good micrometer. Even so, determination of the time when this point was reached was also conditioned by the human factor. Where several observers were gathered at the same station, the problem of influence arose. In some cases, too great uniformity of results would indicate that this obstacle was not surmounted. In other cases where visual signals were given by each observer to recorders, the other observers had no way of knowing when one of their colleagues had reported a contact. A final unexpected difficulty arose when even David Rittenhouse became so excited at seeing the first contact that he failed to report it until some seconds later. His estimate of the passage of time while he was in this state was hardly reliable.[35]

However accurate or careless the observations, they were of no use until they were made known to scientists who could compare them with observations conducted at other points in the world in an effort to work out the sun's parallax. Many of the American observations were not made available in this way. Neither the Newport nor the New York observations got any further than the newspapers. They were copied by papers in other colonies but they never appeared in the journal of a learned society.[36] The observations of Samuel Williams and Tristram Dalton at Newburyport were not

34. *Philosophical Transactions*, 59 (1769), 444-45; American Philosophical Society, *Transactions*, 1 (1771), 13-20, Appendix 5-11.
35. American Philosophical Society, *Transactions*, 1 (1771), 26.
36. *Pennsylvania Chronicle*, June 19, 1769; *New-York Gazette; and Weekly Mercury*, June 26, 1769; *New-York Gazette; or the Weekly Post-Boy*, June 12, 1769.

published in a journal until 1786—much too late to be of any use to the scientists.[37]

In Virginia, John Page put great enthusiasm into several newspaper articles he wrote about the transit.[38] His essay describing the event indicated that he probably understood just what took place, but it also showed that he was completely incapable of explaining it in terms that could be understood by anyone who was not well versed in astronomy. He established his local time with a Hadley's quadrant. He used a pair of Ayscough's perspective glasses with a magnifying power of eighteen which permitted him to see the transit but not to see the tongue of darkness which was the effect of Venus's atmosphere. Years later, Page gave his results to the Virginian Society for the Promotion of Usefull Knowledge and during the war he sent them to the American Philosophical Society.[39] There is no evidence that science was the loser in this failure to publicize his work properly.

The best of the American observations did reach European scientists—most of them through the pages of the *Philosophical Transactions* of the Royal Society. As a matter of course, observations made by Wales and Dymond on Hudson's Bay at the direction of the society were published in its journal.[40] Two sets of observations made by servants of the crown at Quebec were also given space.[41] From the seaboard colonies, four different accounts appeared: that of John Winthrop at Cambridge, that of John Leeds near Annapolis, and the observations made by members of the American Philosophical Society at Norriton and Cape Henlopen.[42] The inclusion of the worthless data of John Leeds was indication enough that the journal printed whatever came to it without any very careful selection. The other three accounts were among the most reliable reported from anywhere in America. From the *Philosophical Transactions* they were picked up and used by English and continental scientists.

At the same time, the first volume of *Transactions* of the American

37. American Philosophical Society, *Transactions*, 2 (1786), 246-49.
38. *Virginia Gazette* (Purdie and Dixon), June 1, 1769, June 29, 1769, Aug. 3, 1769.
39. John Page to James Madison, July 7, 1777, Ms. Communications to American Philosophical Society, Mathematics and Astronomy, I, 13.
40. *Philosophical Transactions*, 59 (1769), 467-88.
41. *Ibid.*, 247-52, 273-80.
42. *Ibid.*, 351-58, 444-45, 289-326, 414-21.

Philosophical Society presented the learned world with a larger variety of American accounts interpreted in more detail. Exactly half of the twenty-two sets of observations reported in America had been carried out under the auspices of the society in the vicinity of Philadelphia. All of the society's observations were presented in elaborate detail in ninety-six pages of the quarto volume. In addition, the volume carried the observations of William Poole at Wilmington, Delaware, William Alexander at Baskenridge, New Jersey, and an abstract of Benjamin West's separately published pamphlet describing the Providence observations.[43] In its second volume (1786), the Newburyport observations were presented.[44] Even at that, the society did not succeed in collecting all of the American observations.

Within the American Philosophical Society, the chief promoters of the transit observations were not driven by a single-minded devotion to science. They continued to act with a sense of expediency but also with a certain lack of finesse. The first newspaper releases erred in reporting one set of contacts to "within a few Seconds"—where every second made a great difference. A questioning reader was informed that the uncertainty had resulted from failure to determine the rate of going of the clock before the transit.[45] Meanwhile, extended discussions over the manner of releasing results rocked the society. Owen Biddle quickly supplied the society with his report on the Cape Henlopen observations. John Ewing's account of the Philadelphia observations came just a little later and by July 20, the Norriton account had also been received from William Smith. At this point, Smith and Dr. Thomas Bond locked horns over the latter's motion that all the reports be sent to the society's president in London.[46] The motion was defeated, but not until high feeling had been generated.

With his usual eye for expediency, William Smith argued that "as the papers of the Society were shortly to be printed, partial pub-

43. American Philosophical Society, *Transactions*, 1 (1771), 97-105, Appendix 20.

44. *Ibid.*, 2 (1786), 246-49.

45. *Pennsylvania Gazette*, June 8, 1769; *Pennsylvania Chronicle*, June 5, 1769, June 12, 1769.

46. American Philosophical Society Minutes, July 20, 1769, [II], 48-71; Evans to Franklin, Nov. 27, 1769, Franklin Papers, II, 201, American Philosophical Society.

lication would lessen the curiosity of the publick, and diminish the demand for the work." [47] Plausible though this statement was, it was negated by Smith's previous dispatch of his own account to Thomas Penn in England. Penn delivered the account to Nevil Maskelyne from whom it found its way into the *Philosophical Transactions* long before the transactions of the Philadelphia society could be published. [48] This strange duplicity did result in an accolade from Maskelyne who described Smith's patron, Thomas Penn, as "first among" those who promoted the undertaking. [49] At the same time, Owen Biddle had given a copy of his account of the Cape Henlopen observations to that inveterate enemy of Smith, Dr. Cadwalader Evans. Evans lost no time in forwarding the paper to Benjamin Franklin who saw that it was published in the same volume of the *Philosophical Transactions* which contained Smith's account of the Norriton observations.

The principal result of all this intrigue was that Ewing's account of the Philadelphia observations was lost in the shuffle. After the society decided not to forward any of the accounts, Ewing held his back until January 1770, when the society finally gave him instructions to send the account to Franklin. In his covering letter, the unhappy clergyman labelled the society's initial decision not to forward the accounts, a "rash Agreement." He explained the reversal in policy as an effort to counter the "possibly inaccurate reports" which had been sent to Europe by private individuals. [50] It did not have that effect because the account arrived too late to be included in the current volume of *Philosophical Transactions* which contained transit observations from other parts of the world. It was not placed in the succeeding volume because by then, it was too old. The Philadelphia observations reached the learned world only through the *Transactions* of the American Philosophical Society which did not have the circulation or the reputation of the Royal Society's *Philosophical Transactions*.

47. Evans to Franklin, Nov. 27, 1769, Franklin Papers, II, 201, American Philosophical Society.
48. American Philosophical Society, *Transactions*, 1 (1771), 40-41.
49. *Ibid.*, 40.
50. Ewing to Franklin, June 14, 1770, Franklin Papers, III, 17, American Philosophical Society.

The accounts which were published in 1771 in the American Philosophical Society *Transactions* were elaborate in scope and almost laborious in detail. The bulk of the transit accounts contrasted sharply with John Winthrop's brief report of his own observations which was published in the *Philosophical Transactions*. Winthrop's figures were accepted without question by European astronomers but this was because Winthrop was already known to them as a competent scientist and because such basic data as the longitude of Cambridge did not have to be determined anew. Yet, William Smith was probably correct when he remarked in the *Philosophical Transactions*, "We thought we should be the easier excused by men of science, for the insertion of twenty superfluous things, than the neglect of anything material." [51] In fact, Jean Bernoulli especially commended the society upon the fullness of its accounts, "a practice," he declared, "well worthy of imitation by those European astronomers who are so sparing of detail and who speak only in general terms of their instruments and their observations." [52] Some of the material included in the volume was a result of European demands. Nevil Maskelyne had specifically requested that cross country surveys of the distances to Norriton and Cape Henlopen be undertaken to link them all with a point in Philadelphia whose latitude and longitude was reliably known. This entailed considerable effort, but the members of the society were more than ready to perform the required tasks after the astronomer royal complimented them, remarking that their transit observations were "*excellent* and *compleat*, and do honour to the gentlemen who made them." [53]

These surveys, the calculations, the writing, and the printing of the transit papers took over a year. When all of the reports, tables, and plates relating to the transit had been printed, it was decided to bind them separately, without waiting for the rest of the volume to be completed and to present this portion to the Pennsylvania Assembly as an acknowledgment of the "Countenance and Encouragement" it had extended. [54] When the whole volume had been completed, a

51. *Philosophical Transactions*, 59 (1769), 326.
52. Jean Bernoulli, *Reçueil pour les Astronomes*, 2 (1772), 307.
53. American Philosophical Society, *Transactions*, 1 (1771), 40.
54. *Votes of Assembly*, VII, 6538.

copy was presented to each member of the assembly.[55] These overtures were a fitting expression of gratitude at the same time that they served as a prelude to requests for still more support. On February 11, 1773, the society sent the assembly a message of distress. It pointed out that "erecting our different Observatories, fitting up Instruments, engraving various Plates, and publishing the different *Transit* Papers, cost us (exclusive of the other Part of our Publication) near *Four Hundred Pounds*, and have involved us in Debt which we are wholly unable to discharge." [56] Once again the assembly responded—this time with a grant of £250. Altogether, the government of the Province of Pennsylvania subsidized the society's transit activities to the tune of nearly £450.

Compared with the £4,000 grant the Royal Society received from the crown for its observations, this was a meager sum, but in the colonies it was unique. In Philadelphia, the entire community took part in the enterprise. In addition to the assembly's benefactions, the proprietor contributed almost £100 and the Library Company gave £10 to improve their telescope which they lent for the purpose. Many individuals contributed heavily of their time while others lent important pieces of apparatus. In time and in skill, David Rittenhouse gave more than any other single person. He not only constructed the observatory and the essential instruments used at Norriton, but he provided an early projection of the transit, carried out the preliminary astronomical observations alone, and aided Provost Smith in calculating the final estimate of parallax from the results.[57]

Elsewhere in the colonies, the observations tended to depend more upon the interest or patronage of a single individual or upon the resources of a collegiate community. One man was solely responsible in the cases of William Alexander, John Leeds, William Poole, and John Page. John Page specifically complained of the failure of the faculty

55. American Philosophical Society Minutes, Feb. 1, 1771, [II], 119.
56. *Votes of Assembly*, VIII, 6948.
57. Rittenhouse's capacities appeared unusual even to some European critics. The Reverend Dr. William Ludlum declared, "*There is not another Society in the world, that can boast of a member such as* Mr. RITTENHOUSE: theorist enough to encounter the problems of determining (from a few Observations) the Orbit of a Comet; and also mechanic enough to make, with his own hands, an Equal-Altitude Instrument, a Transit-Telescope, and a Time-piece." Barton, *Memoirs of Rittenhouse*, 181n.

of the College of William and Mary to make use of the college apparatus for observing the transit.[58] On the other hand, Winthrop's observations were made with college apparatus and within the collegiate community. Most of the New York observations were made by people affiliated with the college. Both of the Rhode Island observations bore some of the marks of community enterprises but in each case, one patron stood out. At Newport, Ezra Stiles—academician without institutional affiliation—assumed more prominence than his patron. Nowhere did the government accept the major burden of support except in Pennsylvania.

The contribution of all this effort to the determination of the solar parallax was never understood by most of the observers. The simplest and most favored method of calculating the parallax depended upon a comparison of the differing times required for the full transit of Venus across the sun at two different observation points. This difference in observed time was a function, primarily, of the displacement of one observer from the other in latitude. It was also possible to calculate the parallax by comparing times of individual contacts between Venus and the sun obtained at observation points where the total duration of the transit had not been visible. These calculations depended more heavily upon exact determination of the difference in longitude between the observers. It was not possible to obtain longitude with the same precision as latitude so the partial transit observations were not favored by most of the men who attempted to work out the parallax.[59]

Of the parallax calculations, the best collections were published in the *Philosophical Transactions* and the *Mémoires* of the Académie des Sciences. The stations most frequently used in the journals for comparison were those in Hudson's Bay, in California, at King George's Isle in the Pacific, and at a few points in northern Europe. At each of these places the full transit had been observed. The European scientists, after scrutinizing their results, presented figures for the mean solar parallax that were within a fairly narrow range. Thomas Hornsby at Oxford calculated it to be 8.78″, Leonard Euler at St. Petersburg put forward 8.82″, and Joseph-Jérôme de La Lande used

58. *Virginia Gazette* (Purdie and Dixon), Aug. 3, 1769.
59. See *Mémoires de l'Académie*, 1770, 558.

the limits 8.53-8.63" to describe it.[60] All of them found individual comparisons which fell far outside this range. When Anders Johan Lexell at St. Petersburg calculated the parallax by comparing individual figures obtained at King George's Isle with the results of the observations at Hudson's Bay, Norriton, and Cambridge, Massachusetts, he found a distressingly wide spread from 8.98" to 10.51".[61]

In America, only one attempt to calculate the parallax was published. This work was undertaken by William Smith and David Rittenhouse. When the Philadelphians compared the Philadelphia, Norriton, and a few other American observations with those obtained at several European observatories, they obtained a myriad of different values ranging from 6.5" to 9.2". The mean value obtained from the comparison of Norriton and Philadelphia observations with ten European observations was calculated to the fourth nonsignificant place at 8.4764". This figure and the 8.805" obtained by comparing Norriton data with that of Greenwich were not far from the figures of Euler and La Lande.[62] The 8.805" estimate, in fact, would yield a distance of approximately ninety-three million miles from the earth to the sun—a value very close to the present accepted figure. That may or may not have been mere luck, but Smith was quite right when he declared, "The final determinations of the Sun's Parallax . . . will not be left to depend on our calculations in America." [63]

As far as actually trading criticism with the European astronomers, John Winthrop operated on a higher level than any other American. He found fault with the assumption made by Bliss and Hornsby that the transit would be accelerated by the equation of light. Winthrop's argument, that the transit would be retarded instead, was published in the *Philosophical Transactions* followed by a paper of the Reverend Richard Price which supported Winthrop's position but pointed out that other factors would have the opposite effect.[64] Price calculated that the net effect would cause the apparent transit to begin five minutes and fifty-five seconds after the instant of its actual beginning.

60. *Philosophical Transactions*, 61, (1771), 397-421; 62 (1772), 69-76; *Mémoires de l'Académie*, 1772, pt. 1, 798, 398.
61. *Philosophical Transactions*, 62 (1772), 75.
62. American Philosophical Society, *Transactions*, 1 (1771), Appendix 54-70.
63. *Ibid.*, Appendix 70.
64. *Philosophical Transactions*, 60 (1770), 358-62, 536-39.

Most of the other Americans who had turned their attention to the transit gave the Europeans raw observational data rather than thoughts of this sort.

The many observations presented in the *Transactions* of the American Philosophical Society brought exclamations of surprise and acclaim from different parts of Europe. They seemed to announce a new stage of maturity in the development of America. There was no indication of any feeling that the Americans were mere observers or that they had neglected anything. The Reverend Dr. William Ludlum wrote, "No astronomers could better deserve all possible encouragement, whether we consider their care and diligence in making the observations, their fidelity in relating what was done, or the clearness and accuracy of their reasonings on this curious and difficult subject." [65] The quality of the work was not much superior to Winthrop's observations of 1761. Now, however, the volume of activity was of a different order of magnitude and the publicity attending it was much more clever. The impact of the observations of 1769 was an entirely different matter.

The American achievement in observing the transit of Venus and in making American work known to European scientists was the major factor in gaining a new recognition for American science. Neither the fact that the American observations were less highly valued than observations of the entire transit nor the wide margin of error attached to the values finally accepted for the sun's parallax was particularly significant. European recognition was the important thing. The observations offered an opportunity to build individual and collective reputations. William Smith was right when he claimed a success that John Winthrop, a better scientist, would never have asserted. Smith declared, "It hath done a Credit to our Country which would have been cheaply purchased for twenty times the Sum!" [66]

65. *Gentleman's Magazine*, 41 (1771), 416.
66. Smith, *An Oration* (1773), 14.

‡‡

Chapter Nine

THE SCIENCES

THE TRANSIT OF VENUS observations were only the most dramatic evidence of a general and continuing interest in astronomy during the decade that preceded the Revolution. Indeed, those observations had assumed such great importance precisely because of the central position assigned to astronomy in the thought of the Enlightenment. Astronomy had been the particular concern of the immortal Newton. Whenever one dreamed of attaining fame in science, he was more than likely to think in terms of astronomy as did Philip Freneau and Hugh Henry Brackenridge when they bade the sons of philosophy:

> ascend with Newton to the skies,
> And trace the orbits of the rolling spheres,
> Survey the glories of the universe,
> Its suns and moons and ever blazing stars! [1]

Astronomy was a study which fused the utilitarian with the utterly useless in a way that could be twisted to confirm the prevailing assumption that all science was useful. The market value of knowledge of the techniques of navigation was evidenced by the frequent advertisements for courses in navigation offered in the coastal cities.[2] The application of astronomy to surveying was equally clear to the

1. Fred L. Pattee (ed.), *The Poems of Philip Freneau* (Princeton, 1902), I, 69.
2. *Pennsylania Gazette*, Sept. 29, 1768; *South-Carolina Gazette*, Oct. 12, 1769; *Virginia Gazette* (Rind), Feb. 23, 1769, March 1, 1770.

Americans who were in the midst of attempting to define their provincial boundaries with some exactness. The English astronomers, Charles Mason and Jeremiah Dixon, were employed to draw a portion of the famous line between Pennsylvania and Maryland. While in America, they also indulged themselves in less obviously useful pursuits, notably in the determination of the length of a degree of latitude in Pennsylvania. The very men who were called upon to complete the survey of Mason and Dixon were the ones who plotted the orbits of comets and speculated upon the properties of their tails. The literate American saw specific uses for astronomy in his almanac and was easily led to feel respect for those who understood the deeper mysteries beyond his ken.

To the thinker, astronomy had an importance beyond utility and beyond the mere satisfaction of intellectual curiosity. In a sense, astronomy seemed the key to the wisdom of the ages. The concept of natural law, the rational religion of the time, the faith in human progress, and the swelling comprehension of infinity all seemed somehow to follow from the ordered nature of the heavens with its precisely predictable events. No better expression of the emotional lift men gained from contemplating the stellar universe could be found than in the statement of David Rittenhouse: "All yonder stars innumerable, with their dependencies, may perhaps compose but the leaf of a flower in the creator's garden, or a single pillar in the immense building of the divine architect. Here is ample provision made for the all-grasping mind of man!" [3]

In this atmosphere, it was to be expected that an attempt would be made to translate the enthusiasm of the transit observations into something of a more permanent nature. It was the Reverend John Ewing who first turned his attention to such an establishment in Philadelphia where the enthusiasm had reached its peak and where the wooden observatory built with funds appropriated by the assembly still stood in the State House Square. Ewing envisaged a permanent public observatory which would make regular observations just as the observatories of Europe did. To this end, he wrote the astronomer royal, Nevil Maskelyne, and Benjamin Franklin in England.

3. David Rittenhouse, *An Oration Delivered February 24, 1775, Before the American Philosophical Society* (Philadelphia, 1775), 26.

Maskelyne, he felt, might be able to find support for the observatory in England, especially if it were placed under the supervision of Greenwich. Without Franklin's active advocacy, he was convinced that it was unreasonable to expect that the Pennsylvania Assembly "would lay out any of the public Money for such a Purpose." [4] In Philadelphia, Ewing remained silent, hoping that Franklin and Maskelyne could be prevailed upon to initiate the proposal. They declined. Franklin's preliminary investigation indicated that the expense would be too heavy.[5]

Five years later, the American Philosophical Society made a concerted effort to bring about the establishment of an observatory. This time the appeal was made to Pennsylvania agencies. It was hoped that the proprietors of the province might be willing to donate a lot upon which the observatory could be erected and that a public subscription would yield the funds needed to construct it. The assembly was counted upon to assume responsibility for maintenance including salaries. An eloquent petition declared to the assembly, "Our Distance from the chief *Observatories* in the World, the Purity and Serenity of our Atmosphere, invite us, nay loudly call upon us, to institute a Series of regular *Astronomical Observations*." [6] Once again, the appeal was made on two grounds: the advancement of learning and utility. Observations would be published annually and sent to the learned societies of Europe. At the same time, the observatory would teach navigation, including the latest method of ascertaining longitude. In addition, a "surveyor of High Roads and Waters," attached to the observatory, would always be available for surveying.

The establishment of an observatory on this plan would serve another very important purpose by providing a satisfactory post for David Rittenhouse—a man "whose Abilities, speculative as well as practical, would do Honour to any Country." As public observer, Rittenhouse would be rescued "from the Drudgery of manual Labour," and given an "Occasion of indulging his bent of Genius." In

4. Ewing to Franklin, Jan. 4, 1770, Franklin Papers, III, 1, American Philosophical Society.

5. Franklin to Ewing, Aug. 27, 1770, Albert Henry Smyth (ed.), *The Writings of Benjamin Franklin* (New York, 1905-7), V, 270.

6. *Votes and Proceedings of the House of Representatives of the Province of Pennsylvania, Pennsylvania Archives*, 8th ser., VIII, 7206.

this respect, the planned observatory was only the most recent of several attempts that had been made to provide adequate patronage for a man whose scientific capacities were rated at the highest level throughout the Middle colonies. The plan was not beyond the capabilities of the colony, but it was brought to an abrupt end by the shooting on Lexington green in April 1775, just one month after the petition was presented. It was after this event that Maskelyne wrote Ewing finally declining to have any part in the establishment of an observatory at Philadelphia "in the present unhappy situation of American affairs." [7] Scientific ambitions which had been nurtured in the stimulating atmosphere of pre-revolutionary thought, were inhibited by the outbreak of actual war.

Before war brought many promising plans to an end, the Americans were able to give evidence of a disposition to honor scientific achievement and to provide for its encouragement. The offer of the presidency of Harvard on two occasions to John Winthrop was evidence of his standing among his colleagues. So was his election to the Royal Society which was accomplished through the efforts of Benjamin Franklin.[8] Similar honor was paid David Rittenhouse when he was granted the degree of master of arts by the College of Philadelphia.[9] For him, however, that was only the beginning.

David Rittenhouse was unusual in his superior understanding of astronomy and mathematics despite very little in the way of formal education. His first important patron was the Reverend Thomas Barton, Anglican clergyman of Lancaster, Pennsylvania, who contributed to his intellectual development and brought him to the attention of important provincial political figures.[10] The Reverend Richard Peters, secretary to the governor, hired Rittenhouse to make some measurements of latitude at the southeastern corner of Pennsylvania preliminary to the surveying of the southern boundary by Mason and Dixon.[11] It was Barton who first urged Rittenhouse to move to Phila-

7. Maskelyne to Ewing, Aug. 4, 1775, Misc. Mss., American Philosophical Society.

8. Raymond P. Stearns, "Colonial Fellows of the Royal Society of London, 1661-1788," *William and Mary Quarterly*, 3d ser., 3 (1946), 249.

9. *Pennsylvania Chronicle*, Nov. 23, 1767.

10. William Barton, *Memoirs of the Life of David Rittenhouse* (Philadelphia, 1813), 104, 150.

11. Edward Ford, *David Rittenhouse* (Philadelphia, 1946), 23.

delphia and seek a sinecure there which would permit him to concentrate his efforts upon creative work. No actual steps to accomplish this move were taken until the Reverend William Smith interested himself in Rittenhouse.

Smith's attention was initially captured by the orrery or mechanical planetarium which Rittenhouse began in 1767. These devices were used in the eighteenth century as teaching aids but that was not the reason they were so highly valued. They also served as monuments to the faith of the Enlightenment in the reasonableness of the world. The same year that Rittenhouse began his mechanical representation of the solar system, Harvard College was billed £92 10s 6d for an orrery constructed by the well-known London instrument maker, Benjamin Martin.[12] Rittenhouse projected an instrument more exact and precise than Martin's, not designed to instruct "the ignorant" but to "astonish the skilful." [13] Barton sent him all the accounts he could find of other orreries while Smith kept promising that if the assembly could not be prevailed upon to purchase the mechanism, he would arrange for the college to do so. Before he made any agreement, however, the new Scottish president of the College of New Jersey, the Reverend John Witherspoon, visited Rittenhouse and purchased the device on the spot.[14] This move by his Presbyterian competitor infuriated Smith, who salvaged something of his expectations by prevailing upon Rittenhouse to construct a second orrery which would go to the College of Philadelphia and would be delivered before the New Jersey orrery. In the end Witherspoon paid £300 for one orrery and Smith advanced £300 for the other. At the same time, the Pennsylvania Assembly made a grant of £300 in recognition of the achievement and offered Rittenhouse another £400 to construct a third orrery "for the use of the Public." [15]

Rittenhouse's orrery had become a symbol upon which substantial funds were lavished when competition required that its value be demonstrated in economic terms. It was a fine piece of craftsmanship,

12. The bill was paid by James Bowdoin who donated the orrery. I. Bernard Cohen, *Some Early Tools of American Science* (Cambridge, 1950), 157.
13. Barton, *Memoirs of Rittenhouse*, 194.
14. *Ibid.*, 214n.
15. *Pennsylvania Gazette*, March 21, 1771, March 28, 1771; *Votes of Assembly*, VIII, 6662, 6716.

presenting the orbits in their proper eccentricity and inclination. It displayed the relative positions of the planets for any time over a five thousand-year period with a margin of error of less than one degree. Its precision was regarded as symbolic of the precision both of God's universe and of man's science. Without restraint the press lauded its builder, "whose singular Genius would have done Honour to the Age and Country in which a Newton lived." [16] The orrery was the best concrete representation of Newton's mechanical world.

Excitement over the orreries helped to sustain the interest in astronomy aroused by the transit of Venus. The telescopes and clocks which had been bought or refurbished for the transit were pressed into recurrent service. Observations of Jupiter's satellites were made to establish geographical positions and, though of little general importance, they were occasionally reported in learned journals.[17] Many American telescopes were trained on a transit of Mercury which occurred just five months after the transit of Venus of 1769, even though it had no importance for determining the sun's parallax. It did serve to check predictions of Mercury's orbit. Accounts of the American observations of the transit of Mercury were published in local newspapers and in the journals of the American Philosophical Society, the Royal Society, and the Académie des Sciences. Again the observations by John Winthrop and the committees of the American Philosophical Society were those most valued.[18]

More spectacular than transits were the awe-inspiring comets which occasionally brought the American astronomers to their telescopes in some numbers. They became particularly interested in the two

16. *Pennsylvania Gazette*, March 28, 1771; James Stokely, "The Rittenhouse Exhibition," *Pennsylvania Magazine of History and Biography*, 56 (1932), 240; Howard C. Rice, Jr., *The Rittenhouse Orrery* (Princeton, 1954), 68-75.

17. *Philosophical Transactions*, 58 (1768), 329-35; 59 (1769), 358; 64 (1774), 171-76, 177-81, 182-83, 190-93.

18. *New-York Gazette; and the Weekly Mercury*, Oct. 30, 1769, Nov. 13, 1769; *Connecticut Courant*, Nov. 13, 1769; *Pennsylvania Chronicle*, Oct. 30, 1769; Nov. 20, 1769; American Philosophical Society, *Transactions*, 1 (1771), 82-88, Appendix 50-54; *Philosophical Transactions*, 61 (1771), 51-52; *Mémoires de l'Académie*, 1772, pt. 1, 445; Isabel M. Calder (ed.), *Letters and Papers of Ezra Stiles* (New Haven, 1933), 22; Stiles, Observations of the Transit of Mercury, 1769, Stiles Papers, Yale University; American Philosophical Society Minutes, [II], 134.

comets that appeared, one in 1769 and the other in 1770, each of which was discovered first in Europe but was then followed closely by the Americans. From Cambridge to Charleston, anxious eyes watched every day for the comet, often reporting its position to local newspapers and sometimes winning front page notice.[19] Most of the observers did no more than note its position but both Winthrop and Rittenhouse attempted to derive the orbit of the comet of 1770. They did not agree in their predictions which unfortunately could not be checked immediately because the comet faded too soon from sight. Confirmation of Rittenhouse's theory was obtained when the *Gentleman's Magazine* for July and August arrived in America. Calculations made by Charles Messier in France agreed very well with those of Rittenhouse, while James Six in England actually observed the comet at a point close to that predicted by Rittenhouse.[20]

Characteristically, Rittenhouse did not indulge in any speculation upon the physical nature of comets. Winthrop did so in his "Cogitata de Cometis" but his ideas were much more restrained and reasonable than those of other Americans who published speculative thoughts on comets.[21] Hugh Williamson wrote an essay in which he went so far as to suggest that comets were cool, solid, opaque bodies and were probably inhabited. This idea was so well regarded in Europe that it won him election to the "Society of Science in Holland."[22] Andrew

19. Frederick P. Bowes, *The Culture of Early Charleston* (Chapel Hill, 1942), 87; *Virginia Gazette* (Purdie and Dixon), Oct. 25, 1770, Sept. 20, 1770, Sept. 13, 1770; *Virginia Gazette* (Rind), Sept. 14, 1769, Oct. 19, 1769, July 26, 1770; Franklin B. Dexter (ed.), *The Literary Diary of Ezra Stiles* (New Haven, 1901), I, 21-27, 57; Stiles, Observations of the Comet, Stiles Papers; *Pennsylvania Gazette*, Sept. 14, 1769, Nov. 9, 1769; *Connecticut Courant*, Sept. 18, 1769, Oct. 30, 1769, Nov. 6, 1769, July 9, 1770; *New-York Gazette; and the Weekly Mercury*, Sept. 11, 1769, Sept. 25, 1769; Franklin to Le Roy, Sept. 22, 1769, Franklin Papers, XLV, 359, American Philosophical Society; William E. Foster, "Some Rhode Island Contributions to the Intellectual Life of the Last Century," American Antiquarian Society, *Proceedings*, 8 (1892), 128.

20. Rittenhouse to Thomas Barton, July 30, 1770, Barton, *Memoirs of Rittenhouse*, 220-22; Rittenhouse to Smith, Dec. 2, 1770, Horace W. Smith, *Life and Correspondence of the Rev. William Smith, D.D.* (Philadelphia, 1879), I, 454-55; Smith to Rittenhouse, [n.d.], *ibid.*, 455; American Philosophical Society, *Transactions*, 1 (1771), Appendix 37-45.

21. *Philosophical Transactions*, 57 (1767), 132-54.

22. Hugh Williamson, "An Essay on the Use of Comets," American Philosophical Society, *Transactions*, 1 (1771), Appendix 27-36; *Pennsylvania Chronicle*, Oct. 11, 1773.

Oliver's *Essay on Comets* attempted to explain the direction of the comet's tail by the pressure of the sun's atmosphere. The idea did not appeal to Joseph Priestley but the French astronomer, Jean-Sylvain Baily, thought enough of Oliver's introduction of electricity into the problem to translate his pamphlet into French.[23] All of these speculations were acceptable to the European world of science, but one newspaper essay went too far. In the *New-York Gazette*, S. Sp. Skinner suggested that comets were absorbed by the sun, serving to replenish its used-up energy and appearing as sunspots.[24]

The sun attracted little attention aside from the observation and calculation of eclipses.[25] However, Humphry Marshall, a cousin of John Bartram and a seedsman and naturalist in his own right, turned his attention to sunspots. Over a period of years he watched the sun through a small telescope, noting and making sketches of the spots he observed. He sent papers on this subject to the American Philosophical Society and to Benjamin Franklin in London, who turned them over to the Royal Society. Franklin acquainted him with current opinions about the nature of sunspots but Marshall had few ideas in this direction. His reports were confined to comments on the apparent closeness of the spots to the sun's surface, their velocity, and their changes of shape. The American Philosophical Society thought well of the observations but never published them while the Royal Society did give them a place in its *Philosophical Transactions*.[26]

Another and more significant error of judgment was demonstrated when the American Philosophical Society failed to publish a paper on the aurora borealis submitted by young Benjamin Thompson of Concord, Massachusetts. Aurorae were perpetual objects of interest. Since one descriptive account had been published in the society's first

23. Andrew Oliver, *Essay on Comets* (Boston, 1772); Priestley to Oliver, Feb. 12, 1775, Massachusetts Historical Society, *Proceedings*, 2d ser., 3 (1886-87), 13-14.

24. *New-York Gazette; and the Weekly Mercury*, Nov. 27, 1769; *Connecticut Courant*, Oct. 16, 1769, Oct. 23, 1769, Oct. 30, 1769.

25. American Philosophical Society Minutes, [II], 130, 131; Dexter (ed.), *Literary Diary of Stiles*, I, 183; *Philosophical Transactions*, 57 (1767), 215-16.

26. *Ibid.*, 64 (1774), 194-95; Franklin to Marshall, May 14, 1774, Franklin Papers, XIV, 19, American Philosophical Society; Marshall to Isaac or Moses Bartram, March 23, 1771, Ms. Communications to American Philosophical Society, Mathematics and Astronomy, I, 9; American Philosophical Society Minutes, Jan. 17, 1772, [II], 133.

volume of *Transactions*, Thompson's letter arrived at a bad time, over a year after the appearance of that volume. There was all the difference in the world between the anonymous published account and Thompson's unpublished letter. Both of them were descriptive, but where the earlier author was general and loose, Thompson was precise, noting the time of each change and the dimensions in degrees of the elements described. Benjamin Thompson was then only nineteen years old; his brilliant scientific career lay well ahead of him, but his meticulous care for details was already apparent in this paper. Perhaps the committee which endorsed it "Doubtful" had just seen too many accounts of aurorae.[27]

Besides the observation of spectacular astronomical events there was another, more utilitarian astronomy. Town lots could be surveyed by men with a knowledge of elementary mathematics but the marking off of provincial boundaries and the laying out of large blocks of land required astronomy. It was for this reason that men of some prominence in science often served as surveyors-general of their province —Cadwallader Colden, James Alexander, William Alexander, John Leeds, and William Parsons among them. The best surveyors available were called upon actually to mark off the boundaries between colonies. In this connection, Captain Samuel Holland of the British Army, former surveyor-general of the Northern District, and some of the members of his company served very capably. Holland played an important part in drawing the New York-New Jersey line, and he began work on the boundary between New York and Pennsylvania just before the outbreak of the Revolution.[28] Members of his company made surveys for New Hampshire, Thomas Wright checking the ends of the Massachusetts-New Hampshire line.[29] By all odds, the most important American surveyor was David Rittenhouse. He

27. *Connecticut Courant*, Sept. 25, 1770; Franklin B. Dexter (ed.), *Excerpts from the Itineraries and Other Miscellanies of Ezra Stiles, D.D., LL.D.* (New Haven, 1916), 476; American Philosophical Society, *Transactions*, 1 (1771), 338-39; American Philosophical Society Minutes, April 2, 1773, [II], 145; Franklin to American Philosophical Society, July 25, 1772, Ms. Communications to American Philosophical Society, Natural Philosophy, I, 4.

28. *Pennsylvania Chronicle*, Oct. 16, 1769; Barton, *Memoirs of Rittenhouse*, 158.

29. Jeremy Belknap, *The History of New Hampshire* (2d ed., Boston, 1813), III, 10-14, 402-3.

was particularly active in connection with Pennsylvania's boundaries, but he served with Holland on the New York-New Jersey line as well as on the New York-Pennsylvania line.[30]

This kind of work advanced the knowledge of American geography in addition to the more practical political and economic functions it served. It also led to occasional scientific by-products of some value. Holland, Thomas Wright, who was his deputy, and Ensign George Sproule published several papers in the *Philosophical Transactions* describing astronomical observations made in the course of their work. Most of them were accounts of eclipses of Jupiter's satellites taken to establish geographical fixes in Canada. Some observations made in New Hampshire and Maine were also included. Copies of most of these papers were laid before the American Philosophical Society but not until they had already been published by the Royal Society.[31]

Certainly the best publicized survey of the entire period was that conducted by Mason and Dixon to determine the boundary between Pennsylvania and Maryland. Between 1763 and 1768, the two astronomers obtained the latitude and longitude at several points close to the line by numerous and lengthy observations. They ran the boundary as far west as possible before Indian troubles stopped them. Although in some respects they were less accurate than the American Philosophical Society's astronomers of 1769, their work long remained a standard.[32] Seeing an important opportunity for scientific work of general importance, Mason and Dixon suggested to the Royal Society the measurement of the length of a degree of latitude. Penn offered to assume the expenses and the society agreed to supervise the measurement. The required apparatus was supplied, including a standard five-foot brass measure and fir rods to be used in terrestrial measurement in place of the less accurate surveyor's chain. Mason and Dixon published their conclusions in the *Philosophical*

30. *Philosophical Transactions*, 64 (1774), 173-74; Barton, *Memoirs of Rittenhouse*, 158.

31. *Philosophical Transactions*, 64 (1774), 171-75, 177, 182, 184-89, 190-93; 58 (1768), 46; Samuel Holland to American Philosophical Society, Read Feb. 7, 1772, Ms. Communications to American Philosophical Society, Mathematics and Astronomy, I, 10.

32. Thomas D. Cope, "Degrees along the West Line," American Philosophical Society, *Proceedings*, 93 (1949), 128.

Transactions. There they also published the results of an experiment to determine the magnitude of gravity at the forks of the Brandy-wine Creek in Pennsylvania, as compared with that at the Royal Observatory at Greenwich, by measuring the rate of going of a clock which had been previously checked at Greenwich.[33] This work was of more scientific value than anything else that grew out of the surveys of provincial boundaries.

The contact maintained by Mason, Dixon, and Holland with the world of science, primarily through the Royal Society, differentiated their work in a striking manner from that of the Americans. They were constantly alert to perform observations of interest to European scientists and to seize every opportunity for publication. Their training, of course, had been superior to that of most of the Americans, but there is no indication that their attainments were superior to those of Rittenhouse. The big difference lay in the intimate ties they had with European scientific circles and the stimulus thus obtained. This difference also characterized a very active group of men who were surveying Florida and charting its waters at this time.

That work began shortly after Britain received Florida from Spain at the conclusion of the Seven Years' War. In March 1764, George Gauld was engaged by the Admiralty to survey the coasts and harbors of West Florida and the west coast of East Florida. In 1771, he began a similar charting of Kingston harbor, Jamaica, and then, with assistance from the navy, the Dry Tortugas and the Florida keys.[34] In June 1764 William Gerard De Brahm, appointed surveyor for the Southern District of North America, began a survey of the Atlantic coast. By June 1772, he was able to return surveys of the coast as far north as 30° 26′ 49″, approximately the latitude of Jacksonville.[35] Bernard Romans, a Dutch engineer, who had been employed privately in Florida, was made principal deputy surveyor for the Southern District under De Brahm. In this capacity, he charted much of the middle and west coast of East Florida. A little later he

33. *Philosophical Transactions,* 58 (1768), 270-335.
34. [George Gauld], *An Account of the Surveys of Florida* (London, 1790), 4-5.
35. William Gerard De Brahm, *History of the Province of Georgia* (Wormsloe, Ga., 1849), 7.

was employed by John Stuart, superintendent for Indian affairs of the Southern District, in surveys of West Florida.[36] Here he met Dr. John Lorimer, a resident of Pensacola who had come to America in 1764.[37] These four men, Gauld, De Brahm, Romans, and Lorimer, added much to the world's understanding of the cartography and hydrography of the Florida region and something to the stock of knowledge of natural history.

All of their initial connections were with London rather than the American colonies. A letter of Lorimer describing a dipping needle he had invented was published in the *Philosophical Transactions* in 1775.[38] He had designed it and had it built before leaving England. The design of the needle permitted its use at sea to yield both azimuth and elevation. Neither Gauld's general *Account of Florida* nor his charts and sailing directions were published until years after the American Revolution and after his own death.[39] De Brahm had only begun his publication by the outbreak of the American Revolution but his *Atlantic Pilot,* a sailing guide based in part on his extensive surveys in Georgia, South Carolina, and Florida, did appear in 1772.[40] It was Bernard Romans who first established intimate contact with American scientific circles.

Romans had passed through a succession of jobs, leaving both friends and a heritage of tension behind him. He asserted that neither De Brahm nor Stuart had paid him for his services; yet he claimed to be on the most intimate of terms with Gauld and Lorimer.[41] When he stopped off at Charleston in 1773, Alexander Garden had an opportunity to observe his enthusiasm for the establishment of a botanical garden in Florida and his tenacity. He had been recommended to John Ellis in London as superintendent of the proposed garden, but Alexander Garden was not much impressed with Romans' knowledge of

36. P. Lee Phillips, *Notes on the Life and Works of Bernard Romans* (Deland, Fla., 1924), 29.
37. *Ibid.,* 27; Lawrence C. Wroth, *Abel Buell* ([New Haven], 1926), 57.
38. *Philosophical Transactions,* 65 (1775), 79-84.
39. [Gauld], *Account of Florida;* George Gauld, *Observations on the Florida Kays* (London, 1796).
40. De Brahm, *History of Georgia,* 11; William Gerard De Brahm, *The Atlantic Pilot* (London, 1772).
41. Phillips, *Works of Bernard Romans,* 29, 30.

botany.⁴² On the other hand, when Ezra Stiles met him a little later, he recorded his opinions on Indians, Eskimos, and the classics with every evidence of esteem.⁴³

Later in 1773, Romans found his way to New York where he presented plans for the publication of charts, sailing directions, and a natural history of Florida and its vicinity. Before making this trip, he had had favorable correspondence with members of the recently formed Marine Society of New York, an organization which proved to be a great help in pushing his plans to completion. The mariners were particularly pleased with his work on the Bahama Banks and were anxious to get the whole of it into print. Romans sought subscriptions to his work in Charleston, New York, Philadelphia, and Boston appealing both to "the Sage in his Cabinet" and to "the Mariner in his Ship." ⁴⁴ In 1775, the charts and *A Concise Natural History of East and West Florida* were published in New York. The book, dedicated to John Ellis, may have been designed to establish a reputation for Romans as a naturalist, but its pages devoted to sailing directions formed its most valuable part.⁴⁵

Romans visited Philadelphia in his quest of pre-publication support shortly after the American Philosophical Society had received communications from Dr. Lorimer and George Gauld. Lorimer sent seeds and botanical specimens while Gauld forwarded specimens of fish, a description by himself of West Florida, and a description by Lorimer of the Chester River.⁴⁶ The last of these communications was dated February 15, 1773, the same month that Romans' work for John Stuart ended and just about the time he left Florida for the North.⁴⁷ The events were not unconnected. Since neither Lorimer nor Gauld mentioned Romans in their letters, they were probably

42. Garden to Ellis, May 15, 1773, Smith (ed.), *Correspondence of Linnaeus*, I, 596.
43. Dexter (ed.), *Literary Diary of Stiles*, I, 524.
44. *Boston Gazette*, Jan. 10, 1774, cited by Phillips, *Works of Bernard Romans*, 26.
45. Bernard Romans, *A Concise Natural History of East and West-Florida* (New York, 1775).
46. American Philosophical Society Minutes, Feb. 21, 1772, July 18, 1773, [II], 134, 146.
47. Phillips, *Works of Bernard Romans*, 30; George Gauld, Height of Catherine's Hill, Feb. 15, 1773, Archives, American Philosophical Society.

anxious that he not assume the sole scientific merit of the Florida group.

At any rate, when Romans attended a meeting of the society on August 20, 1773, the West Florida navigation chart he presented to them was referred to a committee for comparison with the account received from Gauld just the month before.[48] There is no record that Romans received any special commendation from the society for his chart or his papers but he may have learned something from the manner in which one of Gauld's accounts had been handled. In a very brief paper, Gauld had reported the height he had obtained for two hills in Jamaica. The society endorsed it, "Appears to have made very accurate measurement," but "Such and so long a Piece meerly to tell us what is the height of 2 Hills in some Part of the British Dominion could hardly obtain a Place in the Transactions of an American Society." [49] Gauld could not have anticipated the extent of patriotic feeling that had developed by 1773. Romans took it all in, and in 1774 published a poem celebrating the bright destiny of "America." [50]

Army officers played an important part in the surveys of Florida just as Holland and his subordinates did in the North. Captain Phillip Pitman served with this group and Romans himself was a retired captain. The most prominent of the active officers who worked on the Florida surveys was Thomas Hutchins, a lieutenant in the Royal American Regiment and an American by birth.[51] Before his Florida duty, Hutchins had served in several campaigns of the French and Indian War and had had other western assignments which led him to plan a map of the interior of the continent for which he sought subscriptions in 1771 and 1772.[52] It was not published until after the Revolution began when the map was issued with a *Topographical Description* of the areas mapped.[53] From 1772 to 1777 he served as an engi-

48. American Philosophical Society Minutes, Aug. 20, 1773, [II], 148-49.
49. Gauld, Height of Catherine's Hill.
50. *Royal American Magazine*, 1 (1774), 32-33.
51. American Philosophical Society Minutes, July 20, 1769, [II], 47; *Maryland Gazette*, April 18, 1765.
52. *Virginia Gazette* (Purdie and Dixon), May 28, 1772; *Pennsylvania Gazette*, Dec. 5, 1771; Ruth Henline, Travel Literature of Colonists in America, 1754-1783 (Doctoral Dissertation, Northwestern University, 1947), 70.
53. Thomas Hutchins, *A Topographical Description of Virginia, Pennsylvania, Maryland, and North Carolina* (London, 1778).

neer in West Florida where he became well acquainted with George Gauld. It took many years longer to get the results of these surveys into print but in 1775, he did publish in the *Philosophical Transactions* a series of experiments on the dipping needle, "Made by desire of the Royal Society." [54] To the American Philosophical Society he sent an account of the Illinois country.[55]

Surveying, exploring, map-making, and the charting of navigable waters were carried on with much more energy in the newer and more remote colonies than in the settled parts of America. Charts and sailing directions were already in print for the most frequented ports and the best of the charts did not have to be done again; Joshua Fisher's *Chart of Delaware Bay* of 1756 remained standard until the end of the century.[56] In cartography too, earlier maps of the settled regions often continued to stand against later efforts. John Henry's *Map of Virginia* did not in any way challenge the Fry-Jefferson *Map of the Inhabited Part of Virginia* of 1754.[57] Nothing superseded the Mitchell map or the Evans map of the Middle colonies despite the numerous more limited maps which did appear.[58] It was with respect to the frontier regions that the greatest advances were made.[59] Yet, even in the busiest ports, more accurate information—more precise fixes for instance—continued to increase.[60] This was a process of refinement.

Most of the surveyors and map-makers avoided the formation of theoretical hypotheses, but with respect to one problem they were anxious to find general law. They wanted to reduce the apparent vagaries of the earth's magnetic field to a predictable pattern. The variation of the compass from true north was not constant but seemed

54. Thomas Hutchins, *An Historical Narrative and Topographical Description of Louisiana and West-Florida* (Philadelphia, 1784); *Philosophical Transactions*, 65 (1775), 129.

55. American Philosophical Society Minutes, Dec. 20, 1771, [II], 132.

56. Lawrence C. Wroth, *Some American Contributions to the Art of Navigation* (Providence, [n.d.]), 22-23.

57. *Virginia Gazette* (Rind), Sept. 27, 1770.

58. *Archives of the State of New Jersey*, 1st ser., *Documents Relating to the Colonial History, 1631-1800* (Newark, etc., 1880-1906), XXVI, 187-89; *Pennsylvania Gazette*, Feb. 2, 1769; see Henry Monzon, *An Accurate Map of North and South Carolina with their Indian Frontiers* (London, 1775).

59. Samuel Langdon to Stiles, [n.d.], Dexter (ed.), *Itineraries of Stiles*, 524.

60. *New-York Gazette; and the Weekly Mercury*, Oct. 23, 1769.

to change in a haphazard manner from point to point on the surface of the earth. John Winthrop gave some thought to this matter and Hugh Williamson submitted a plan to the American Philosophical Society for collecting data on variation.[61] It was generally recognized that the earlier attempt of William Mountaine and James Dobson in England to construct a variation chart had been unsatisfactory because they possessed inadequate information. The Philosophical Society, therefore, used the public press to invite "the Ingenious and Learned, through the several colonies," to collect data on compass variation.[62] The results were very disappointing but, even so, William Alexander felt that he had enough information to formulate the general law that on the eastern side of the line of no variation, the variation increased uniformly as one moved to the northeast.[63] All of these men sought a simplicity that did not exist.

In the broad field of exploring, surveying, and map-making, the astronomers and the men trained in mathematics met the naturalist. Sometimes the two were united in one man, an ambition which Romans and De Brahm sought to fulfill. Rittenhouse, who made no real attempt to be a naturalist, and Bartram, who left mathematics and astronomy alone, were perhaps wiser; certainly they were more typical. The naturalist and the surveyor could work on different aspects of the same problem. Just as De Brahm, Gauld, and the others were set to work surveying Florida shortly after its acquisition, so John Bartram, appointed King's botanist, was permitted to study the area in a different way. His trip resulted in the introduction of many new plants to England and in a journal describing the country. This book was published in England because of the new and growing importance of Florida to land speculators as well as to naturalists.[64]

Bartram's Florida trip demonstrated anew the vitality of the natural history circle. Peter Collinson expressed himself well satisfied

61. Clap to Stiles, June 26, 1765, Dexter (ed.), *Itineraries of Stiles*, 452; American Philosophical Society Minutes, Dec. 20, 1771, [II], 131.

62. *Philosophical Transactions*, 50 (1757), 329-49; *Pennsylvania Gazette*, Feb. 20, 1772.

63. William Alexander, Variation of the Compass, March 27, 1773, Archives, American Philosophical Society.

64. It appeared as a separately paged section in William Stork, *An Account of East-Florida, with a Journal, kept by John Bartram of Philadelphia, Botanist to his Majesty for the Floridas* (London, [1767]).

with John's southern excursion, urging him to send journals of all his trips to the King, his patron.[65] The King himself was pleased with the seeds and other items Bartram sent back from Florida.[66] During this trip Bartram had an opportunity to visit Alexander Garden and his other Charleston friends. It was on this occasion, too, that he first introduced his son, William, to Florida.[67] Then, just a few years later, William returned under the patronage of John Fothergill for an extended investigation of the interior. It was through a new member of the circle, Dr. Lionel Chalmers of Charleston, that Fothergill's money was advanced to the younger Bartram.[68]

Like Lining and Garden, Chalmers was an Edinburgh-trained Scot with energy and ambition. His primary interest was medicine. Among the five papers he had submitted to the old American Society at Philadelphia, the most important tried to relate weather and disease. Most eighteenth-century physicians gave some thought to this problem; John Lining had even written on it. None of them studied the relationship in America so closely as Chalmers.[69] His carefully collected records of weather and disease were published in London in a two-volume work entitled *An Account of the Weather and Diseases of South-Carolina*.[70]

Interest in meteorology extended well beyond the medical group to include academic men and naturalists. One reason for the appeal of weather study was the development of instruments for the quantitative measurement of temperature, atmospheric pressure, humidity, and wind force. In the eighteenth century, for the first time, it became possible to weigh and measure the weather and then to classify

65. Bartram to Franklin, April 10, 1769, Franklin Papers, II, 169, American Philosophical Society; Collinson to Bartram, Feb. 10, 1767, William Darlington, *Memorials of John Bartram and Humphry Marshall* (Philadelphia, 1843), 286.

66. Collinson to Bartram, Aug. 21, 1766, Darlington, *Memorials*, 282.

67. Garden to Ellis, July 15, 1765, Smith (ed.), *Correspondence of Linnaeus*, I, 536-38.

68. Chalmers to Bartram, April 7, 1773, Gratz Collection, Historical Society of Pennsylvania.

69. Chalmers to the American Society, Nov. 2, 1768, American Society Minutes, 144, American Philosophical Society; *Pennsylvania Chronicle*, Dec. 12, 1768.

70. Lionel Chalmers, *An Account of the Weather and Diseases of South-Carolina* (2 vols., London, 1776).

it in a manner so dear to the mind of the Enlightenment. The faith
that such observation and classification would yield important scientific results was unclouded. Numerous projects developed in Europe
for systematic study of meteorology, and the general interest of the
literate classes easily spilled over into America. American newspapers
and magazines were full of meteorological observations on a day-by-
day basis in addition to occasional reports of unusually severe snow
storms, dark days, and similar phenomena.[71] Meteorological diaries
were kept by John Winthrop, Ezra Stiles, Samuel Williams, Edward
A. Holyoke, John Page, and Phineas Pemberton, among others.[72]
Winthrop sent two reports to the Royal Society but much of the
data on American weather received by that body came from individuals working under its direction or for some branch of the government. Thus Wales and Dymond returned meteorological records
while on their trip to observe the transit of Venus, and Thomas
Hutchins sent in others.[73] In manuscript notebooks and in print, extensive records of daily variations of temperature, pressure, wind,
and general weather conditions were piled up.

A few Americans sought to interpret these data in terms other than
in relation to the incidence of disease. Ezra Stiles was convinced that
urban temperatures were higher than those of the surrounding countryside, and Hugh Williamson wrote a paper trying to establish
general changes in climate as a function of the settlement of the
country.[74] Benjamin Franklin was also interested in this problem but
his speculations went much deeper. In attempting to account for
various meteorological phenomena, Franklin postulated air particles

71. *New-York Weekly Mercury*, Feb. 4, 1765; *Maryland Gazette*, Feb. 21,
1765; *Pennsylvania Chronicle*, May 15, 1769, July 10, 1769, Aug. 28, 1769; *Penn-
sylvania Magazine*, 1 (1775), 53, 99, and each successive month; *Royal American
Magazine*, 1 (1774), 40, and each successive month.
72. John Winthrop, Meteorological Journal, 3 vols., American Academy
of Arts and Sciences; Stiles, Meteorological Journal, Stiles Papers; Edward
Holyoke, Meteorological Journal, American Academy of Arts and Sciences;
Samuel Williams, Account, Jan. 21, 1774, American Philosophical Society
Minutes, [II], 163; David James and John Page, Meteorological Observations,
Ms. Communications to American Philosophical Society, Natural Philosophy,
I, 3, 4; Phineas Pemberton, A Meteorological Register, *ibid.*, 8.
73. *Philosophical Transactions*, 60 (1770), 137-78; 56 (1766), 291.
74. Stiles, Meteorological Journal, I, not paged, Stiles Papers; American
Philosophical Society Minutes, Aug. 17, 1770, [II], 111.

which were in the form of equilateral triangles. Published in the *Philosophical Transactions*, this idea received considerable attention even though in the long run it did not help much.[75] In fact, those who wrote on such simple topics as tornadoes, hurricanes, and water-spouts were probably wiser although here too Franklin stumbled in accepting the common idea that waterspouts were composed of ocean water.[76]

Everyone could talk about the weather but more esoteric subjects were introduced by some of the explorers. The learned world became particularly excited about the big bones occasionally brought back from the interior of America. Such bones had been turning up from a very early date and had been interpreted quite variously—even attributed to human giants. It was not until George Croghan, the Indian agent, sent back some bones from the deposit known as Big Bone Lick on the Ohio River in 1766 that interest became general.[77] Some were sent to Lord Shelburne and others to Benjamin Franklin but soon they became known to all of the London literati. Peter Collinson became excited enough to publish an account of them in the *Philosophical Transactions*. Their mere size aroused attention, for they were larger than elephant bones to which they were immediately compared because of the tusks found among them. Dr. William Hunter pointed out, however, that the molars were those of a carnivorous animal while the elephant was known to be herbivorous.[78] The facts became clearer and more numerous even though many years elapsed before the mastodon could be satisfactorily described.

Indian mounds and relics were often found in the West by explorers, but not exclusively there. The most celebrated object of this sort, the Dighton rock, lay in the Taunton River in eastern Massachusetts. It was a large boulder with inscriptions on its face so indistinct that they led to many divergent interpretations. The academic

75. *Philosophical Transactions*, 55 (1765), 182-92; *Pennsylvania Chronicle*, July 6, 1767.
76. *Pennsylvania Chronicle*, July 6, 1767; *Philosophical Transactions*, 55 (1765), 192.
77. George Croghan, "Journal," in Reuben G. Thwaites (ed.), *Early Western Travels, 1748-1846* (Cleveland, 1904), I, 135; Collinson to Colden, Feb. 10, 1768, *The Letters and Papers of Cadwallader Colden*, VII, New-York Historical Society, *Collections for 1923*, 56 (1923), 132.
78. *Philosophical Transactions*, 57 (1767), 464-67; 58 (1768), 34-45.

group was ready with explanations of the origin and nature of the writing but few could be found who would agree. Cotton Mather, Dean Berkeley, Isaac Greenwood, John Winthrop, Stephen Sewall, and Ezra Stiles all visited and attempted to study the rock. Stiles, who made three drawings of the inscriptions in 1767 and one in 1768, pronounced them Phoenician. It was a copy of the life-size drawing made by Stephen Sewall, Hancock Professor of Hebrew and other Oriental Languages at Harvard, that was used by the French scholar, Court de Gebelin, when he asserted in his *Monde Primitif* that the writing was Punic or Carthaginian. John Winthrop, answering an inquiry of Timothy Hollis in 1774, came closer to the truth when he spoke of them as Indian inscriptions, though very ancient.[79]

The literary character of this Indian relic led the academic group to interest itself in the Dighton rock, and by the same token, these men became interested in the literary culture of the Indian. Some of the clergy had from the beginning concerned themselves with Indian languages in their efforts to convert the natives to Christianity. They were led to the necessity of translating parts of the Bible and prayer book into the Indian languages.[80] A few clergymen who maintained continued contact with the Indian became interested in his whole culture, as did the Reverend Eleazar Wheelock.[81] On the whole, however, it was the Indian traders and military men who had the most intimate contact with the Indians in this period and it was they who wrote the most comprehensive accounts of Indian life. Most important were Lieutenant Henry Timberlake's *Memoirs* (1765) on Cherokee life and the *History of the American Indians* (1775) by James Adair, who had spent thirty years among the southern tribes. Both books were published in London, although Adair first tried to get enough support in Georgia, South Carolina, Pennsylvania, and

79. See Edmund B. Delabarre, "Early Interest in Dighton Rock," Colonial Society of Massachusetts, *Publications*, 18 (1916), 235-99.

80. Bernhard Adam Grube (trans.), *A Harmony of the Gospels Translated in the Language of the Delaware Indians* (Friedenstahl bei Bethlehem, 1763); Bernhard Adam Grube, *Delawaerisches Gesang-Buchlein* (Friedenstahl bei Bethlehem, 1763); William Andrews, Henry Barclay, and John Ogilvie, *The Order for Morning and Evening Prayer . . . Collected and Translated into the Mohawk Language* (New York, 1769).

81. Eleazar Wheelock to Stiles, [n.d.], Calder (ed.), *Letters of Stiles*, 27-34.

New York for publication in America.[82] Beside these works, such general travel accounts as that of Jonathan Carver which relied very heavily upon earlier travel literature were distinctly inferior.[83] Andrew Oliver, of Salem, Massachusetts, published in the *Philosophical Transactions* an arresting little report upon epidemic sickness among the Indians of Martha's Vineyard and Nantucket.[84] The first attempt to write an account of medicine among the Indians was made by Benjamin Rush using literary sources and interviews rather than direct experience.[85]

The great breadth of interest demonstrated by the physicians did not lead them to neglect their own professional work. As yet they had no medical periodicals in America despite the organization of medical schools and societies between the Stamp Act and the Revolution. Their medical papers were published in London or Edinburgh medical journals, in the *Philosophical Transactions*, the *Transactions* of the American Philosophical Society, the general magazines, or as pamphlets. The medical schools at New York and Philadelphia began to turn out theses, too, and Americans continued to publish medical theses at Edinburgh. Most American medical papers related specific—often unusual—case histories, describing new methods of treatment or telling of surgical operations. Pervading much of this work was a growing appreciation of the fact that "natural Philosophy" was "the grand Basis of true medical Knowledge," as Oliver Fuller told the Medical Corporation of Litchfield County.[86] James McClurg, of Virginia, published a study of bile and liver in 1772 which was a model of experimental research.[87] Smallpox continued to demand much attention in many quarters. The best publicized paper on this

82. Henry Timberlake, *Memoirs, 1756-1765*, ed. by Samuel C. Williams, (Johnson City, Tenn., 1927); James Adair, *History of the American Indians* [1775], ed. by Samuel C. Williams (Johnson City, Tenn., 1930), xxi-xxvi.

83. J[onathan] Carver, *Travels through the Interior Parts of North America* (London, 1778).

84. *Philosophical Transactions*, 54 (1764), 386-88.

85. Benjamin Rush, *An Oration...Containing an Enquiry into the Natural History of Medicine among the Indians in North America* (Philadelphia, 1774).

86. Address by Oliver Fuller to Medical Corporation of Litchfield County, *Connecticut Courant*, Sept. 14, 1767.

87. James McClurg, *Experiments upon the Human Bile* (London, 1772).

subject was a study of the results of the use of mercury in inoculation by Benjamin Gale of Killingworth, Connecticut, published in the *Philosophical Transactions* and extracted by the general magazines.[88] In Philadelphia, a Society for Inoculating the Poor was founded with the full support of leading physicians, but in most colonies, inoculation was still a debatable procedure.[89]

Some physicians continued their interest in botany but their activity was more pronounced in that other annex of medicine—chemistry. With the appointment of Benjamin Rush as professor of chemistry at the College of Philadelphia and James Smith—soon succeeded by Peter Middleton—at King's College, the study attained academic recognition. The physicians were interested in chemistry because of its relation to the materia medica rather than to physiology. As in Europe, some attention was paid to chemical analysis of mineral waters which were being discovered throughout the country and were held in high repute. Newspaper controversies even developed from claims and counterclaims about the virtues of mineral springs.[90] Dr. John De Normandie's careful chemical analysis of the mineral water found near Bristol, Pennsylvania, served a real purpose.[91] The announcement of the discovery of a mineral spring in Philadelphia shortly after De Normandie's work led Benjamin Rush to compare the efficacy of Bristol, Philadelphia, and Abington water. In 1773, he published a long series of tests on water specimens and a comparison of the medical effects of the different waters.[92]

The concern of the academic group with physics did not produce such significant work as was done outside college halls. The teachers were more important for diffusing this knowledge than for advancing it. Ebenezer Kinnersley at the College of Philadelphia did experiment with the compressibility of water and continued his work in elec-

88. *Philosophical Transactions*, 55 (1765), 193-204; *Universal Museum*, new ser., 2 (1766), 333; *American Museum*, 5 (1789), 242-44, 495-96. See also John Duffy, *Epidemics in Colonial America* (Baton Rouge, 1953), 63.

89. *Pennsylvania Gazette*, Feb. 2, 1774.

90. See dispute in *New-York Mercury*, Jan. 5, 1767, Jan. 26, 1767, Feb. 16, 1767.

91. American Philosophical Society, *Transactions*, 1 (1771), 303-14.

92. Benjamin Rush, *Experiments and Observations on the Mineral Waters of Philadelphia, Abington and Bristol* (Philadelphia, 1773), 6, 7, 15; American Philosophical Society Minutes, July 18, 1773, [II], 146.

tricity.[93] More important certainly, were Franklin's speculations and experiments on the use of oil in calming waves and his letters on heat and smoky chimneys.[94] Even Cadwallader Colden was able to fit the phenomena of light into his analysis of matter more success-fully than he had handled earlier phases of his work.[95] While in the Hudson's Bay region, Thomas Hutchins duplicated the experimental freezing of mercury.[96] No one did anything to advance mathematics.

The fame of the Philadelphia experiments in electricity and the success Franklin continued to win insured attention to that spec-tacular study not only in Philadelphia, but almost everywhere in America where men sought the fame of science. Such academic figures as Kinnersley, Winthrop, and Hugh Williamson turned their thoughts to electricity. So did the medical naturalists, Garden and Lining. David Rittenhouse the astronomer, David Colden the civil servant, William Henry the gun manufacturer, and many others sought to contribute something to this most popular of all the physical sciences. Franklin himself, by now the foremost "electrician" of the world, lived intellectually in Europe rather than America. It was the other men who sent their experiments to the American Philosophical So-ciety and who wrote and talked in America. Electricity had a broad and continuing appeal because it appeared to have applications of many kinds. It not only proved immediately useful in saving buildings from destruction through the lightning rod but many thought it was of use as a medical cure as well. In fact, there was at least one report of a cure by lightning.[97]

In electricity, the sensation of the early 1770's was the electric eel, a fish which could transmit an electrical shock to anything it touched. Philosophers on both sides of the Atlantic were amazed and delighted; naturalists as well as academic men and physicians began

93. Barton, *Memoirs of Rittenhouse*, 155-56.
94. *Philosophical Transactions*, 64 (1774), 445-60; Franklin to Lord Kames, Feb. 28, 1768, Albert Henry Smyth (ed.), *The Writings of Benjamin Franklin* (New York, 1905-7), V, 108-10.
95. Colden, Some Remarks on some obvious Phenomena of Light, [1771], Ms. Communications to American Philosophical Society, Natural Philosophy, I, 3.
96. *Philosophical Transactions*, 66 (1776), 174-78.
97. Franklin to [?], [about 1773], Franklin Papers, XLVI, 23, American Philosophical Society; *Pennsylvania Chronicle*, Sept. 17, 1770.

to study this curiosity. William Walsh, one of the leading students of electricity in England, ran a series of experiments on the torpedo, as it was named, in order to determine whether the shock it delivered was electrical in nature.[98] The surgeon, John Hunter, happily dissected one, absorbed in its anatomy.[99] Henry Cavendish, brilliant experimental chemist, also turned his talents upon the problem.[100] Letters to Franklin indicated that there was high interest on the continent, too, but the Americans were at least as excited as anyone. The best source for the fish was Surinam, and since American trade routes ran to that area it was easy to import them. In fact, some of the electrical eels sent to England went by way of American ports.

When one of the fish was brought to Philadelphia, the American Philosophical Society appointed a committee to arrange with the owner for a set of experiments. As a result, Ebenezer Kinnersley, David Rittenhouse, Isaac Bartram, Owen Biddle, and Levi Hollingsworth carried out several experiments designed to test the nature and extent of the shock. As subjects they used eight or ten people, and some live fish. They also brought into play pith balls and other apparatus of "electricians." [101] The experiments impressed Henry Cavendish, who ran a similar series with a different source of electricity and succeeded in answering most of the questions raised by the Philadelphia experimenters. Hugh Williamson carried out another series of experiments on an electrical eel from Guiana which he felt to be different from the torpedo used by Walsh.[102] Head and shoulders above Williamson's account was the paper by Alexander Garden which also was published in the *Philosophical Transactions* of 1775. This was not a series of experiments, but an accurate and detailed description of the fish with some informed comments upon its habits and life. It was the report of a naturalist.[103]

98. *Philosophical Transactions*, 63 (1773), 461-77; 64 (1774), 464-73.
99. *Ibid.*, 65 (1775), 395.
100. *Ibid.*, 66 (1776), 196-225.
101. *Philadelphia Medical and Physical Journal*, 1 (1805), 96-160; American Philosophical Society Minutes, July 30, 1773, [II], 148.
102. Williamson to Rush, Feb. 19, 1774, Rush Papers, XLIII, 12; *Philosophical Transactions*, 65 (1775), 94-101.
103. *Ibid.*, 102-10.

Chapter Ten

AMERICAN IMPROVEMENT

I T WAS ONE of the articles of faith of the Enlightenment that science could be applied to the improvement of the material conditions of life. Sir Isaac Newton was not worshipped because of the simple beauty of his laws alone but because he had provided a key which promised to unlock the wisdom of the ages and to permit man to put that wisdom to work. Men became convinced that they might remake the world according to the image of their dreams. It was no accident that Francis Bacon, Lord Verulam, was held in such high esteem despite the dubiousness of his contributions to logic and despite the limited value of his science. Bacon remained the greatest prophet of the application of science to life—to useful purposes. "Science," he had written, "must be known by its works. It is by the witness of works rather than by logic or even observation that truth is revealed and established. It follows from this that the improvement of man's lot and the improvement of man's mind are one and the same thing." [1]

Not all thinkers and scientists were ready to accept Bacon's identification of truth and utility, but the utilitarian goal was widely acknowledged. The Americans frequently met Bacon's ideas in their press.[2] The Royal Society, which played such an important part in colonial intellectual development, was in some measure a monument to the Baconian ideals, its *Philosophical Transactions* showing a con-

1. Benjamin Farrington, *Francis Bacon, Philosopher of Industrial Science* (New York, 1949), 68.
2. See Academicus in *Virginia Gazette* (Purdie and Dixon), Aug. 5, 1773.

tinuing attention to useful knowledge. The London Society of Arts was interested in nothing except immediately and demonstrably useful knowledge. Even the high priest of the natural history circle, Linnaeus, advised his friends to "study that Botany may always be turned to some Beneficial purposes." [3] The historical record did not give a clear indication that science had produced utilitarian results, but there was ample faith that it would do so.

This utilitarian emphasis was particularly welcome in America where youth and necessity gave added prestige to all that was useful. Praise of useful science echoed and re-echoed through the colonies. One magazine declared, *"Rome* was never wiser or more virtuous, than when moderately learned, and meddled with none but the useful Sciences. *Athens* was never more foolish than when it swarmed with Philosophers.... Use is the Soul of Study." [4] The physicians, whose study must inevitably be concerned with useful results, were particularly responsive to this emphasis. Dr. John Kearsley prefaced his influential *Observations on the Angina Maligna* with a declaration that must have appeared a mere truism: "It is the duty of every man to aim at being useful." [5] Thomas Paine was quick to express the aspiration as if it were a description of historical process, remarking, " 'Tis by the researches of the virtuoso that the hidden parts of the earth are brought to light, and from his discoveries of its qualities, the potter, the glassmaker, and numerous other artists, are enabled to furnish us with their productions. Artists considered *merely* as such, would have made but a slender progress, had they not been led on by the enterprising spirit of the curious." [6]

The attitude of the colonies' outstanding creative scientist upon this problem is significant. Benjamin Franklin was a product of his time. He became one of the foremost representatives of the Enlightenment in Europe but he attained this eminence without ever losing his American coloration—without ever wanting to. Useful

3. Linnaeus to Ellis, Dec. 8, 1758, James Edward Smith (ed.), *A Selection of the Correspondence of Linnaeus and Other Naturalists* (London, 1821), I, 111.

4. *The Instructor,* 1 (1755), 34.

5. [John Kearsley], *Observations on the Angina Maligna* (Philadelphia, 1769), 3.

6. Philip S. Foner (ed.), *The Complete Writings of Thomas Paine* (New York, 1945), II, 1021-22.

knowledge was important to Franklin as it was to most of his philosophical friends in Europe and to almost all his fellow Americans.
When he wrote, "What signifies Philosophy that does not apply to
some Use?" he meant just that.[7] He had the faith of a good Baconian
that genuine science would ultimately yield useful results. He felt
apologetic about the time he had spent upon the useless construction
of magic number squares and he felt chagrined until he was able
to find something in his electrical experiments "of use to mankind."[8]
Franklin's most important scientific investigations were not begun
with any specific utilitarian objective. He sought only the truth
but he sought it with the conviction that the truth was useful.[9]

There was a wide gulf between the faith of Franklin and Paine
that scientific researches would lead to the betterment of mankind
and the demand that philosophers focus their efforts upon material
improvement. Charles Thomson had expressed the latter view when
in 1768 he sought to define "the one end" of the American Society as
"the Advancement of useful Knowledge and improvement of our
Country."[10] Thomson, however, was not a scientist; he spoke for
the merchant and the average literate man. When the American Philosophical Society came to frame a public petition, it revealed a more
hazy view of the relationship between the promotion of science and
the general welfare. "The experience of ages shews," their petition
announced, "that by such institutions, arts and sciences in general are
advanced; useful discoveries made and communicated; many ingenious artists, who might otherwise remain in obscurity, drawn forth,
patronized and placed in public usefulness; and (what is of great con-
sequence to these young countries, especially in their present situation) every domestic improvement, that may help either to save or
acquire wealth, may by such means be more effectually carried on."[11]
Whether material improvement could be planned as Thomson seemed

7. Franklin to Mary Stevenson, [Sept. 20, 1761], Albert Henry Smyth (ed.),
The Writings of Benjamin Franklin (New York, 1905-7), IV, 115.
8. I. Bernard Cohen (ed.), *Benjamin Franklin's Experiments* (Cambridge,
1941), 199.
9. Cf. I. Bernard Cohen, "How Practical was Benjamin Franklin's Science?"
Pennsylvania Magazine of History and Biography, 69 (1945), 284-93; Carl
and Jessica Bridenbaugh, *Rebels and Gentlemen* (New York, 1942), 325.
10. American Society Minutes, Jan. 1, 1768, 64.
11. American Philosophical Society Minutes, Feb. 7, 1769, [II], 114.

to hope or only spurred by the general promotion of arts and sciences, it was clear that utilitarian results were anticipated from the advance of science.

At the same time, some of the academic men were ready to recognize values in science which did not depend upon its ultimate application to material welfare. William Smith called for the promotion of "all those Branches of *Literature* and *Science*, whereby the Mind may be humanized, the Spirit of Religion and Liberty supported, and the Genius of our Country exalted." At the same time that he placed science with the humanities, Smith recognized another equally important category of activity which included "such Mechanic Arts, Inventions and useful Improvements, as tend to shorten Labor, to multiply the Conveniences of Life, and inrich the Community." [12] David Rittenhouse went even further than Smith in picturing astronomy as a humanity, yet at the same time he could remark that he despised the kind of mathematical "juggle, where no use is proposed." [13] Even these men who could see in science values that did not depend upon material betterment, still held high their utilitarian expectations.

The yield of all these exalted hopes was disappointingly meager. Science had not yet reached a stage at which it could offer much help to the farmer, the merchant, or the artisan in the ordinary pursuit of his business. Advances in technics still came by trial and error methods rather than through the application of externally constructed patterns of theory. Except in a few isolated cases, the best that could be done was to attempt to study the most successful practice found in any part of the world and to experiment with new methods as freely and frequently as possible. Knowledge thus attained might then be placed at the disposal of the trade or business in question—though persuading its practitioners to adopt the better methods was still another problem. It was not yet a question of applying scientific knowledge to economic pursuits as Justus Liebig was able to do so successfully with respect to agriculture in the next century. All that could be done was to study practice with the detachment that had

12. Smith, *An Oration* (1773), 8.
13. David Rittenhouse, *An Oration Delivered February 24, 1775, Before the American Philosophical Society* (Philadelphia, 1775), 26; William Barton, *Memoirs of the Life of David Rittenhouse* (Philadelphia, 1813), 234n-35n.

been useful in the study of science and to attempt to disseminate the knowledge thus acquired.

This approach was being made in England, particularly in agriculture and the mechanic arts. The Americans learned of it from acquaintance with individual reformers, from their newspapers and magazines, through the efforts of the London Society of Arts, and through the pages of that society's unofficial journal, *Memoirs of Agriculture*. In America, however, despite widespread approval of this process there were only spotty efforts to support it. The American Philosophical Society, with its heavy merchant and nonscientific membership, did turn some of its energies to "the Several Supports of Mankind at large, Agriculture, Manufactures and Commerce." [14] Three of its six committees, Trade and Commerce, Mechanics and Architecture, and Husbandry and American Improvements, dealt with utilitarian pursuits. Most members clearly felt themselves less equipped to do justice to natural philosophy, natural history, and medicine than to "such subjects as tend to the improving of their country, and advancement of its interest and prosperity." [15] This principle of action combined patriotism with personal gain to produce a feeling of exhilaration.

Agriculture was not only the chief support of the American people but it was in dire need of improvement. As early as 1758, the *American Magazine* had suggested that "township companies or societies" be formed to discuss agricultural improvement and to make new experiments.[16] Little was done to introduce the new English practices: the Norfolk system of crop rotation, the marling and pulverizing of the soil, purposeful breeding, and the use of new and better implements. As late as 1775, the perceptive author of *American Husbandry* still saw wasteful practices on every hand and again urged the need for a society specifically devoted to the encouragement of agriculture. Such a society might "settle a plan of operations, which would, in a few years, by means of an annual subscription, given in bounties and premiums, alter the face of things. They might reduce

14. American Society Minutes, Sept. 18, 1767, 56.
15. *Ibid.*
16. Quoted from *American Magazine and Monthly Chronicle* by Winthrop Tilley, The Literature of Natural and Physical Science in the American Colonies (Doctoral Dissertation, Brown University, 1933), 54.

David Rittenhouse's First Orrery, 1767-1771.
Photographed after its reconstruction in 1952-1953.

This mechanical planetarium was designed with unusual precision, relative speeds and distances being accurate except where the scale makes that impossible. All the planets then known are represented by small ivory balls which circle around the brass sun in the center of the device. The longest geared arm, extending toward the upper left, supports Saturn, its distinctive ring, and five circling satellites. Jupiter and four satellites are held by the shorter arm extending toward the lower right. All the minor planets can be seen too, the earth and its accompanying moon being above and to the left of the sun—just beyond Venus. The eccentricity of the planetary orbits is apparent in the construction of the central wheels.

NATURALISTS: *Upper left*, Peter Collinson, detail from an engraving by T. Miller after a portrait by Thomas Gainsborough in John Fothergill, *Some Account of the Late Peter Collinson* (London, 1770). *Upper right*, Cadwallader Colden, from a portrait by John Wollaston (Metropolitan Museum of Art). *Lower left*, Benjamin Waterhouse, from a portrait by Gilbert Stuart (Redwood Library and Athanaeum). *Lower right*, Gotthilf Henry Ernest Muhlenberg, from a portrait by Charles Willson Peale (National Park Service).

NATURAL PHILOSOPHERS: *Upper left*, Benjamin Franklin, from a pastel portrait by Joseph Siffred Duplessis (New York Public Library). *Upper right*, John Winthrop, detail from a portrait by John Singleton Copley (Harvard University). *Lower left*, David Rittenhouse, from a portrait by Charles Willson Peale (University of Pennsylvania). *Lower right*, William Smith, detail from a portrait by Gilbert Stuart (From the original owned by Mr. John Brinton; photograph courtesy of the Frick Art Reference Library).

The Franklin Tree (Franklinia Altamaha), discovered by John and
William Bartram, 1765. From a water color by William Bartram.

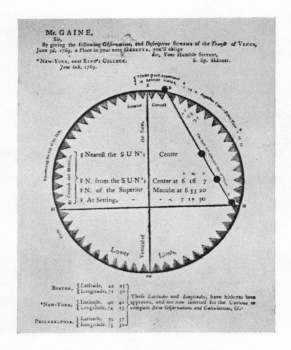

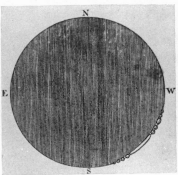

Above, Newspaper Projection of the Transit of Venus, *New-York Gazette, and Weekly Mercury*, June 12, 1769. *Below*, First Representation of Baily's Beads, discovered by Samuel Williams, 1780. From the American Academy of Arts and Sciences, *Memoirs*, 1 (1785).

PHYSICIANS: *Upper left,* Thomas Bond, from a miniature by an unknown artist (Pennsylvania Hospital). *Upper right,* John Morgan, from a portrait by Angelica Kauffman (From the original owned by the Washington County (Pennsylvania) Historical Society; photograph courtesy of the Frick Art Reference Library). *Lower left,* Benjamin Rush, detail from a portrait by Charles Willson Peale. (From the original owned by Mrs. T. Charlton Henry; photograph courtesy of the Frick Art Reference Library). *Lower right,* John Warren, from a portrait by Rembrandt Peale (Boston Medical Library and the Society for the Preservation of New England Antiquities).

PROPHETS OF GLORY: *Upper left*, Charles Thomson, from a portrait by Charles Willson Peale (National Park Service). *Upper right*, Thomas Jefferson, from a portrait by Charles Willson Peale (National Park Service). *Lower left*, Ezra Stiles, a detail from a portrait by Samuel King (Yale University Art Gallery). *Lower right*, David Ramsay, from a portrait by Charles Willson Peale (From the original owned by Mrs. F. Barnard O'Connor; photograph courtesy of the Frick Art Reference Library).

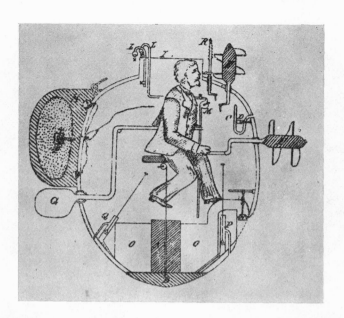

Above, David Bushnell's Submarine, a restoration, used against the Royal Navy, 1776. From Henry L. Abbot, *The Beginning of Modern Submarine Warfare* (1881). *Below*, John Fitch's Steamboat. A later version was demonstrated to members of the Constitutional Convention, 1787. From the *Columbian Magazine*, I (1786-87).

these doubtful points to certainty; they might introduce a better system of rural oeconomy, and be in a few years of infinite service to their country." No such effort was made. The only experimentation was done by enlightened individuals: by Thomas Jefferson in Virginia, William Allen in Pennsylvania, Benjamin Gale in Connecticut, and a handful of others.[17]

Through their magazines and newspapers, the Americans were given a frequent diet of new techniques and new implements, often copied from English and occasionally from French sources. Sometimes they appear to have been inserted merely as space fillers.[18] The *New American Magazine* published in 1760 an article on planting trees that it extracted from the *Philosophical Transactions* of 1669! [19] Yet evidences of the new agriculture were to be found in newspaper advertisements offering new implements for sale and offering to locate marl and instruct in its use.[20] Americans tried new plows and seed drills. They invented harvesting machines, threshing machines, and seed drills of their own.[21] Benjamin Gale of Killingworth, Connecticut received a gold medal from the Society of Arts for the completion of a superior seed drill which his father-in-law, Jared Eliot, had begun.[22] Even so, the practice of the dirt farmer was not much affected.

17. Lewis B. Walker (ed.), *Extracts from Chief Justice William Allen's Letter Book* ([n.p.], 1897), 78; Edwin M. Betts (ed.), *Thomas Jefferson's Garden Book* (Philadelphia, 1944), 48-50; Edwin M. Betts (ed.), *Thomas Jefferson's Farm Book* (Princeton, 1953), 88-89; George C. Groce, Jr., "Benjamin Gale," *New England Quarterly*, 10 (1937), 708.

18. *Connecticut Courant*, June 15, 1767; *Pennsylvania Chronicle*, Feb. 23, 1767; *Massachusetts Gazette*, April 23, 1767.

19. *New American Magazine*, 2 (1760), 24-25.

20. *Maryland Gazette*, June 11, 1772; *Pennsylvania Gazette*, July 21, 1768.

21. *Connecticut Courant*, June 15, 1767; *Virginia Gazette* (Purdie and Dixon), Nov. 19, 1772; *Pennsylvania Magazine*, 1 (1775), 163-64; John Jones to [Owen Biddle], Oct. 26, 1771, Archives, American Philosophical Society; *Votes and Proceedings of the House of Representatives of the Province of Pennsylvania*, Pennsylvania Archives, 8th ser., VII, 6453; William Franklin to Benjamin Franklin, May 11, 1769, Bache Collection, American Philosophical Society.

22. Groce, "Benjamin Gale," 708; *Public Records of the Colony of Connecticut*, ed. by J. H. Trumbull and C. J. Hoadly, (Hartford, 1850-90), XIII, 566; Jared Eliot, *Essays upon Field Husbandry in New England and Other Papers, 1748-1762*, ed. by Harry J. Carman and Rexford G. Tugwell (New York, 1934), 116-20.

One of the most serious difficulties was the fact that English experience was not always applicable to the American environment. The extent of land in the colonies, the climate, and the soil differed markedly from England. Intensive farming would have meant economic suicide in America; even the root crops which had proven so beneficial in England did not fare well in America.[23] The problem was recognized in many quarters but the experiments necessary to test the applicability of the English methods were carried out only in a very small part. To the end of the colonial period, Jared Eliot's *Essays upon Field Husbandry* remained the best treatise based upon a knowledge of English methods but written in terms of American experience and experiment.

In the case of crop pests, experience suggested that the insects which attacked American plants were different from those known in Europe. When Thomas Gilpin wrote his paper on the seventeen-year locust, he called it the American locust.[24] The American Philosophical Society published three articles which offered methods of fighting pests.[25] In one of them, Landon Carter, Virginia planter, dealt with the most serious problem, the wheat fly. Much had already been written in England, on the continent, and in America on the control of this or similar insects attacking wheat.[26] Carter was acquainted with some of these writings but he made his own experiments and came up with a method of control by heat that seemed to work well. He was so pleased with the reception of his article that he went on to make more experiments which he reported in the *Virginia Gazette*.[27]

To most of those concerned, agricultural improvement was not primarily a question either of adapting English techniques to America or of attacking specific problems. The widest support was given to

23. Rodney C. Loehr, "The Influence of English Agriculture on American Agriculture, 1775-1825," *Agricultural History*, 11 (1937), 14.

24. "Memoir of Thomas Gilpin," *Pennsylvania Magazine of History and Biography*, 49 (1925), 301.

25. American Philosophical Society, *Transactions*, 1 (1771), 205-17, 241-44; *Pennsylvania Gazette*, March 16, 1769.

26. "Memoir of Thomas Gilpin," 296; Charles Vancouver, *General Compendium or Abstract of Chemical, Experimental, and Natural Philosophy* (Philadelphia, 1785), I, 37; *Pennsylvania Chronicle*, May 2, 1768; Franklin to Jan Ingenhousz, Sept. 30, 1773, Franklin Papers, XLV, 72, American Philosophical Society.

27. *Virginia Gazette* (Rind), Nov. 19, 1772.

AMERICAN IMPROVEMENT

the introduction of new crops that offered economic advantage. The natural history circle had been founded in considerable measure upon the desire to introduce American plants to Europe, but a reversal in emphasis was effected when the American Philosophical Society reprinted John Ellis's pamphlet on foreign plants that might be profitably introduced into the colonies. Three other articles in the *Transactions* dealt with new, American sources of table oil, two of them relating experiments in the extraction of oil from sunflower seeds and the third, by John Morel of Georgia, advocating the use of the bene seed.[28] A contemporary letter by George Brownrigg of North Carolina, urging the use of the ground nut or arachis for the same purpose, was published in London in the *Philosophical Transactions*.[29] The American society also published a recipe for currant wine, along with Isaac Bartram's old paper on the distillation of persimmons. Bartram was aware of the use of persimmon beer in the southern back country but what he sought to develop was a rum-like hard liquor. More widespread efforts were made to encourage viniculture, silk culture, and the raising of hemp and flax.[30] These crops had been encouraged from almost the beginning of settlement in America. They were not distasteful to the English nor did they conflict with imperial policy. At the same time, they promised to increase the riches and potentialities of America. Americans of all shades of political opinion could join in supporting the cultivation of new products, convinced that they were acting in the best interests of patriotism.

The culture of grapes for the purpose of making wine had been advocated from an early time in the hope of decreasing the large importation of foreign wine into Britain and her colonies. Some early local success was attained, particularly by the Huguenots who had been settled for that purpose in South Carolina by the British government.[31] After the French and Indian War, encouragement be-

28. American Philosophical Society, *Transactions*, 1 (1771), 234-35, 235-39, 239-40.
29. *Philosophical Transactions*, 59 (1769), 379-83.
30. American Philosophical Society, *Transactions*, 1 (1771), 117-97, 198-204, 224-30, 339-40.
31. Arthur H. Hirsch, "French Influence on American Agriculture in the Colonial Period with Special Reference to Southern Provinces," *Agricultural History*, 4 (1930), 5.

came more general: a bounty was offered by the British government, a premium was announced by the Society of Arts, and in 1769, a vineyard was established by the Province of Virginia.[32] Individuals began to attempt the raising of grapes and the manufacture of wine on a wider scale. One of the most successful was Edward Antill of New Jersey, who was rewarded for his activity by the Society of Arts. In 1768, Antill proposed a most ambitious project through the columns of the *New-York Gazette* and the *Maryland Gazette*. Antill sought subscriptions toward the establishment of a public vineyard from which each province would be able to draw free cuttings.[33] The project failed as did John Leacock's attempt to establish a similar vineyard in 1773 by means of a public lottery. Antill's most lasting contribution was in the form of an eighty-page treatise on viniculture, the best article on the subject to come out of the colonies.[34]

Viniculture was basically a question of skill rather than learning or study, and the men who came closest to success were foreigners who had acquired the techniques in Europe. The two most successful vineyards were established by Frenchmen, each with the aid of Lord Hillsborough. Lewis St. Pierre made his plantings along the Savannah River with such success that he was awarded a gold medal by the Society of Arts. Pierre LeGaux began with a much smaller vineyard on the outskirts of Philadelphia. His enterprise lasted longer despite the withdrawal of Hillsborough's support and the later opposition of the French government.[35] Another considerable vineyard was begun in Virginia by Philip Mazzei who brought with him seeds, cuttings, and plants as well as laborers from Italy. He was enabled to establish himself principally through the assistance of Thomas Jefferson, who induced him to settle on land bordering his own. Mazzei's project

32. *The Statutes at Large of Virginia*, ed. by W. W. Hening (Richmond, 1820), VIII, 364; IX, 239.

33. John Leacock to Thomas Coombe, Dec. 29, 1772, Ms. Communications to the American Philosophical Society, Trade, 5; *New-York Gazette; and the Weekly Mercury*, Feb. 1, 1768.

34. American Philosophical Society, *Transactions*, 1 (1771), 117-97; Francis James Dallett, Jr., "John Leacock and the Fall of British Tyranny," *Pennsylvania Magazine of History and Biography*, 78 (1954), 456-75.

35. Hirsch, "French Influence," 3-5; *Virginia Gazette* (Purdie and Dixon), July 22, 1773.

failed when his laborers' terms ran out and he himself became involved in the developing Revolution.[36]

Another crop long advocated for mercantilist reasons was hemp, used for cordage, bags, and sails, and even urged in some quarters for clothing. As early as 1735, an Irish pamphlet on flax and hemp was reprinted in Boston, but it was not until the imperial tensions of the 1760's that general interest rose.[37] Parliament established a bounty on hemp in 1765 and Virginia and North Carolina soon followed suit with bounties of their own.[38] While viniculture had attracted the greatest attention in the southern colonies, the raising of hemp was recommended especially to the attention of the northern colonies.[39] It was by order of the Massachusetts Assembly that Edmund Quincy's *Treatise of Hemp Husbandry* was published in 1765. Admittedly a collection from the best European and American accounts, it also contained a description of the extensive experiment in hemp production then being undertaken at Salem. The following year, Boston saw the reprinting of Marcandier's valued work on hemp in an English translation. Even this contained an American element in its "Plan of the *Pennsylvania* Hemp Brake." [40] Hemp was widely attempted, but although its virtues were generously extolled, it did not supplant flax which the hemp advocates considered much less generally useful. Hemp production continued on a limited basis. Madder and tea which also received encouragement failed altogether.[41]

Of all the attempts to introduce new crops, the struggle to estab-

36. Jefferson to Albert Gallatin, Jan. 25, 1793, Andrew A. Lipscomb and Albert E. Bergh (eds.), *The Writings of Thomas Jefferson* (Washington, 1903-4), IX, 14-15; *Virginia Gazette* (Rind), July 28, 1774; Howard R. Marraro (trans.), *Memoirs of the Life and Peregrinations of the Florentine, Philip Mazzei, 1730-1816* (New York, 1942), 192-94.

37. Lionel Slator, *Instructions for the Cultivation and Raising of Flax and Hemp* (Boston, 1735).

38. Edmund Quincy, *A Treatise of Hemp Husbandry* (Boston, 1765), 31; *The Statutes at Large of Virginia*, VIII, 363; Carl R. Woodward, "Agricultural Legislation in Colonial New Jersey," *Agricultural History*, 3 (1929), 18.

39. Edward Antill, "Observations on the Raising and Dressing of Hemp," *Select Essays* (Philadelphia, 1779), 76.

40. *An Abstract of the Most Useful Parts of a late Treatise on HEMP, translated from the French of M. Marcandier* (Boston, 1766), 31.

41. Aaron Loocock, *Some Observations and Directions for the Culture of Madder* (Charleston, 1775), 1; David Ramsay, *The History of South-Carolina* (Charleston, 1809), II, 220; *Boston Gazette*, Nov. 9, 1767.

lish silk culture was the most dramatic. Like wine and like hemp, silk was a product that Britain and her colonies had to import from foreign lands. For that reason attempts had been made ever since 1621 to establish its production in the colonies. Initial failures in Virginia did not discourage later enthusiasts in Georgia, South Carolina, New York, Connecticut, and Pennsylvania. The pattern of encouragement was extensive, involving premiums for the growth of mulberry trees, bounties as well as premiums on raw silk, favorable instructions to royal and proprietary governors, financial inducements to stimulate the immigration of skilled workers, the establishment of public filatures where silk could be wound from the cocoons, and the printing of numerous accounts, advertisements, and appeals. First and last, it meant the investment of considerable sums of private as well as public money. The whole effort was accelerated during the Seven Years' War when the Society of Arts began to offer premiums in Georgia, Pennsylvania, and Connecticut.[42] After 1765, imperial tensions influenced encouragement, too.

The greatest measure of success was met in Georgia where the proprietors had early fastened upon silk as the most desirable product for that colony. Joseph Ottolenghe, a native of Piedmont, was appointed to direct the silk culture, a filature was built at Savannah, and Italian workers were brought in to do the winding. Maximum production of 1,084 pounds of silk was attained in 1766. In 1767, after having paid out a total of £1,370 in silk premiums nearly all of which went to Georgia, the Society of Arts became discouraged with the results and abandoned its efforts to stimulate silk production in America. In 1769, the parliamentary bounties on silk were reduced. Production had so declined by 1771 that the filature itself was given up.[43]

42. Sir Henry T. Wood, *A History of the Royal Society of Arts* (London, 1913), 84-85; Lyman Carrier, *The Beginnings of Agriculture in America* (New York, 1923), 133; Lewis C. Gray, *History of Agriculture in the Southern United States to 1860* (Washington, 1933), I, 184; Robert Dossie, *Memoirs of Agriculture* (London, 1768), I, 239; *Premiums by the Society, Established at London for the Encouragement of Arts, Manufactures, and Commerce* (London, 1758), 26.

43. Gray, *History of Agriculture*, I, 186-88; James W. Holland, "The Beginning of Public Agricultural Experimentation in America; the Trustees' Garden in Georgia," *Agricultural History*, 12 (1938), 294; Wood, *Society of Arts*, 84.

South Carolina never succeeded as well as Georgia, despite the whole-hearted cooperation of the provincial assembly in providing a bounty and appropriations toward the erection of two filatures and two spinning factories.[44] Failure was undeniable in both colonies before the Revolution, but in other quarters hope still persisted.

Production of silk in the Northern colonies never equalled that of South Carolina and Georgia. In most of the North, silk culture was promoted only by private individuals.[45] It was in Pennsylvania that the best organized, most literate effort was made in association with America's one learned society. The very men who were most anxious to use science to advance the material welfare of the colonies were in the forefront of the Pennsylvania movement. Ignoring previous failures, a small group of men led by the Reverend Francis Alison began in April 1769 to offer free silk worm eggs to all who would raise them. In May, the *Pennsylvania Journal* and Lewis Nicola's *American Magazine* began to run extracts from Samuel Pullein's standard *Culture of Silk* in the hope that it would "contribute to the Improvement of a Country in an infant State." During the next month, the *Pennsylvania Chronicle* reprinted selections from a French treatise on silk culture. When Dr. Cadwalader Evans wrote to Benjamin Franklin in London for advice, the enterprise was lifted to a new level.[46]

The attention of the American Philosophical Society was turned to the culture of silk when Franklin's reply to Evans was read before it in January, 1770.[47] Franklin transmitted a recent French pamphlet on the subject, promised to get a copy of the act of Parliament establishing the bounty on silk, and outlined the steps to be taken in establishing silk production. He urged that a public filature be established

44. Hirsch, "French Influence," *Agricultural History*, 4 (1930), 7; *Pennsylvania Gazette*, July 17, 1766; Gray, *History of Agriculture*, I, 185; Frederick P. Bowes, *The Culture of Early Charleston* (Chapel Hill, 1942), 89; Ramsay, *History of South-Carolina*, II, 220-21.

45. *Boston Gazette*, April 10, 1769, June 12, 1769.

46. *Pennsylvania Chronicle*, April 3, 1769, April 10, 1769, June 5, 1769, June 12, 1769, June 19, 1769; *American Magazine*, 1 (1769), 147-51, 183-85, 214-16; Franklin to Evans, Read Jan. 5, 1770, American Philosophical Society Minutes, [II], 89-90.

47. Franklin to Evans, Read Jan. 5, 1770, American Philosophical Society Minutes, [II], 89-90.

and that the Pennsylvania Assembly be induced to promote the growth of mulberry trees. The society disagreed with the need for mulberry trees since they were native to the province but it did immediately prepare a petition to the assembly asking for aid in establishing the filature and in providing annual premiums to be offered for the ensuing five years.[48]

When the assembly demonstrated a reluctance to appropriate the required £500, even the first year, another approach was made. Two hundred fifty pounds proved easy to collect by way of private subscription. The subscribers then elected a twelve-man board of managers, which was made up exclusively of members of the American Philosophical Society. It at once set to work, established the filature, hired an experienced silk reeler, set the prices it would pay for cocoons at 25 per cent above the market, and offered various premiums to residents of Pennsylvania, Delaware, New Jersey, and Maryland. A rising enthusiasm led to increasing press attention which occasionally did more harm than good. One account had it that if a calf fed on mulberry leaves were killed and the maggots found on the meat were then set on mulberry leaves, they would turn out to be silkworms! Despite such old wives' tales, the first year's operations were reckoned successful when 150 pounds of silk could be exported. Encouraging reports received from England on the quality of the exported silk led the managers to make bigger plans for the following year. The Pennsylvania Assembly proved willing to make a conditional grant of £1,000 to the silk society to be paid as soon as it could be demonstrated that £1,000 had been raised from other sources. The qualification was soon met and the future appeared bright.[49] It even looked as if the New Jersey Assembly might soon grant some encouragement.

The silk society continued its activities right into the war period, offering premiums and exporting silk each year. Franklin handled the affairs of the organization in London, even presenting a dress of

48. American Philosophical Society Minutes, [II], 86, 87; *Votes of Assembly*, VII, 6496-98, 6513.

49. List of Subscribers, 1770, Philadelphia Miscellaneous Box 7B, Historical Society of Pennsylvania; *Pennsylvania Gazette*, March 8, 1770, March 15, 1770, June 14, 1770, Jan. 24, 1771, April 30, 1772; *Pennsylvania Chronicle*, April 9, 1770; *Votes of Assembly*, VIII, 6685, 6846.

American silk to the Queen. At one point, he urged the Americans to protect themselves against future developments by preparing to manufacture the silk cloth themselves, but without effect. The society was responsible for bringing Joseph Ottolenghe to Philadelphia from Georgia and encouraging him to write a little handbook on raising silkworms.[50] Comprehensive directions for breeding silkworms were put together by four of the managers who offered extracts from the works of Samuel Pullein and Pierre de Boissier.[51] Actually, the most original paper to emerge from this whole period was Moses Bartram's study of the life cycle of "native American Silk worms" which had been written before the establishment of the silk society.[52] When newspaper essays and unpublished manuscripts were included, the total literary output was surprising, but it did not save the enterprise from collapse. Humphry Marshall expressed the proper degree of skepticism when he reported: "Our people Seem to make a great Noise about raising Silk, how it will turn out I know not." [53] When a silk depression hit England in 1772, it turned out rather badly. It was a lost cause when the last list of premiums was offered in May 1776.

There were probably many people throughout the country who wrote some notes on their experiments in silk culture, but it may be doubted that anyone was more careful in this respect than Ezra Stiles at Newport, Rhode Island. He produced a manuscript entitled "Observations on Silk Worms and the Culture of Silk," a narrative account of his experiments with three thousand worms beginning in the summer of 1763. He carried it through to his participation in unsuccessful efforts after the Revolution to revive the culture of silk in Connecticut. The account was embellished with remarks from

50. Joseph Ottolenghe, *Directions for Breeding Silk-Worms* (Philadelphia, 1771).

51. Franklin to Evans, Feb. 6, 1772, Smyth (ed.), *Writings of Franklin*, V, 388-89; [l'Abbé Pierre de Boissier and Samuel Pullein], *Directions for the Breeding and Management of Silk-Worms* (Philadelphia, 1770).

52. Moses Bartram, "Observations on Native Silk Worms of North America," American Philosophical Society, *Transactions*, 1 (1771), 224-30.

53. Humphry Marshall to Franklin, May 28, 1770, Franklin Papers, III, 16, American Philosophical Society; Franklin to Abel James and Benjamin Morgan, Feb. 10, 1773, Smyth (ed.), *Writings of Franklin*, VI, 10-11; *Pennsylvania Journal*, May 15, 1776.

various Italian and Chinese authorities. It revealed relationships with Philadelphia silk cultivators and with Franklin in London who sent him Chinese prints of silk culture scenes. On one occasion Franklin agreed to sell him enough Pennsylvania silk to permit a dress to be made for his wife entirely of American silk—the Connecticut and Pennsylvania strains no doubt being "indistinguishable." Stiles recorded that the only thing that led him momentarily to neglect his silk worms was his greater absorption with the transit of Venus of 1769.[54]

The effort to establish various manufactures in the colonies was a counterpart of the attempts to alter patterns of agriculture. Manufactures represented only a small part of the American economy, but in the years of tension after the peace of Paris they loomed very large. The encouragement of manufactures had two distinct phases. In the first, the aim was to produce such goods as pig iron, bar iron, potash, and pearl ash which would supplement the economy of the empire. Just as England was anxious to have the colonies produce the agricultural exotics, wine and silk, so it welcomed partly manufactured raw materials that would otherwise have to be imported from foreign nations. The second phase of the American enthusiasm for manufactures involved finished products which were not encouraged in England because they competed with English goods. The Americans made attempts to rediscover the techniques and processes that would permit them to compete with the mother country especially in the manufacture of textiles.

The encouragement of manufactures in the colonies was closely coordinated with the developing Revolutionary movement. Before the imperial reforms that followed the Seven Years' War, the appeal for manufactures had usually been limited to a desire to benefit the poorer elements of the cities which would provide the needed labor. At least the arguments were couched in those terms.[55] After 1764, a series of non-importation movements cut down the supply of British manufactures. The Sugar Act, Stamp Act, Townshend Acts, and

54. Stiles, "Observations on Silk Worms and the Culture of Silk," Stiles Papers, Yale University.
55. William Bailey, *A Treatise on the Better Employment and more Comfortable Support, of the Poor in Workhouses* (London, 1758); Thomas Bernard, *A Sermon Preached in Boston, New England, Before the Society for Encouraging Industry and Employing the Poor* (Boston, 1758), 22-23.

finally the Coercive Acts introduced a succession of efforts to stimulate manufactures and to create a preference for American goods. In several colonies, societies offered premiums to aid specified manufactures, town and provincial governments extended encouragement, and home markets flourished.[56] Harvard students demonstrated their patriotism by appearing at commencement in clothes entirely of American manufacture.[57] The use of "foreign Geegaws" was deprecated as Lynn boasted that it produced forty thousand pairs of shoes in a year and Germantown six thousand dozen pairs of stockings.[58] England had early shown its attitude toward this sort of manufacture in the Woolens Act, the Hat Act, and the Iron Act but it did not at this time reply by similar restrictive legislation. In 1766 and 1768 the Board of Trade required reports on manufactures established since 1734 from all colonial governors and in 1774 Parliament prohibited the export of all tools and machinery used in textile manufactures.[59]

Textile manufacture was the most serious problem. Textiles, the foundation of Britain's trade, were at the same time the product the Americans found most desirable to manufacture. The Americans already provided the bulk of their own clothing although little of that got into "trade." Beginning with the New York Society of Arts of 1764, several organizations were formed to manufacture cloth. The most successful of them was the United Company of Philadelphia for Promoting American Manufactures which, soon after its establishment in 1775, had seven hundred people at work making cloth.[60] There was notably less literary activity in connection with cloth manufacture than there was relating to agricultural improvement. There was less interest on the part of the American Philosophical Society and the Royal Society. Occasionally a new textile machine was pictured in a

56. Victor S. Clark, *History of Manufactures in the United States* (Washington, 1929), I, 215; *New-York Mercury*, Oct. 28, 1765; *Connecticut Courant*, Oct. 28, 1765; April 7, 1766, June 4, 1770, Nov. 16, 1767; *Georgia Gazette*, July 19, 1770; *Boston Gazette*, Feb. 29, 1768.

57. *Boston Gazette*, July 23, 1770.

58. *Massachusetts Gazette*, Nov. 5, 1767; Andrew Burnaby, *Travels through the Middle Settlements in North-America* (Dublin, 1775), 103.

59. Arthur C. Bining, *British Regulation of the Colonial Iron Industry* (Philadelphia, 1933), 101; *Virginia Gazette* (Purdie and Dixon), Oct. 6, 1774.

60. *Pennsylvania Journal*, April 3, 1776; *Votes of Assembly*, VIII, 7390.

magazine and one plan for a water-powered spinning mill was presented to the American Society. Assembly grants to encourage inventors or to help men duplicate English inventions had some effect.[61]

Less essential products received attention, too. Paper, for example, had been manufactured in the colonies for a long time but the Townshend duty on the importation of paper led to a new effort to increase its production. In New York, the collection of rags was publicized as a patriotic duty. In Boston, rags were bought to sustain the paper mill at Milton; in Connecticut, an assembly bounty kept at least one paper mill in operation; and in Philadelphia, the American Philosophical Society offered premiums for rags with which the manufacture of paper could be extended.[62] The society also gave to William Henry Stiegel a kind of scientific stamp of approval for the flint glass he manufactured. This certificate he was able to use to advantage in his advertising and in his successful appeal for an assembly grant.[63] Another attempt by the Philosophical Society to encourage American manufactures was its request that clays be submitted from all over the country so that the deposits most suitable for pottery could be located. One such specimen, sent from Charlestown in 1772, was adjudged to be particularly good for making crucibles.[64]

Abel Buell of Killingworth, Connecticut, made at this time the first type fonts ever cast in America. When he worked out his plan for producing them in 1769, all printing type had to be imported from England. He won the support of Benjamin Gale, who sent samples of his type to the American Philosophical Society and wrote to the Reverend Ezra Stiles and the Reverend John Devotion of the achievement. The Philosophical Society was pleased with the specimens but it was the Connecticut Assembly that gave Buell support in the form of a loan of £100 without interest. Even with this aid Buell was not able to place his enterprise on a commercial basis nor was a Scottish immigrant in Boston who attempted type founding just a

61. *Votes of Assembly*, VIII, 7463; *Pennsylvania Magazine*, 1 (1775), 137.

62. *New-York Gazette; and the Weekly Mercury*, Aug. 15, 1768; *Public Records of Connecticut*, XIII, 212; *Pennsylvania Gazette*, March 31, 1773; *Massachusetts Gazette*, March 19, 1767.

63. American Philosophical Society Minutes, [II], 125; *Pennsylvania Gazette*, June 27, 1771; *Votes of Assembly*, VIII, 6857.

64. American Philosophical Society Minutes, Oct. 6, 1769, [II], 78; *Pennsylvania Gazette*, Aug. 5, 1772.

little later. Although Christopher Sower did establish type foundries in Germantown which produced principally gothic type for his German publications, they were not of general importance. Franklin also planned to establish a foundry in Philadelphia with materials he brought back from Europe in 1775, but that undertaking was prevented by the war. Successful production did not begin until after the Revolution when Philadelphia and Hartford became the centers of the trade.[65]

Unlike many of the manufactures encouraged by the Americans in this period, iron production was welcomed by the British—to a degree. The Iron Act of 1750 encouraged the production of pig and bar iron in the colonies by dropping the duty on those products, but at the same time, it prohibited the erection of plants for the production of finished iron products. Iron as a raw material was very much needed in Britain where pig and bar iron were considered semi-finished basic products because they had to be further processed before reaching the consumer. When Jared Eliot found a means of extracting iron from "Black Sea Sand," his discovery was welcomed by the Society of Arts with a gold medal because it looked toward increased production of the American raw material.[66] When Connecticut granted legislative subsidies to sustain steel production that was another matter.[67] Steel was a finished iron product which the English did not want the Americans to make.

Potash, like pig and bar iron, was welcomed by the British. Potash was one of the very limited number of industrial chemicals then in considerable demand. It was used not only in the manufacture of glass and soap but also in bleaching and dyeing which were becoming increasingly important in England with the unfolding of the industrial revolution. The substance was imported by England from the Baltic and the forested regions of Europe. There seemed no good reason

65. Isaiah Thomas, *The History of Printing in America* (Worcester, 1810), I, 213-19; Lawrence C. Wroth, *Abel Buell* ([New Haven], 1926), 34-40; Lawrence C. Wroth, "The First Work with American Types," *Bibliographical Essays: a Tribute to Wilberforce Eames* ([Cambridge], 1924), 129-30; Benjamin Gale to Thomas Bond, April 25, 1769, Archives, American Philosophical Society.

66. Franklin B. Dexter (ed.), *Extracts from the Itineraries and Other Miscellanies of Ezra Stiles, D.D., LL.D., 1755-1794* (New Haven, 1916), 214.

67. *Public Records of Connecticut,* XIII, 617.

why it could not be produced in quantity in America where an abundant supply of timber existed. With that view in mind, Parliament had removed the duty from American potash in 1751, and very shortly after its foundation the Society of Arts began to encourage production by establishing premiums and publicizing methods of manufacture.[68]

Here, if anywhere, was an opportunity for turning science to useful ends, for potash was the product of a series of chemical processes. Potash was an impure form of potassium carbonate. Relatively pure potash went under the name of pearl ash and commanded a higher market price. The processes of manufacture and refinement had been developed on an empirical rather than a theoretical basis; indeed, chemical knowledge did not even permit a satisfactory description of the processes that were involved. Nevertheless, much knowledge had accumulated which could be organized and used by those conversant with it.

In the encouragement of potash there was an endeavor to be as scientific as—in the nature of things—it was possible to be. As early as 1755, an Englishman, Thomas Stevens, was granted £3,000 for a new method of manufacture. Two years later in the Southern colonies he was selling a pamphlet that described that method. On the same trip he succeeded in inducing the Virginia legislature to build a potash furnace at Williamsburg.[69] The Society of Arts sent James Stewart to New England in 1763 and to Maryland in 1771 with the mission of promoting the production of potash and similar alkalis. For years, by paying generous premiums for American potash the society gave some color to its claim to responsibility for the establishment of potash manufacture in the colonies.[70] The best treatises on potash were two published in London, one by W. M. B. Lewis and the other by Robert Dossie, one of the leaders of the Society of Arts. Both of them gave precise descriptions of the methods of production and suggested chemical tests to determine alkalinity. Both were primarily concerned

68. [John Mascarene], *The Manufacture of Pot-Ash in the British North American Plantations Recommended* (Boston, 1757), 3; *New-York Mercury,* March 31, 1766.

69. [Mascarene], *Manufacture of Pot-Ash,* 3; Carl Bridenbaugh, *The Colonial Craftsman* (New York, 1950), 105.

70. *Ibid.;* Robert Dossie, *Memoirs of Agriculture* (London, 1768), I, 24.

with the American production of potash.[71] The New York Society of Arts also published directions on production and paid out premiums of its own.[72] In 1766, a pamphlet published in Boston described the process for manufacturing pearl ash.[73] The newspapers were full of efforts to promote this manufacture and full of optimism. The *Virginia Gazette* reported that in 1772 the Northern colonies alone exported potash to the value of £200,000 sterling.[74] There is no evidence that the Americans contributed to the knowledge of the chemistry involved in manufacturing and testing potash but the knowledge available was placed at their disposal and used by them to advantage.

Since the improvement of agriculture and manufactures was promoted primarily by the merchant element in the colonies, it was to be expected that specifically commercial demands would also receive some attention from this quarter. Interest was demonstrated in the building of wharfs, the dredging of channels, and general harbor improvement in all of the port cities. At the same time, improvement in ship design was evolving in the colonial trade. More concerted effort, however, was lavished upon internal improvements than upon the elements of foreign trade. Moreover, it was the merchant class— rather than the West which has been traditionally associated with this demand—that was most active in press agitation, legislative lobbying, and the formation of organizations to complete specific improvements. Roads, bridges, and river improvement were all involved, but the most striking attempt to apply science and to use men of science was the proposed construction of canals.

Although there was no canal in any of the colonies, the Americans were kept well informed of the building of many British and continental canals to answer needs which could not have seemed so imperious as their own.[75] After 1765, projects were set on foot in several

71. Robert Dossie, *Observations on the Pot-Ash Brought from America* (London, 1767); W. M. B. Lewis, *Experiments and Observations on American Potashes* (London, 1767).

72. *New-York Mercury*, March 31, 1766.

73. *Directions for Making Calcined or Pearl-Ashes* (Boston, 1766).

74. *Connecticut Courant*, June 3, 1765, June 16, 1765; *Boston Gazette*, Feb. 15, 1768, March 30, 1772; *New York Post Boy*, Aug. 5, 1771; *Virginia Gazette* (Rind), Jan. 28, 1773.

75. *New American Magazine*, 2 (1759-60), 52. Lester J. Cappon and Stella F. Duff, *Virginia Gazette Index, 1736-1780* (Williamsburg, 1950), I, 177-78, list twenty-eight different European canals reported in the *Virginia Gazettes*.

of the colonies to provide canals where they seemed urgently needed. In North Carolina, an act of assembly provided for opening subscriptions to build a canal that would connect Beaufort and the Neuse River.[76] In Virginia, both legislative and popular support was given to projects for cutting canals around the falls of the James and the Potomac and for connecting the James and York rivers by means of a canal.[77] New England's terrain was less attractive to such projects, although Connecticut did devote much attention to the improvement of the Connecticut River and some thought was given to a canal around the falls of the Merrimack River in New Hampshire.[78] Nowhere was there so much enthusiasm for canal construction or so extensive an effort to apply science to the problem as in Philadelphia.

It was Thomas Gilpin, landed Quaker merchant, who began the campaign for a canal to connect the Delaware with the Chesapeake. Gilpin's broad interest in science resulted in papers on the wheat fly, the seventeen-year locust, a hydraulic wind pump, and the migration of herrings.[79] He had twice visited Europe before he purchased a tract of land at the head of Chester Creek on the Maryland eastern shore. There, the utility of a canal of the sort he had seen in his travels became immediately apparent. A relatively short cut would connect the waters of the Delaware with the Chesapeake. Sometime in 1766 or 1767, he began a survey of a canal route from Head of Chester to Duck Creek on the Delaware and subsequently surveyed other routes with the help of some of his Maryland neighbors. In February 1769, Gilpin presented the canal project to the American Philosophical Society of which he was a member.[80]

The strong merchant membership of the Philosophical Society was particularly noticeable in the two committees to which the canal

76. C. Christopher Crittenden, "Inland Navigation in North Carolina, 1763-1789," *North Carolina Historical Review*, 8 (1931), 145.

77. Hening (ed.), *Statutes at Large of Virginia*, VIII, 556, 564, 570; *Virginia Gazette* (Dixon and Hunter), Jan. 7, 1775; *Virginia Gazette* (Purdie and Dixon), Jan. 16, 1772.

78. *Public Records of Connecticut*, XII, 319-20; XIII, 383-84, 503, 643; XIV, 96-97.

79. See especially, "Memoir of Thomas Gilpin," *Pennsylvania Magazine of History and Biography*, 49 (1925), 289-328.

80. American Philosophical Society Minutes, Feb. 17, 1769, II, 17; Thomas Gilpin, Chester-Duck Creek Surveys, Gilpin Papers, Historical Society of Pennsylvania.

plan was referred. To these men, the canal was no mere academic question. The threat of Baltimore to Philadelphia's control of the rich trade of the interior of Pennsylvania was becoming increasingly clear. The committee sought an intercolonial approach, arguing that the canal would benefit the Marylanders as much as the Philadelphians by the great increase in trade it would promote, but Baltimore merchants were not convinced, charging that the real aim of the whole project was to bring the trade of the Susquehanna to Philadelphia instead of Baltimore.[81] Philadelphia merchants were so attracted to the proposed canal that they quickly subscribed £140 toward a survey of routes.[82]

The initial survey was begun by a committee, four members of which had been appointed by the Philosophical Society and five by the merchants. The best results were assured by the inclusion of two capable surveyors, John Lukens and John Sellers. John Ewing and Joel Bailey were among those who served later in connection with surveys of other routes. Including Gilpin's Chester-Duck Creek survey, estimates were prepared on the cost of five possible routes by which the Delaware and Chesapeake could be united. In its report, the committee favored the building of a barge canal rather than one which could accommodate seagoing ships because the cost of the former was estimated to be only about one-third as great. It also preferred one of the more northerly routes: either the one from Elk River on the Chesapeake to Christiana Creek on the Delaware or that from Elk River to Red Lion Creek on the Delaware. Barge canals along these routes could be constructed at an estimated £19,396 10s and £14,426 respectively, sums not considered beyond the capabilities of the people of that region. At the same time, the committee recommended that the Susquehanna River be improved and that a road be opened between Peach Bottom on the Susquehanna River and the tide waters of Christiana Creek on the Delaware at an estimated cost of £1,500.[83]

81. Gilpin to Franklin, Oct. 10, 1769, [copy], Gilpin Papers; *Pennsylvania Chronicle*, March 23, 1772.
82. American Philosophical Society Minutes, April 21, 1769, May 3, 1769, [II], 21-22, 24.
83. *Ibid.*, Dec. 1, 1769, May 19, 1769, [II], 81, 27; American Philosophical Society, *Transactions*, 1 (1771), 297, 299, 300.

The American Philosophical Society accepted the work of the committee and published in its first volume of *Transactions* a summary of its recommendations illustrated by an excellent map showing the routes surveyed. The society labelled the surveys "public-spirited undertakings," having a "tendency to advance the landed and commercial interest of the British Colonies in general, and particularly of those Middle Colonies with which they are more immediately connected." [84] Many letters to the Philadelphia newspapers in 1769 and 1770 indicated that there was much support in Philadelphia and the back country for the fulfillment of the proposals but that the Marylanders were distinctly less enthusiastic.[85] When the Pennsylvania Assembly took hold of the problem in 1771, no real hope prevailed that it would be handled in a way to benefit any province but Pennsylvania. A survey under the direction of David Rittenhouse and Samuel Rhoads was instituted for the purpose of finding practical and reasonable routes for navigation or land carriage between the Susquehanna, the Schuylkill, and the Lehigh rivers. One of the most important recommendations of this rough survey called for "Procuring from *Europe* such Assistance as the Importance of the Work may require." [86] This had been Franklin's advice, too, in a letter to Samuel Rhoads, and after Rhoads and Rittenhouse completed the survey, they fully agreed that "Some Experience in works of this kind" was necessary before it would be possible to "estimate the Expence with any Degree of precision." [87] Rittenhouse had great competence in science, but science and practical engineering were widely separated.

For the time, that was the end of the matter. The Susquehanna-Schuylkill Canal was not built until 1828, the Chesapeake-Delaware Canal was not completed until 1829. Throughout the colonies roads were improved, rivers cleared of obstructions, ferries established, and bridges built, but such things had been done for years. Like canals, great bridges also had to wait for future developments although

84. *Ibid.*, 293.
85. *Pennsylvania Chronicle*, May 16, 1768, Jan. 8, 1769, Dec. 25, 1769, March 23, 1772; *Pennsylvania Gazette*, Dec. 28, 1769, Jan. 4, 1770, Jan. 18, 1770, Feb. 15, 1770, March 5, 1772; List of newspaper articles by Thomas Gilpin, Gilpin Papers.
86. Report of Samuel Rhoads and David Rittenhouse, Jan. 30, 1773, *Votes of Assembly*, VIII, 6934.
87. Franklin to Rhoads, Aug. 22, 1772, Historical Society of Pennsylvania.

Thomas Gilpin projected a bridge with a three-hundred-foot arch and John Jones of Indian River actually built a model of a suspension bridge designed to support an arch of four hundred feet.[88]

Perhaps Christopher Colles, an immigrant of 1771, was the man best prepared by his knowledge of mathematics and science to apply learning to practical engineering. He called himself an "Engineer and Architect" in an early newspaper advertisement in which he offered to design mills and hydraulic engines at the same time that he proclaimed himself ready to instruct "young Gentlemen, at their Houses, in the different Branches of Mathematics and Natural Philosophy." [89] In 1772, Colles gave one series of lectures on hydraulics in the Hall of the American Philosophical Society in Philadelphia and another on pneumatics. The following year, he offered a more general course in natural philosophy and mechanics from the second floor of the Pennsylvania State House.[90] Colles also demonstrated before the American Philosophical Society a steam engine he had constructed for raising water at a Philadelphia distillery. Despite the fact that he was well considered by the Philadelphia scientists, he drifted to New York in 1774. There he succeeded in getting his plan adopted for a water system using hollow wooden logs as conduits.[91] Except for the war, this ambitious project might have been completed at a time when piped water in America was limited to unusual local situations. In a part of Providence, for example, water was conveyed from an artesian well to the interior of a group of houses where it was available on tap.[92]

The first medal awarded by an American organization for a practical invention came from the Virginia Society for Promoting Useful Knowledge which was founded in Williamsburg, Virginia, on May 13, 1773. It awarded a gold medal to John Hobday in recognition

88. See Joshua Gilpin, *Memoir on the . . . Chesapeake and Delaware Canal* (Wilmington, 1821); Chester-Duck Creek Surveys, Gilpin Papers; John Jones to American Philosophical Society, May 27, 1773, May 19, 1773, Archives, American Philosophical Society.

89. *Pennsylvania Chronicle,* Aug. 26, 1771.

90. *Pennsylvania Gazette,* Sept. 26, 1771, Feb. 13, 1772; *Pennsylvania Chronicle,* Aug. 26, 1771; *Pennsylvania Packet,* March 2, 1772, March 13, 1773.

91. American Philosophical Society Minutes, Aug. 20, 1773, [II], 149; *New-York Gazette; and the Weekly Mercury,* Aug. 1, 1774, Sept. 5, 1774.

92. *Boston Gazette,* Sept. 7, 1772.

of his threshing machine, a "cheap and simple" device which was credited with a capacity for beating out 120 bushels of wheat in three days. The inventor claimed that it would cost only £15 to duplicate.[93] The curious thing about this award was that the threshing machine was invented before the society was established; it may even have been instrumental in turning the attention of the founders of the society toward the promotion of such useful knowledge. John Page, the most active founder of the society, was among those who in 1772 had sought to get up a subscription to encourage John Hobday to distribute models of his machine through the country.[94]

Indeed, the decision to establish the Virginian Society for the Promotion of Usefull Knowledge was taken by a group of eight men, including John Page, on November 20, 1772.[95] This was just one day after the original announcement of Hobday's machine was made in Purdie and Dixon's *Virginia Gazette*. The meeting was held in Williamsburg during the session of the legislature—the only time that it was possible to count on finding a large group of intelligent, informed men in any one place in Virginia. Inevitably, the meetings of the new society were thereafter tied to the sessions of the legislature. The establishment of this society came at a peculiarly frustrating time of the Revolutionary struggle, just after the opposition to the Townshend Acts had collapsed. Many Virginians proved ready to support the promotion of useful knowledge, particularly when it promised to benefit their sagging economy.

For a time, the Virginia Society commanded much attention. John Clayton, the now aged author of *Flora Virginica*, was elected president; John Page, vice-president; and the royal governor assumed the post of patron. Some one hundred Virginians accepted membership in an organization which, in a sense, was conceived "in humble imitation of the Royal Society" although from the beginning utilitarian objectives seemed paramount.[96] One early advocate sought to use

93. *Virginia Gazette* (Purdie and Dixon), June 16, 1774.

94. *Ibid.*, Nov. 19, 1772.

95. Plan of Virginian Society for the Promotion of Usefull Knowledge, Nov. 20, [1772], Colonial Williamsburg; see also [Lester J. Cappon], *Williamsburg's First Institute* ([Williamsburg], 1954).

96. Francis Hargreaves' Manuscript on the Society for the Advancement of Useful Knowledge, quoted in *William and Mary Quarterly*, [1st ser.], 4 (1896), 201.

the writings of Francis Bacon as a guide, pointing to his opinion that the practical arts were based upon the sciences. He lamented the lack of cities in Virginia but felt that planters could, at least, "make Experiments in Agriculture, without Detriment to the usual Course of their Business." [97]

Some papers, meteorological journals, and observations were collected during the ensuing two years when meetings were held with some regularity. After Clayton's death, John Page became president and most of the other offices were given to people associated with the College of William and Mary. A few corresponding members were elected, most of them from among the leadership of the American Philosophical Society. Thought, too, was given to the publication of a volume of scientific papers, but after the granting of the medal to John Hobday, very little of a positive nature was accomplished. With the approach of war, the society declined, never to be revived effectively again. [98]

97. *Virginia Gazette* (Purdie and Dixon), Aug. 5, 1773.
98. *Ibid.*, June 16, 1774.

PART THREE

The New Nation

1775-1789

Chapter Eleven

THE WAR

W ITH THE OUTBREAK of war on April 19, 1775, the conditions under which science had been encouraged in America were altered in many significant ways. The approach of Revolution had been intellectually stimulating. It had led men to see visions, including the prospect of a richer scientific life on this side of the Atlantic. Few of their aspirations had been realized, although many attempts had been made to establish institutions in which science—especially useful science—could flourish. The war brought much of this activity to an end. Indeed, the mounting tensions of 1774 and 1775 so occupied the thoughts of many men that their efforts to promote science began to decline even before the fatal shot on Lexington green.

War affected the promotion of science much more than it did the life of the average American who continued to do the same things he had always done in much the same way he had always done them. The life of science was a much more fragile plant. The men, the institutions, and the interrelationships that sustained science were badly disturbed and disrupted by the war. Whereas the independent farmer could still work his land and find a ready market for his products, those who were scientists or sought to promote scientific endeavor found nothing quite the same. To begin with, the countryside, in which nine-tenths of the population lived, was much less altered by war than the cities in which most of the attempts to advance science were centered. The major cities, Boston, New York, Newport, Philadelphia, and Charleston were all occupied for a time by the British, while New Haven, Richmond, and many of the small towns were subjected to military raids. Even under American control, the cities, all of which were commercial in character, found their complex activities disordered by the cessation of trade with Great Britain and the difficulties imposed in the way of all other trade. The many imperatives of war relegated the pursuit of science to a distinctly secondary consideration.

The men of science of the eighteenth century had a tendency to consider their efforts above and beyond the distractions of national wars but even that attitude did not serve to keep the international scientific community inviolate. Arthur Lee, of Virginia, took an unusual stand when he resigned from the Royal Society on the grounds that membership was inconsistent with his duty to his country, now that she was at war with Great Britain. In reply to his letter, Sir Joseph Banks stated the generally accepted position of the philosopher, that there was a great difference between scientific societies and political associations, the objects and interests of the former being universal, "belonging to the republic of letters, and to the community of man and mind." [1] Even so hotheaded a patriot as Arthur Lee could sympathize with this viewpoint. Indeed, despite the great difficulties involved, Lee maintained occasional contact throughout the war with his English scientific friends, the Reverend Richard Price and the Reverend Joseph Priestley. [2]

Several other Americans also maintained a correspondence with the two nonconforming clergymen and their circle. Priestley was able to get his own publications to Benjamin Franklin at Paris without much trouble but found it very difficult to send them to John Winthrop at Cambridge. [3] Winthrop sought to send his letters for English friends through Franklin. [4] In fact, Franklin's diplomatic post at Paris

1. Richard Henry Lee, *The Life of Arthur Lee* (Boston, 1829), I, 17.
2. Arthur Lee to Richard Price, April 20, 1777, "Letters of Richard Price," Massachusetts Historical Society, *Proceedings*, 2d ser., 17 (1903), 310.
3. Price to Charles Chauncy, Dec. [1775], *ibid.*, 307; Joseph Priestley to Franklin, May 8, 1779, Franklin Papers, XIV, 88, American Philosophical Society.
4. Winthrop to John Adams, June 21, 1775, "Correspondence between John Adams and Prof. John Winthrop," Massachusetts Historical Society, *Collections*, 5th ser., 4 (1878), 293.

permitted him to serve as the most effective link between science in America and science in England. On one occasion, Benjamin Vaughan made use of the Spanish ambassador as a courier for transmitting letters on science to Franklin.[5] Even the French ministry was ready to countenance a scheme for importing, through Franklin, American seeds desired by English gardeners. Franklin wrote to his friend John Bartram in this connection, but before the program could be worked out, France and England were at war and John Bartram was dead.[6] Such contacts and plans were a monument to the enlightened attitudes of men on both sides of the Atlantic, but they were not a substitute for the untrammelled traffic of peacetime.

Franklin's unremitting efforts to keep open the channels of communication were more important but less spectacular than the instructions he issued for the protection of the last expedition of Captain James Cook. Without real authority to do so, Franklin wrote and circulated a paper "To all Captains and Commanders of armed Ships acting by Commission from the Congress of the United States of America, now in war with Great Britain" directing them not to consider Cook's men as enemies but "as common friends to mankind." [7] Sir Joseph Banks, as president of the Royal Society, thanked him for this gesture—but not publicly, for that would have been "willfully misconstrued." He pointed out that the Royal Society was going to present one of the medals struck in honor of Captain Cook to the King of France because of similar orders he had issued. Banks hoped that he might be informed of the Congress having taken such action so that a medal might also be presented to them.[8]

It was a little odd that the prevailing faith in the ultimate utility of science did not lead men to oppose scientific intercourse as injurious to the war effort of their nation, but such was seldom the case. The Reverend Joseph Willard, president of Harvard College, expressed his own feelings very clearly when he sent an account of

5. Benjamin Vaughan to Franklin, April 9, 1779, Franklin Papers, XIV, 21, American Philosophical Society.

6. Franklin to Bartram, May 27, 1777, William Darlington, *Memorials of John Bartram and Humphry Marshall* (Philadelphia, 1843), 406.

7. Franklin order dated March 10, 1779, Albert Henry Smyth (ed.), *The Writings of Benjamin Franklin* (New York, 1905-7), VII, 242-43.

8. Banks to Franklin, March 29, 1780, Bache Collection, American Philosophical Society.

astronomical observations to Nevil Maskelyne. "I hope, Sir," he saluted the astronomer royal, "no umbrage will be taken at my writing to you on account of the political light in which America is now viewed by Great Britain. I think political disputes should not prevent communications in matters of mere science; nor can I see how any one can be injured by such an intercourse." [9] For his part, Maskelyne, through the Royal Society, presented Harvard College with observations made at Greenwich at a time when they could not have been obtained "but by the favourable interposition of that very learned and illustrious Society." [10]

Governmental recognition was given to this principle of the internationalism of science in the wartime charter of the American Philosophical Society. That document provided, "It shall and may be lawful for the said Society, by their proper officers, at all times, whether in peace or war, to correspond with learned societies, as well as learned men of any nation or country upon matters merely belonging to the business of the said Society." [11] A little later, Benjamin Rush expressed the feeling as well as it could be expressed when he said, "In science of every kind men should consider themselves as citizens of the whole world." [12]

This favorable attitude did much to mitigate the worst effects of war but it could not create a favorable atmosphere for individual scientific activities. Most of the men who had given evidence of a capacity and desire to make creative contributions to science found their support threatened, their environment drastically altered, or their attention diverted to other objects. The experience of America's most celebrated scientific figure, Benjamin Franklin, was particularly enlightening. From the time that Franklin had first begun his electrical experiments, nothing was able to command his interest so fully as scientific work. There is no reason to imagine he spoke less than the whole truth when he said he would rather talk of science with his

9. Willard to Maskelyne, Read July 5, 1781, *Philosophical Transactions*, 71 (1781), 507.

10. Harvard Corporation Records, III, 107.

11. American Philosophical Society, *Transactions*, 1 (2d ed., 1789), xvi.

12. Rush to Price, April 22, 1786, "Letters of Price," Massachusetts Historical Society, *Proceedings*, 2d. ser., 17 (1903), 347.

"philosophic Friends" than be with "all the Grandees of the Earth." [13] Yet despite this devotion to science, no man realized better than he that there were times when the pursuit of science must be dropped for more important things. Long before the Revolution he had written Cadwallader Colden, declaring, "Had Newton been Pilot but of a single common Ship, the finest of his Discoveries would scarce have excus'd, or atton'd for his abandoning the Helm one Hour in Time of Danger; how much less if she carried the Fate of the Commonwealth." [14] From the beginning of the War of Independence, Franklin was prepared to throw all that he had to give into the struggle.

Not the least of the acquirements that Benjamin Franklin placed at the service of the Americans was his scientific reputation. He had had long and successful experience as the agent of several of the colonies in London, but even this was less valuable to the new United States when he represented them at the French court than his reputation as a learned philosopher. At its base, his reputation rested upon the electrical experiments but it had come to flower during the many years he spent in England among the savants of the Royal Society displaying qualities of character and mind that made him appear the personification of wisdom and wit. According to John Adams, who had reservations of his own regarding the stature of his compatriot, "His reputation was more universal than that of Leibnitz or Newton, Frederick or Voltaire; and his character more beloved and esteemed than any or all of them." [15] French admiration of Franklin was the best possible setting for a successful mission to that country. Thomas Jefferson reported that "there appeared to be more respect and veneration attached to the character of Doctor Franklin in France than to that of any other person in the same country, foreign or native." [16]

In France, Franklin was feted by the scientific community almost as much as he was by the ladies of the French court. He was appointed to scientific commissions by the Académie des Sciences and by

13. Franklin to Banks, Sept. 9, 1782, Smyth (ed.), *Writings of Franklin*, VIII, 593.

14. Franklin to Colden, Oct. 11, 1750, *The Letters and Papers of Cadwallader Colden, IV*, New-York Historical Society, *Collections for 1920*, 53 (1921), 227.

15. Charles F. Adams (ed.), *Works of John Adams* (Boston, 1850-56), I, 660.

16. Jefferson to William Smith, Feb. 19, 1791, Etting Collection, Signers of the Declaration of Independence, 50, Historical Society of Pennsylvania.

Louis XVI.[17] He was elected to the Royal Academy of Medicine and escorted to its meetings by the King's personal physician.[18] At Passy, he gathered about himself a stimulating little community of intellectuals of which he became the patriarch. Scientists of every sort as well as those who sought to storm the citadels of science asked eagerly for Franklin's support.[19] Despite all this, despite the fact that the scientific capital of the world almost prostrated itself before him, Franklin did not enjoy an atmosphere favorable for creative science. His diplomatic, political, and social activities absorbed his time to the exclusion of any opportunity for experiment or quiet thought. The two papers written on the long homeward voyage to the United States were more valuable scientific contributions than anything he wrote during the years he spent in France.

Whether as a direct result of the war or not, the whole configuration of American science was strikingly altered very shortly after hostilities began. The natural history circle was particularly affected. Three of the most distinguished American botanists died within a few years of the outbreak of the war: John Clayton in 1774, Cadwallader Colden in 1776, and John Bartram in 1777, just four days before the British occupied Philadelphia. It was Bartram's death that seemed most to disturb visiting foreigners. One British officer who maintained a considerable garden of his own had hoped to purchase seeds and to get advice on American plants from him.[20] François de Barbé-Marbois, a member of the French legation, made a kind of pilgrimage to the home of "the American Linnaeus" only to be shocked at finding the garden "in a state of neglect which caused us actual pain."[21] As for Alexander Garden, the only remaining naturalist

17. Jean B. Le Roy to Franklin [1778?], Franklin Papers, XLIV, 143, American Philosophical Society; [Jean-S.] Bailly to Franklin, June 17, 1784, *ibid.*, XXXII, 19.

18. Membership Certificate, Royal Society of Medicine of Paris, June 17, 1777, American Philosophical Society, Portfolio; Vicq d'Azyr to Franklin, [1777?], *ibid.*, XL, 127.

19. Lavoisier to Franklin, [Circa 1780], American Philosophical Society, XLIV, 277b; Court de Gebelin to Franklin, May 6, 1781, *ibid.*, XXII, 7; Mesmer to Franklin, Dec. 1, 1779, *ibid.*, XVI, 138.

20. Captain Fraser to [?John Bartram], Dec. 15, 1777, Darlington, *Memorials*, 465.

21. François de Barbé-Marbois, *Our Revolutionary Forefathers: The Letters of François de Barbé-Marbois* (New York, 1929), 132.

who enjoyed the international esteem of Clayton, Colden, and Bartram, some attempt was made to introduce him to younger men in the field. David Ramsay opened a correspondence between the Charleston physician and Benjamin Rush in Philadelphia but it was never developed. Garden, a decided loyalist, emigrated to England where he spent the balance of his life.[22]

With the removal of the leading naturalists of the colonial period, other men found new opportunities. The seed trade which had played a large part in forming the prewar natural history circle was largely discontinued, but the war introduced some peculiar demands of its own. Dr. Thomas Bond, for example, decided that it would be scientifically as well as diplomatically desirable to exchange seeds with the botanically-inclined French minister to the United States, Conrad Alexandre Gérard. With this in mind, Bond turned to Humphry Marshall, who at John Bartram's death became the leading seedsman in the Philadelphia area. Marshall willingly presented a collection of seeds for the minister, but he soon found that this was only the beginning. Other requests followed, for trees and plants to be used in the Royal Garden at Paris. Some French seeds were returned by way of thanks and Bond and Marshall had the satisfaction of being informed that "the King of France examined every article of our collection, and was extremely pleased with it." [23] Such donations were not unusual. In a similar manner, John Page presented the Marquis de Chastellux with a hollow amethyst containing some fluid and a bubble of air. The curiosity had been given to Page in order that it might be donated to the Virginian Society for the Promotion of Usefull Knowledge but with the declining fortunes of that organization, Page decided it might better be used as a gift to a noble ally. The amethyst, too, found its way to the cabinet of the King of France.[24]

Some naturalists became less productive while others were given their first encouragement in the unusual circumstances that accompanied the armed conflict. In the early war years William Bartram

22. Rush to David Ramsay, Nov. 5, 1778, L. H. Butterfield (ed.), *Letters of Benjamin Rush* (Princeton, 1951), I, 218.

23. Bond to Marshall, Aug. 7, 1779, Nov. 3, 1779, March 16, 1781, Darlington, *Memorials,* 536, 537, 539; Barbé-Marbois to Marshall, Dec. 9, 1780, Marshall Correspondence, Dreer Collection, I, 6, Historical Society of Pennsylvania.

24. Page to Jefferson, April 28, 1785, Julian P. Boyd (ed.), *The Papers of Thomas Jefferson* (Princeton, 1950–), VIII, 119.

completed his famous travels, little affected by marching armies, but after 1778 when he returned to Philadelphia, he remained very quiet. Besieged by debt and bereft of the bounty of Dr. Fothergill and the sustaining hand of his father, Bartram's talents seemed for a time to sleep.[25] On the other hand, it was during the unsettled years of war that the Reverend Henry Ernest Muhlenberg first began the systematic study of botany. Muhlenberg was the youngest son of the patriarch of the Lutheran Church in America and brother of two other clergymen. Having graduated from the University of Halle, he had scholastic equipment enabling him to study botany profitably as soon as his interests led him to that occupation. Moreover, although he was the most religious of the Muhlenberg boys, botany was a study which he could follow with a clear conscience, for his father had told him "not to buy a large library, but to read the book of Nature." It was not until the enforced leisure of a year spent at New Hanover, Pennsylvania, during the British occupation of Philadelphia that he actually began the study and collection of plants.[26] Once embarked upon this hobby, he pursued it so determinedly that his father remonstrated, declaring that his parish work would give him "infinitely more blessing and reward than all this research into hidden variants and the plants of Linnaeus." [27] Henry Muhlenberg continued his studies, nevertheless, and before the war had ended, he became a very respectable botanist.

The war disrupted the usual channels of communication between American and European naturalists but at the same time it established new connections with individuals among the many soldiers who came to America to fight. Some of these contacts proved very important because of the direct nature of the relationship. Several French officers had some interest in natural history although only the Marquis de Chastellux was a member of the Académie. Chastellux could usually recall some reference from his reading to objects of conversation even

25. *Pennsylvania Packet*, April 22, 1785.
26. George W. Corner (ed.), *The Autobiography of Benjamin Rush* (Princeton, 1948), 184; Henry M. M. Richards, "Gotthilf Henry Ernest Muhlenberg, D.D.," *Pennsylvania-German*, 3 (1902), 149-50.
27. Paul A. W. Wallace, *The Muhlenbergs of Pennsylvania* (Philadelphia, 1950), 193.

when he was not a master of the topic himself.[28] It was fashionable to have such interests, as the Chevalier D'Anmours indicated when he introduced the Chevalier de Ternant to Thomas Jefferson with the comment, "His hobby-horse is like mine, natural history." [29] Important work was done by some of the officers attached to the German mercenary troops fighting for Great Britain. Friedrich von Wagenheim, a Hessian captain, got back to Germany so quickly that before the war was over he was able to publish a description of several American trees and shrubs he thought might be suitable for German forests.[30] His own field work was largely limited to the Middle states but he made use of the writings of Colden and Gronovius on the subject. Wagenheim's book was distinctly less valuable than the several writings of the Beyreuth surgeon, Johann David Schöpf, upon American geology, fish, turtles, frogs, climate, and the materia medica, but the war saw only the introduction of Schöpf to the American scene. The bulk of his investigations and their fruition belong to a later period. This was true also of the work of another learned German who came to America as chaplain to a regiment of Brunswick dragoons and remained to become an American citizen, Frederick V. Melsheimer. In later years, Melsheimer became a professor at Franklin College and the leading entomologist in the country.[31] Such men provided a new stimulus for work in natural history which offset in some measure the dislocations and interruptions of the war.

The same kind of European influence was felt in the academically-housed physical sciences but the total effect was less impressive. David Rittenhouse and John Ewing demonstrated the celebrated orrery to the Marquis de Chastellux just as eagerly as John Page had given him his hollow amethyst but the Marquis seemed less impressed with Rittenhouse's accomplishments than with the empty curiosity.[32] A part of the difficulty was that American academic accomplishments were

28. Chastellux to Jefferson, June 10, 1782, Boyd (ed.), *Papers of Jefferson*, VI, 190.

29. Chevalier D'Anmours to Jefferson, Feb. 27, 1782, *ibid.*, 161.

30. Friedrich A. J. von Wagenheim, *Beschreibung einiger Nordamerican-ischen Holz und Buscharten* (Gottingen, 1781).

31. Herbert H. Beck, "Henry E. Muhlenberg, Botanist," Lancaster County Historical Society, *Papers*, 32 (1928), 101.

32. Marquis de Chastellux, *Travels in North-America in the Years 1780-81-82* (New York, 1827), 112.

THE PURSUIT OF SCIENCE

not on a level that would impress learned Europeans, whereas much of the natural history they encountered was entirely unknown and absorbing. In addition, physics, mathematics, and astronomy suffered from their institutional associations. Most of the men who cultivated those sciences were attached to the colleges and the colleges were badly disrupted by the war.

The war profoundly altered the life of John Winthrop who was the best known and most productive college scientist of this type. Winthrop was wholeheartedly devoted to the cause of the Revolution, serving as a member of the provincial council and concerning himself from an early date with the problem of munitions production. The early disorganization of Harvard College robbed his life of whatever calm it then had. In May 1775, the buildings of Harvard College were taken over by provincial troops to be used as barracks. In June, the provincial congress ordered the college to pack up the library and apparatus for removal to the country for safety. In November, the General Court finally decided that the college was to send to Concord the books and apparatus immediately required for instructing students at that town.[33] Most of this equipment found its way to Winthrop's Concord home but some was so dispersed that much effort was required to collect it once again when the British evacuated Boston.[34] Harvard was fortunate, nevertheless, for when the British left the soil of Massachusetts in 1776, they left for good; thereafter, neither Harvard nor any other institution in the area was threatened with direct assault.

With some interruptions, John Winthrop continued to teach classes despite the political demands upon him, but he did not write anything of a scientific character in this period. Winthrop spent time inspecting powder mills, advising on the production of sulphur, and trying to stimulate saltpeter production.[35] Harvard's vicissitudes and his own preoccupation with war activities, however, did not prevent his attention to astronomical observations; they only made such work more

33. Winthrop to Adams, June 21, 1775, "Correspondence between Adams and Winthrop," 292; Samuel E. Morison, *Three Centuries of Harvard* (Cambridge, 1936), 148-49; Lawrence S. Mayo, *The Winthrop Family in America* (Boston, 1948), 186.

34. Harvard Corporation Records, II, 439, 448.

35. Winthrop to Adams, May 23, 1776, June 1, 1776, "Correspondence between Adams and Winthrop," 304, 306.

difficult. In this period he made use of his wife's assistance to observe an eclipse of the sun and a transit of Mercury across the sun's face.[36] In 1779, John Winthrop died—as serious a blow to science at Harvard as anything that befell the college as a result of the war. Mercy Warren stated only the truth when she remarked, "I fear it will be long before Harvard sees a successor that will fill the Chair of the professor with Equal Honour and Ability." [37] For a time, his son, James Winthrop, and Caleb Gannett delivered his lectures but the chair was then given to one of his students, the Reverend Samuel Williams, who ultimately proved himself inferior to Winthrop in scientific attainment and incompetent in his personal relations.[38]

Even more than Harvard, the College of Philadelphia had been deranged by the war. Its halls were used by the Americans as a barracks and by the British as a hospital. After the British evacuated the city, the state government in 1779 was led to reorganize the college as the University of the State of Pennsylvania with a new faculty and a board of trustees whose devotion to the Revolution could not be suspected. David Rittenhouse was immediately appointed a trustee and shortly afterward was made a vice-provost of the institution. Enjoying nothing about his faculty post, he soon resigned it, but other faculty changes proved more lasting. Because of the personal and political enemies he had made and because of his equivocal attitude toward the Revolution, the Reverend William Smith was replaced as provost by the Presbyterian clergyman, John Ewing. Ewing, who took up the courses in natural philosophy, and Robert Patterson, who was appointed to the faculty just a little later, both became effective leaders in the scientific life of the city.[39]

David Rittenhouse was the only man in America whose capacity in mathematics, physics, and astronomy could be compared with John Winthrop's. Rittenhouse's devotion to the cause of the Revolution was as complete as that of Winthrop, and it resulted in his rapid absorption in a multitude of tasks only a few of which required any

36. Hannah Winthrop to Mercy Warren, Jan. 14, 1777, *Warren-Adams Letters*, I, Massachusetts Historical Society, *Collections*, 72 (1917), 283-84.
37. Mercy Warren to Abigail Adams, July 29, 1779, *ibid.*, II, 115.
38. See William Bentley, *Diary* (Salem, 1905), I, 100.
39. Edward P. Cheyney, *History of the University of Pennsylvania* (Philadelphia, 1940), 132-33; Edward Ford, *David Rittenhouse* (Philadelphia, 1946), 116-17.

scientific capacity. He served as engineer of the Pennsylvania Committee of Safety, he was elected vice-president of the Council of Safety, and he succeeded to Franklin's seat in the state assembly. His most important offices were treasurer and Loan Office trustee of Pennsylvania which he held to the end of the war. He helped to write the Pennsylvania Constitution of 1776, served on the Board of War, and accepted other occasional tasks so occupying his time that there was little opportunity left for scientific activity. Rittenhouse, however, lived longer than Winthrop, and before the war was over, he was able to return to science.[40]

His scientific activities fell into two categories: the first embracing attempts to contribute to the war effort and the second being a continuation of basic scientific work. Rittenhouse helped to select strategic points for fortification about Philadelphia, surveyed a part of the Delaware River, experimented in rifling cannon, supervised the manufacture of other munitions, and with Charles Willson Peale, experimented upon the use of telescopic sights for rifles.[41] The biggest obstacle in the way of scientific work of a fundamental character was the want of leisure—particularly after he accepted the appointment as treasurer. It was Hannah Rittenhouse, his wife, who finally solved this difficulty. "For a time Dr. Rittenhouse managed the business of his office with the utmost attention and assiduity; but his all-capacious mind could no longer be restrained from its native pursuits; his money and his counter, therefore, he resigned into the hands of his beloved wife, who, although possessed of all the feminine virtues, performed the arduous duties of the office with a masculine understanding, with accuracy and unwearied attention."[42] In company with William Smith, and sometimes with John Lukens and Owen Biddle, Rittenhouse was enabled to observe a transit of Mercury and two eclipses of the sun.[43] In 1780 and 1781 he was free enough of the restraints of war to write his important papers on magnetism and the cameo-

40. Ford, *Rittenhouse*, 71 ff.; *Pennsylvania Journal*, Nov. 27, 1776.
41. Charles C. Sellers, *Charles Willson Peale* (Philadelphia, 1947), I, 128.
42. Col. Francis Johnson quoted by William Barton, *Memoirs of the Life of David Rittenhouse* (Philadelphia, 1813), 463n-64n.
43. Ms. Notes in Tobias Meyer, *Tabulae Motuum Solis et Lunae* (London, 1770), flyleaf, American Philosophical Society; Horace W. Smith, *Life and Correspondence of the Rev. William Smith, D.D.* (Philadelphia, 1879), I, 570; Barton, *Memoirs of Rittenhouse*, 587, 589, 591.

THE WAR

intaglio illusion.[44] The deleterious effect of war was finally reduced to manageable size.

Science in the other American colleges had been largely a pedagogical affair before the war, but everywhere it was further weakened by the actual conflict. King's College was taken over by American troops in 1776 and used by British troops during the occupation, until 1783.[45] The College of Rhode Island was used by the American military and by the French between 1776 and 1782.[46] Both the College of New Jersey and Queen's College, situated in the "cockpit of the Revolution," experienced interruptions in their activities, but of shorter duration.[47] Yale was not so harshly handled, although Edmund Fanning asserted that he barely prevented the "utter destruction" of the college during Governor Tryon's raid upon New Haven in 1779.[48] Even so, Thomas Clap's manuscripts were carried off and, while Tryon was ready to return them upon demand, they could never be found.[49] Indirectly, it was the war that caused the college to fall behind in its salary payments to its rather ineffective professor of mathematics and natural philosophy, Nehemiah Strong. He finally resigned in 1781. With the election of Ezra Stiles to the presidency of the college in 1777, science at Yale was placed upon a firmer and more enthusiastic base than it had been for some years.[50]

The sequence of events at the College of William and Mary was rather the reverse of that at the other colleges. The first effect of the war was to change the composition of the faculty by eliminating the Tories, thus bringing to the presidency the Reverend James Madison, who was eagerly devoted to the cultivation of the sciences. In 1779, as a part of a general educational reform, Thomas Jefferson sought to convert the school into a state university in which science would be

44. American Philosophical Society, *Transactions*, 2 (1786), 37-42, 178-81.
45. Horace Coon, *Columbia, Colossus on the Hudson* (New York, 1947), 51.
46. Walter C. Bronson, *The History of Brown University* (Providence, 1914), 63, 67-68.
47. Thomas J. Wertenbaker, *Princeton, 1746-1896* (Princeton, 1946), 59; William H. S. Demarest, *A History of Rutgers College, 1766-1924* (New Brunswick, 1924), 118.
48. Amos Botsford to Stiles, July 27, 1789, Isabel M. Calder (ed.), *Letters and Papers of Ezra Stiles* (New Haven, 1933), 107.
49. Abiel Holmes, *The Life of Ezra Stiles* (Boston, 1798), 262-63.
50. Louis W. McKeehan, *Yale Science, The First Hundred Years, 1701-1801* (New York, 1947), 46-52.

strenuously cultivated by the establishment of professorships of mathematics, anatomy and medicine, and natural philosophy and natural history. Jefferson also proposed to purchase a Rittenhouse orrery for the school. Very little in the direction of this reform could actually be accomplished before war engulfed the Yorktown peninsula and the college was converted into a hospital. When it was finally turned back to use as a college, most of its assets were in a declining state.[51]

Other institutions which encouraged the development of science suffered just as the colleges did. The philosophical societies did not have so much in physical assets to lose but the war just as effectively interrupted their meetings. No recorded meeting of the Virginian Society for the Promotion of Usefull Knowledge was held after June 15, 1774 although John Page made occasional observations and continued to act as if the society still existed.[52] As for the American Philosophical Society, its meetings and vitality began to decline long before the outbreak of war. When no meeting was held for nine months in 1774, the patriotic secretary explained it as a result of the Coercive Acts which, having "alarmed the whole of the American colonies, the members of the philosophical Society partaking with their countrymen in the distress and labors brought upon their country were obliged to discontinue their meetings for some months until a mode of opposition to the said acts of parliament was established." [53] The three-year lapse in activities between January 1776 and March 1779 was similarly explained as a result of "The Calamities of War and the Invasion of this City by the Enemy." [54] This, however, was not the whole story, for David Rittenhouse indicated that the meetings

51. Committee of Revisors Report, Boyd (ed.), *Papers of Thomas Jefferson*, II, 540 (Bill No. 80); Wythe to Jefferson, Dec. 31, 1781, *ibid.*, VI, 144; Philip A. Bruce, *History of the University of Virginia, 1818-1919* (New York, 1920), I, 65; James Madison to Stiles, June 19, 1782, Calder (ed.), *Letters of Stiles*, 50-51; *Virginia Gazette* (Dixon and Hunter), April 4, 1777; Theodore Hornberger, *Scientific Thought in the American Colleges, 1628-1800* (Austin, 1945), 62.

52. *Virginia Gazette* (Purdie and Dixon), June 9, 1774; *Virginia Gazette* (Purdie), May 16, 1777; Rittenhouse to Page, Aug. 18, 1777, [Photographic Copy], Historical Society of Pennsylvania; Page to Arthur Lee, March 12, 1778, Lee, *Life of Arthur Lee*, II, 324.

53. American Philosophical Society Minutes, [III], following entry for Feb. 4, 1774, not paged.

54. *Ibid.*, before entry for March 5, 1779, [III], 31.

were discontinued "rather through the disputes between Whig and Tory than any public necessity." [55]

In any event, the patriot wing seems to have controlled the property of the American Philosophical Society before the British occupied the city. The Declaration of Independence was first read publicly in Philadelphia from the platform of the observatory built by the society to observe the transit of Venus in 1769. It was in the rented hall of the society that the official committee of inspection assembled for this occasion.[56] Then, when the Supreme Executive Council of the state had the great seal of the colony of Pennsylvania broken up, the remnants were placed in the society's hall, only to be stolen soon afterward.[57] During the British occupation, some of the society's books also were stolen, not from its building but from the home of Benjamin Franklin, by the celebrated Major John André.[58] In 1779, meetings were revived, papers were received, and a future began to appear at least possible.

A similar course of decline and recovery during the war was demonstrated by the publishing media which served to disseminate knowledge. Newspapers throughout the colonies encountered the greatest difficulty in maintaining continuous publication. It was not only a question of military occupation and political antagonism, although few editors were as adroit as Benjamin Towne of the *Pennsylvania Evening Post* who managed to put out a patriot paper at the beginning of the war, a loyalist paper during the British occupation of Philadelphia, and a patriot paper after the city was evacuated. Even the physical problems of obtaining paper, ink, and type were often too much for the editors.[59] Magazines ceased publishing altogether except for one short-lived effort. During the first three years of the war, almost the only separate titles published in fields related to science

55. Rittenhouse to Page, Aug. 18, 1777, [Copy], Historical Society of Pennsylvania.
56. Charles H. Hart, "Colonel John Nixon," *Pennsylvania Magazine of History and Biography*, 1 (1877), 196.
57. Philip S. P. Conner, Memoir of Edmund Physick, 65, American Philosophical Society.
58. [Memorandum of Deborah Logan], *Pennsylvania Magazine of History and Biography*, 8 (1884), 430.
59. Clarence S. Brigham, *History and Bibliography of American Newspapers, 1690-1820* (Worcester, 1947), I and II, *passim*.

were concerned with fighting the war: military medicine, munitions manufacture, and military engineering. There was hardly any publishing at all in 1779; then in 1780 and 1781 more general titles in the sciences and in medicine began to appear.[60]

The French added a new element in publication that was altogether a wartime phenomenon. A printing press carried by the French fleet was used to strike off several items during the war, the most scientific of which was an almanac in French—the first Roman Catholic almanac printed in the United States.[61] Not particularly accurate, it was distinctly less valuable than occasional flashes of interest in science demonstrated by some of the French officers. Major General de Grauchain observed an eclipse of the sun and an eclipse of the moon in 1780 while stationed at Newport. He sent his observations and remarks to the American Philosophical Society and to the American Academy of Arts and Sciences and both societies published them.[62]

Of the groups which sustained the advance of science in America, the physicians were the most directly and permanently affected by the war. The New Jersey Medical Society was suspended for the duration, the Philadelphia Medical School was broken up for a time, and the medical school in New York could not be put back together again for many years after the Revolution. The normal flow of correspondence, books, journals, and freshly graduated doctors from Britain was drastically reduced, although some of the most patriotic Americans did continue their medical studies on the continent and even in Britain during the war years. Some of the best educated and most successful physicians remained loyal to the crown and left the country. Yet, the most profound effect upon the medical profession —for good and for ill—followed from the medical services that were immediately required by the American army.

The erection of a military medical service was very difficult despite the existence throughout the country of a reservoir of variously

60. Lyon N. Richardson, *A History of Early American Magazines, 1741-1789* (New York, 1931), 163; Charles Evans, *American Bibliography* (Chicago, 1903-34), V and VI, *passim*.

61. Howard M. Chapin, *Calendrier Français pour l'Année 1781 and the Printing Press of the French Fleet* (Providence, 1914), not paged.

62. American Philosophical Society Minutes, Feb. 6, 1781, [III], not paged; American Philosophical Society, *Transactions*, 2 (1786), 239-46; American Academy of Arts and Sciences, *Memoirs*, 1 (1785), 151-55.

trained physicians. At the outset, several different agencies employed medical men, wherever found, to take care of the sick and wounded.[63] Throughout the war, the individual states continued to provide medical services for their troops but what coordination was accomplished followed from the efforts of the Continental Congress to create a continental establishment.

The Congress established the table of organization and filled numerous subordinate positions but its appointment of the director general and chief physician was the step that had the most far reaching and dramatic consequences. Its first director general was Dr. Benjamin Church, who was nominated by Massachusetts and confirmed by Congress in July 1775. Before three months were out, Church had to be removed for treasonable activities which might never have come to light but for the jealousy of the regimental surgeons. These men resented his perfectly justifiable efforts to have all the seriously ill sent to his general hospital and otherwise to limit their autonomy.[64] The appointment to Church's post of Dr. John Morgan, who had never been noted for tact or organizing capacity, did not solve the problems which increased in magnitude as the war was extended. The nearly disastrous reverses of the New York campaign served to reveal the great inadequacy of the medical services. They also gave to Morgan's twice-insulted enemy, Dr. William Shippen, the opportunity to plot his rival's downfall. It proved easy enough to collect affidavits detailing shortages of supply, inadequate medical care, and even poor direction by Morgan. When Morgan was removed and replaced by Shippen, he in turn found it easy to collect enough information to win a vindication of his own conduct and, with the help of Dr. Benjamin Rush, to build up a case against Shippen's "mal-practice and misconduct in office." [65] Despite this success, Morgan did not

63. William O. Owen, *The Medical Department of the United States Army During the Period of the Revolution* (New York, 1920), 1; Gurdon W. Russell, *Early Medicine and Early Medical Men in Connecticut* ([n.p.], 1892), 133.

64. Worthington C. Ford (ed.), *Journals of the Continental Congress* (Washington, 1904-37), II, 76-78, X, 128-31; Allen French, *General Gage's Informers* (Ann Arbor, 1932), 147-201; Warren to Adams, Oct. 1, 1775, *Warren-Adams Letters*, I, 121-22.

65. *Journals of the Continental Congress*, especially XIV, 724, 733; James Thacher, *A Military Journal during the American Revolutionary War* (2d ed., Boston, 1827), 39; Rush to Julia Rush, March 17, 1780, Butterfield (ed.), *Letters of Rush*, I, 248.

recover his pride or become as effective in the promotion of science as he had formerly been. Moreover, the appointment of Dr. John Cochran to Shippen's resigned post in January 1781 did not end supply shortages or needless death, although the worst of the crises were then past.

More important than the dramatic feud between the two Philadelphia doctors were the positive accomplishments that flowed from the medical necessities of the war. The mere association of physicians on a scale never before approached in America was one of the most important factors in bringing about later organization in medicine and the establishment of standards for practice. The war presented the specific necessity for examining surgeons, physicians, and mates. As early as May 1775, the provincial congress of Massachusetts established a medical examining board which gave rigorous four-hour examinations in anatomy, physiology, surgery, and medicine even to those seeking appointment as surgeon's mates.[66] Morgan gave examinations of his own to many of those in the service when he became director general. By this time examinations had become the accepted means of entering the medical service, but they could not remedy defects in training.[67] They did no more than uncover them.

Efforts to make the medical men who served with the army more effective led to the publication of medical handbooks. As early as 1775 in New York, Dr. John Jones published his very useful *Plain, Concise, Practical Remarks on the Treatment of Wounds and Fractures*. This was based largely upon the standard English works and various articles in the London and Edinburgh *Medical Observations* but also in some measure upon Jones's own experience in the Seven Years' War. It contained a pertinent section on military hospitals. A second and much larger edition was issued the following year in Philadelphia, and a third as late as 1795.[68] A book that was a pure compendium was the Philadelphia edition in 1776 of Baron Van Swieten's much used *Diseases Incident to Armies*, which also contained related treatises

66. Owen, *Medical Department*, 2; Thacher, *Military Journal*, 31.
67. Thacher, *Miltary Journal*, 40; Joseph Warren to Dr. Lemuel Hayward, June 3, 1775, Warren Papers, Massachusetts Historical Society.
68. John Jones, *Plain, Concise, and Practical Remarks on the Treatment of Wounds and Fractures* (New York, 1775), 3; James Mease, *The Surgical Works of the Late John Jones* (Philadelphia, 1795), v; James Thacher, *American Medical Biography* (Boston, 1828), I, 53.

by John Ranby and William Northcote. The only thing American about it was the introduction, but it was nonetheless useful and was reprinted in Boston in 1777.[69] A brief but original treatise was Benjamin Rush's *Directions for Preserving the Health of Soldiers,* which he issued first as a newspaper article and in 1778 as a pamphlet. It was a very sensible essay upon public health, treating dress, diet, cleanliness, encampment, and exercise. It was used not only during the Revolution, but in later editions even during the American Civil War.[70]

In 1778, Dr. William Brown, a Virginian with an Edinburgh medical degree, was appointed physician general of the Middle Department and in the same year his *Pharmacopoeia* appeared. While admittedly a compilation based largely on the Edinburgh *Pharmacopoeia,* it was highly selective and was the first ever published in the United States. In a brief thirty-two pages Brown tried to present "such formulae as it is always in our power to obtain" leaving interleaved blank pages for the owner to insert "favourite or more useful formulae." The pamphlet served the author's aim of introducing a degree of uniformity throughout the several hospitals.[71]

John Morgan wrote one pamphlet aimed primarily at the civilian population of the Boston area but designed to protect the health of the troops. A smallpox epidemic struck the town in 1776, communicated, according to Morgan, by the British soldiers. In response to the threat, general inoculation of the troops and citizenry was undertaken by the medical men attached to the army. In this connection and for the benefit of civilian practitioners, Morgan wrote his *Recommendation of Inoculation According to Baron Dimsdale's Method,* which seemed as much calculated to quiet fears of inoculation as it was to detail any particular method.[72] Benjamin Rush's *New Method of Inoculation for the Small Pox* (1781) was much more concerned with

69. Baron Girard Van Swieten, *Diseases Incident to Armies* (Philadelphia Printed, Boston Reprinted, 1777), 3.

70. Benjamin Rush, *Directions for Preserving the Health of Soldiers* (Lancaster, 1778); Butterfield (ed.), *Letters of Rush,* I, 146n.

71. William Brown, *Pharmacopoeia Simpliciorum et Efficaciorum* (Philadelphia, 1778).

72. John Morgan, *Recommendation of Inoculation According to Baron Dimsdale's Method* (Boston, 1776), 3.

technique, but it was a college lecture without any direct relationship to the army.[73]

The war experience conferred benefits upon medical men associated with the army not only by giving them many new cases to handle, but also by exposing them to techniques and systems of medicine that were better than they had previously known. Much could even be learned from the enemy, as James Thacher discovered when he observed the great speed and skill of the British surgeons operating upon their own wounded after the battle of Saratoga.[74] The French ally had more to offer because the gentler French medical tradition was less well known in America than the British pattern and the Americans were inferior to the French in certain techniques, notably in hospital administration. The chief physician of Rochambeau's army, Jean-François Coste, played a particularly important role while in America. He published a brief pharmacopoeia for use in the French hospitals. He willingly wrote a pamphlet for the Philadelphia Humane Society on the subject of asphyxia. Then in 1782, he gave an address on adapting ancient medical philosophies to the New World which was marked with good advice. He asserted that as the United States was now independent, medicine in the country must be free from subjection to authority, even that of the most celebrated masters. Though later published in the Netherlands, the address was known in America only to those who attended the occasion of it, the granting of an honorary degree of doctor of medicine to Coste by the College of William and Mary.[75]

The war experience provided opportunities for publication which some medical officers seized. One French surgeon made chemical and physical tests upon Boston water, publishing his results in the *Memoirs* of the American Academy of Arts and Sciences in 1785.[76] Another French officer published a circumstantial account of a partridge that

73. Benjamin Rush, *The New Method of Inoculating for the Small Pox* (Philadelphia, 1781).

74. Thacher, *Military Journal*, 112.

75. John E. Lane, "Jean-François Coste, Chief Physician of the French Expeditionary Forces in the American Revolution," *Americana*, 22 (1925), 2-10. See also, J. F. Coste, "An Account of some Experiments with Opium in the Cure of the Venereal Disease," *London Medical Journal*, 9 (1789), 7-27.

76. American Academy, *Memoirs*, 1 (1785), 55-64.

had been found with two hearts.[77] More important work was done by a physician attached to the Hessian mercenaries operating with the British Army, Dr. Christian Friedrich Michaelis, but it was in European journals that his articles appeared. The most significant of these were his experiments on the regeneration of nerves which were written up in *Chirurgische Bibliotek*. He also published reports upon specific cases he had observed in America.[78]

Some Americans improved their opportunities too. Dr. Barnabas Binney wrote of "A Remarkable Case of a Gun-Shot Wound" in which a very serious internal wound healed without any surgical intervention, leading Binney to conclude that "Surgeons may be too officious, as well as too tardy." This was well enough considered in London to be reprinted by the *London Medical Journal* after it had first appeared in the *Memoirs* of the American Academy.[79] Dr. Ebenezer Beardsley wrote of the experience of the Twenty-second Regiment of the American Army with dysentery in which it appeared that overcrowding in stuffy basements and garrets was the cause of the epidemic. Companies that had been quartered in tents remained healthy and all of those who were sick recovered when removed from the close atmosphere.[80]

In later years, Benjamin Rush frequently looked back to his war experience for ideas and facts. His "Observations on the Cause and Cure of the Tetanus" followed from wartime experiences.[81] He also summarized his medical experiences during the war in a very brief paper entitled "Result of some Observations made by Benjamin Rush, . . . during his Attendance as Physician General of the Military Hospitals of the United States in the late War." [82] More important

77. American Philosophical Society, *Transactions*, 2 (1786), 333-34.
78. Whitfield J. Bell, Jr., "A Box of Old Bones," American Philosophical Society, *Proceedings*, 93 (1949), 172n; *Medical Communications*, 1 (1784), 307-58, 407-8.
79. *London Medical Journal*, 7 (1786), 294-97; American Academy, *Memoirs*, 1 (1785), 545.
80. American Academy, *Memoirs*, 1 (1785), 542-43.
81. American Philosophical Society Minutes, March 17, 1786, [III], not paged; American Philosophical Society, *Transactions*, 2 (1786), 225-31; Medical Society of London, *Memoirs*, 1 (1787), 65-76; *London Medical Journal*, 7 (1786), 424-34.
82. *London Medical Journal*, 7 (1786), 76-77; under slightly different title in Literary and Philosophical Society of Manchester, *Memoirs*, 2 (1785), 506-9.

still, was the effect of the war and the Revolution generally upon Rush's whole pattern of thought regarding medicine. He was more than ready to agree with Coste that the United States should be as independent in medicine as it was politically independent. Years later, he declared that it was the Revolution that led him to discard his old medical system and to try to build an American system.[83] From this point, he went on to erect a theory of medicine that was an impressive intellectual edifice, hard as it may be to condone.

Specific positive advances coming out of the war were much more difficult to find than they have been in recent wars, and some of the assertions of advance were certainly spurious. Joshua Clayton, for example, succumbed to the prevailing efforts to find ersatz products for items that could no longer be easily obtained. Clayton worked out what he thought was a substitute for Peruvian bark, compounded of poplar bark, the bark of dogwood root, and the bark of white oak. Presented as a remedy for gangrene and mortifications, it was certainly not the equivalent of Peruvian bark though it was probably not particularly harmful.[84] On the other hand, some improvements were genuine. There was general agreement with Dr. John Jones's proposal of an improved hospital system free of overcrowding but it was hard to find the means to implement the idea. Dr. James Tilton did work out a satisfactory solution at Princeton where he built a series of log huts instead of one large hospital building. They provided unusual isolation, were cheap and well-ventilated, and with a fireplace in the center had some warmth.[85]

Medical advance was limited, but at least the Americans possessed the medical skills required by the war in more ample measure than the corresponding military and engineering skills needed for fighting the war. To supply this deficiency, a halting attempt at the literate approach was made by publishing several European works on the art of war. Lewis Nicola, versatile Philadelphia bookseller, made the most sustained effort of this sort. In 1776, his *Treatise of Military Exercise* attempted to present the manual of arms and drill move-

83. Corner (ed.), *Autobiography of Rush*, 89.
84. Thacher, *American Medical Biography*, I, 225-26; on substitute tea, see *United States Magazine*, 1 (1779), 116, 117 [actually 216, 217].
85. L. P. Bush, *The Delaware State Medical Society and Its Founders in the Eighteenth Century* (New York, 1886), 5.

THE WAR

ments insofar as they would be useful to the Americans, eliminating "such Manoeuvres, as are only for Shew and Parade."[86] In this, he made no effort to strip down the innumerable commands in the manual of arms as Steuben later did, but he was clearly thinking of that need. As steps beyond this elementary piece, he published translations of the Chevalier de Clairac's *Field Engineer* and Major General de Grandmaison's *Treatise on . . . Light Horse, and Light Infantry*.[87] In 1779, John Muller's *Treatise of Artillery* was reprinted, an English work which supplemented the Nicola publications very well.[88]

It was clear from the outset that this approach was a very limited aid, particularly in the matter of engineering which required a specialized knowledge beyond the reach of a manual or two. In December 1775, Congress instructed its Committee of Correspondence to find and engage not more than four skillful engineers, but only when Franklin reached France in December 1776 was progress made in this effort. Four officers headed by Louis Lebeque de Presle Duportail were finally engaged and in 1777 commissioned by the Congress. Duportail, Radière, Gouvion, and Laumoy were all commissioned French officers who were permitted to accept Continental commissions on a sort of loan basis. Duportail accepted appointment as commandant of the Corps of Engineers and Companies of Sappers and Miners with the rank of brigadier general and the French officers became the nucleus of the corps. Duportail came to play a part in Washington's counsels and, after much wrangling, to command the bulk of the engineering activities of the Continental Army. The corps came to include several other foreign officers recruited in various ways. Some had been attached to the staff of General Tronson du Coudray before he died—among them Pierre Charles L'Enfant. Others had had altogether independent assignments, the most important of these being Colonel Thaddeus Kosciuszko who, although brought under the technical supervision of Duportail, continued to

86. [Lewis Nicola], *A Treatise of Military Exercise, Calculated for the Use of the Americans* (Philadelphia, 1776), [i].
87. Chevalier de Clairac, *L'Ingenieur de Campagne, or Field Engineer*, trans. by Lewis Nicola (Philadelphia, 1776); Major General de Grandmaison, *A Treatise on the Military Service, of Light Horse, and Light Infantry, in the Field, and in Fortified Places*, trans. by Lewis Nicola (Philadelphia, 1777).
88. John Muller, *A Treatise of Artillery* (Philadelphia, 1779).

THE PURSUIT OF SCIENCE

be given separate commands. In assuming the engineering functions of
an army, these officers made a vital contribution to the winning of the
war. As late as April 1782, when the corps contained fourteen officers
only one was an American, a Captain Niven, ranking thirteenth in
the list.[89]

The Engineer Corps never took care of all the military engineering
carried out by the Americans; many different individuals made scattered contributions in the field, particularly before the formation of
the corps but afterwards as well. In line with the close connection
then obtaining between engineering and artillery it was reasonable
to expect engineering work from Henry Knox, who became colonel
in charge of artillery as early as 1775. He directed the construction of
fortifications in the vicinity of Boston, planned Washington's crossing
of the Delaware, and started the arsenal at Springfield. The Dutch
engineer, Bernard Romans, who had left British employ in Florida
before the war, planned fortifications in New York, and Arthur
Donaldson, a clever mechanic, built river defenses on the Delaware.
Even line officers planned fortifications and exercised engineering
talents on occasion.[90]

The war revealed the striking lack of adequately trained American
engineers. Knox was conscious of the need that some provision be
made for training technical officers and Duportail made an effort to
give his companies of sappers a broad engineering training. He required that their officers be "skilled in the necessary branches of
mathematics," picking them by a process of examination. As a permanent solution, Duportail pointed out that since the preliminary knowledge in mathematics and natural philosophy required of artillerists
and engineers was the same, the Department of Artillery ought to be
united with the Corps of Engineers. Training could then be carried
out at a single academy erected for the purpose. "The necessity of an
Academy, to be the Nursery of the Corps, is too obvious to be insisted

89. Elizabeth S. Kite, *Brigadier-General Louis Lebeque Duportail* (Philadelphia, 1933), 12, 27-29, 91, 226, 247, 295; Anon., "General Thaddeus Kosciuszko," *American Catholic Historical Researches*, new ser., 6 (1910), 132-
216; Martin I. J. Griffin, "General Count Casimir Pulaski," *ibid.*, 4-128.
90. Petition of Arthur Donaldson, Jan. 23, 1784, *Minutes of the General
Assembly of the Commonwealth of Pennsylvania, Eighth* (Philadelphia, 1781-
92), 94; Phillips, *Romans*, 54, 56, 60; Francis S. Drake, *Life and Correspondence of Henry Knox* (Boston, 1873), 18, 20-21, 35.

upon," he declared. In addition to men teaching military subjects its faculty should include a master of mathematics and of natural philosophy, a master of chemistry, and a master of drawing. Had such an institution been erected at this time, it would have had a great influence upon the development of science in the country, but the aspiration could not be realized. It is not clear that the French engineers had very much influence upon the later development of American engineering.[91]

A still more specific requirement of the war was good, usable maps. There were many difficulties. Prewar cartography had often favored very large area mapping which was inadequate for military needs. Yet, even what had been done was often not available to American field commanders unless they were lucky enough to obtain reissues of the maps by the British, sometimes in the form of atlases.[92] Some of the best cartographers were not available to the army when most needed: Thomas Hutchins, for instance, retained his commission in the British Army until 1779 and even published a map of London in 1778; and Bernard Romans spent most of his time executing maps designed to be sold to the general public. Robert Erskine, who had come to America in connection with Peter Hasenclever's great iron enterprise, became first geographer to the Continental Army, performing very capably in that post as did his successor, the American-born Simeon DeWitt.[93] The Americans had been better prepared by their colonial experience to do an acceptable job of mapping and charting than to provide for military engineering.

The war brought many imaginative designs of weapons intended to win the war, although Franklin's experience at Paris indicated that the Americans were no more ingenious in this respect than the French.[94]

91. Kite, *Duportail*, especially 269.

92. Douglass S. Freeman, *George Washington, A Biography* (New York, 1948-54), V, 524; *American Military Pocket Atlas* (London, [1776]).

93. Freeman, *Washington*, IV, 259; Erwin Raisz, "Outline History of American Cartography," *Isis*, 26 (1937), 375; P. Lee Phillips, *A List of Maps of America in the Library of Congress* (Washington, 1901), 859-61; Maps by Bernard Romans and Robert Erskine, Map Room, N.Y. Public Library; Ruth Henline, Travel Literature of Colonists in America (Doctoral Dissertation, Northwestern University, 1947), 173.

94. The Franklin Papers at the American Philosophical Society are full of letters from inventive Frenchmen.

Most of the ideas, even Thomas Paine's incendiary arrows shot from steel crossbows and Sion Seabury's rolling earthworks, were never realized.[95] Some wartime innovations, such as the iron chain thrown across the Hudson to prevent British ships from sailing beyond West Point, were ingenious in execution rather than concept—the iron-master, Peter Townsend, completing this work in six weeks at the Sterling iron works.[96] Undoubtedly the greatest of the wartime inventions—a remarkable achievement even in its failure—was the submarine.

The submarine was the accomplishment of a Yale student, David Bushnell, who built it in 1775 to "pulverize the British navy." [97] The American Turtle, as its inventor dubbed it, was a cask made of oak planks bound with iron, caulked and sealed with tar. It was higher than it was wide, and looked much like two turtles clamped together. At its top it had a glassed-in conning tower which was the only part of the craft above the surface when afloat except for two ventilating tubes. It was submerged by admitting water into a tank at the bottom which could then be pumped dry by a foot pump. Another foot-actuated treadle was used to propel the submarine, and it was moved vertically and horizontally by hand-operated cranks all of which actuated screw propellers. A barometer to measure depth and a compass were illuminated by phosphorus when submerged. It was an ingenious device and—as a submarine—it worked.[98]

Its failure came when it was applied to the purpose for which it had been constructed: the destruction of enemy ships. It was designed to approach its victim underwater and attach to its hull a time bomb that would explode after the submarine had made good its escape. Although on its first trial, it reached a hostile ship in New York harbor, the bomb could not be screwed to the hull. Later trials were still less successful and Bushnell gave up the submarine to experiment with captive floating mines in the Hudson and free floating mines in the

95. T[homas] P[aine] to Franklin, July 9, 1777, Bache Collection; Sion Seabury to Col. John Hancock, Nov. 11, 1774, Franklin B. Dexter (ed.), *The Literary Diary of Ezra Stiles* (New Haven, 1901), I, 497.

96. Benson J. Lossing, *The Pictorial Field Book of the Revolution* (New York, 1852), II, 137-38.

97. Franklin B. Dexter (ed.), *Excerpts from the Itineraries and Other Miscellanies of Ezra Stiles, D.D., LL.D., 1755-1794* (New Haven, 1916), 530.

98. Gale to Franklin, Aug. 7, 1775, Franklin Papers, IV, 61, American Philosophical Society; Thacher, *Military Journal*, 121-23.

Delaware. The whole episode was important, not in the winning of the war, but as the first step in the realization of submarine navigation.[99]

More sustained ingenuity and persistence were required in the effort to adapt the American colonial economy to the fighting of a major war. In the late phases of the Revolutionary movement, before the war broke out, attempts had been made to encourage the production of silk, wine, and other exotics which might benefit British trade but would not be much help during a war. In 1776, John Beale Bordley asserted that this whole process had been encouraged by the British with the intention of drawing the Americans "by degrees, into an humble state of dependence and submission absolute, by such employments as may divert them from a prudent and due attention to the raising of necessaries. Bread being wanting," he declared, *"submission must follow."* [100] This was a strained interpretation of attitudes but it was clear enough that the Americans had done little thinking about how to adapt their economy to war.

Their failure to provide adequate supplies of munitions was a glaring illustration of the lack of planning. Although efforts had been made to collect arms and powder before the war, the Continental Association in 1774 prohibited the importation of gunpowder from the mother country. Later, the Privy Council not only prohibited the exportation of powder from Britain but as John Adams complained, the ministry took "Every Step that human Nature could devise to prevent the Americans obtaining so essential an Article." [101] When Congress opened American ports to world trade in 1775, it was too late to obtain gunpowder through trade channels. After the war broke out, a furious effort was made to repair this deficiency by manufacturing the most critical component of gunpowder—saltpeter or potassium nitrate. Exhortations and printed directions went out from the Congress, bounties and instructions came from the provincial governments, and the newspapers and almanacs were full of the effort.[102]

99. *Public Records of Connecticut*, V, 395; American Philosophical Society, *Transactions*, 4 (1799), 303-12.

100. [John B. Bordley], *Necessaries Best Product of Land, Best Staple of Commerce* (Philadelphia, 1776), 4, 8; Carl Bridenbaugh, *Myths and Realities* (Baton Rouge, [1952]), 50.

101. Adams to Warren, Oct. 12, 1775, *Warren-Adams Letters*, I, 135.

102. Evans, *American Bibliography*, IV, *passim*; *Pennsylvania Journal*, March 22, 1776; *Virginia Gazette* (Dixon and Hunter), Feb. 18, 1775; *Connecticut*

Probably the best treatise on the subject was Benjamin Rush's *Process of Making Salt-Petre* which gained credit because of the author's position as professor of chemistry. It recounted recipes and processes of manufacture used in Europe.[103] Thomas Paine and a Captain Pryor varied conditions of manufacture experimentally to see what circumstances yielded the greatest quantity of saltpeter crystals.[104] John Adams even heard of one recipe for extracting saltpeter from the air but the process was hardly a satisfactory means of nitrogen fixation! [105] The most serious effort to produce gunpowder within the country was made in the first year and a half of the war. It was a failure. Out of the 2,347,455 pounds of gunpowder used by the Americans before 1777, only 115,000 were manufactured in America from components produced in America. Later there was practically none.[106] The Americans were much more successful in the manufacture of muskets and even in the casting of cannon which had not been done in the colonial period.[107] Gunpowder remained the most critical shortage.

Although the most imperative need was for munitions, the war also brought encouragement for a variety of manufactures. An extensive effort was made in 1775 to establish the manufacture of cotton, wool, and linen textiles through the formation of the United Company of Philadelphia for Promoting American Manufactures. It employed as many as seven hundred people at one time but its success was temporary and by 1779 it had ceased to exist.[108] Still less successful was the manufacturing society established in Williamsburg in 1777.[109] The manufacture of textiles for the market in this way was desirable

---

Courant, March 7, 1768; *Journal of the Continental Congress*, II, 218-19; III, 296, 345-48; *Statutes at Large of Virginia*, IX, 72; *Votes and Proceedings of the House of Representatives of the Province of Pennsylvania, Pennsylvania Archives*, 8th ser., VIII, 7471.

103. *Essays upon the Making of Salt-Petre and Gun Powder* (New York, 1776), 3 ff.

104. *Pennsylvania Journal*, Nov. 22, 1775.

105. Adams to Warren, Oct. 23, 1775, *Warren-Adams Letters*, I, 163.

106. Orlando W. Stephenson, "The Supply of Gunpowder in 1776," *American Historical Review*, 30 (1924-25), 277.

107. *Virginia Gazette* (Purdie), June 21, 1776; Peter Force (ed.), *American Archives* (Washington, 1837-46), 4th ser., IV, 495; Robert E. Gardner, *American Arms and Arms Makers* (Columbus, 1938), 41.

108. *Pennsylvania Journal*, April 3, 1776; *Pennsylvania Magazine*, 1 (1775), 140-41; *Votes of Assembly*, VIII, 7390-91.

109. *Virginia Gazette* (Purdie), April 18, 1777, Aug. 15, 1777.

but not so essential as increasing the production of such an item as salt which had been imported in large quantity in the colonial period. Methods of producing salt were, as a result, much publicized.[110] The inventive capacity of American mechanics made possible continuing production in some manufactures injured by the war restrictions. Abel Buell's types were used in 1781 by both the *Connecticut Journal* and the *Connecticut Gazette* when imported type became unavailable. In Wilmington, Oliver Evans mechanized the manufacture of the wire-toothed leather cards used for carding wood by inventing a machine to cut and bend the wire and set it in the leather. In Rhode Island, Jeremiah Wilkinson made the first cold cut iron nails from iron plate.[111]

The disruptive influence of the war upon the whole pattern of science in America was much more serious than the limited number of beneficial influences it provided. Yet in some respects, the deleterious effects of the war were compensated by corresponding gains. This was notably true of the loss suffered by science in America from the emigration of the loyalists. The counter currents of temporary visitors from France, Germany, and England who established new ties resulted in a permanent gain. When Robert Erskine, Bernard Romans, and William Gerard De Brahm decided to remain in the country, they, too, represented a permanent gain. So did the desertion of Frederick Melsheimer and the Americanization of such French soldiers as Pierre Charles L'Enfant. Even a few civilians who later promoted scientific work managed to get into the country—among them, Albert Gallatin. More important was the large tide of American friends enabled to come to this country when normal immigration could be resumed at the end of the war. By then, many of the injurious effects the war had had upon science began to appear less significant in the bright dawn of hope that flooded over the new nation.

110. *The Art of Making Common Salt* (Boston, 1776); *The Art of Making Common Salt* (Philadelphia, 1776); *Virginia Gazette* (Dixon and Hunter), Dec. 9, 1775.
111. Greville and Dorothy Bathe, *Oliver Evans* (Philadelphia, 1935), 7; Wroth, *Abel Buell*, 47; Israel Wilkinson, *Memoirs of the Wilkinson Family in America* (Jacksonville, Ill., 1869), 471.

Chapter Twelve

THE PROPHETS OF GLORY AND THEIR TEMPLES OF SCIENCE

THE REVOLUTION caused no more of a break with the past in science than in other phases of American life: in large measure, the achievements of the national period were the fulfillment of colonial aspirations. On the other hand, the enthusiasm of the Revolution did awaken a cultural nationalism that took its color from the ideals of the Enlightenment and placed the highest value upon attainment in science. The emotional faith of the Revolution, added to the continually increasing maturity of the country, led to demands for scientific achievement which found expression primarily in the formation and expansion of institutions designed to encourage science—secondarily in actual creative science.

The concept of a special and glorious American destiny had flourished during the course of the imperial contest that began in 1763, reaching a peak just before the War for Independence broke out. Newspapers recalled Bishop Berkeley's famous verses, written nearly fifty years earlier, in which he had referred to America as "Time's noblest offspring." [1] At each imperial crisis it was pleasant to contemplate the frequently voiced prediction that the colonies would "soon be more Flourishing and Populous than the Mother Country." [2] A deep faith that time was on the side of the colonies made it possible to see many evidences of America's "growing greatness, to which

1. *Virginia Gazette* (Rind), Aug. 25, 1774.
2. Gale to Stiles, April 17, 1767, Stiles Papers, Yale University.

every day seems more or less to contribute." [3] This vision of "the rising Grandeur of America" usually implied and sometimes expressed the expectation that time would be particularly kind toward "the Arts and Sciences planted and springing up towards Maturity." [4]

In 1771, the Americans had been told that their "Rising Glory" in science depended upon the maintenance of the American heritage of liberty:

> This is a land of ev'ry joyous sound
> Of liberty and life; sweet liberty!
> Without whose aid the noblest genius fails
> And science irretrievably must die.[5]

By 1774, Benjamin Rush was still reflecting that "the sciences" were "fond of the company of liberty and industry," but he was much more acutely conscious of the threat to conditions which encouraged science. He tied patriotism and literature together, remarking that "a man cannot neglect the one, without being destitute of the other." [6] With the assembling of the Continental Congress, David Rittenhouse became still more specific, declaring that it was to that body that "the Future Liberties, and Consequently the Virtue, Improvement in Science and Happiness, of America are Intrusted." [7]

The vision did not alter with the outbreak of war but it was proclaimed in a more aggressive and decisive tone. After entering the military service himself, Dr. John Warren of Massachusetts read an address to the troops in which he sketched America in 2000 A.D. with a population of one and a half billion. He hoped that this phenomenal increase would be accompanied by "proportionable Improvements, in Agriculture, the military and polite Arts and Sciences together with that Increase of Wealth and Importance" that would make us "the most powerful and august Empire which the Annals of History can

3. Jacob Duché, *Observations on a Variety of Subjects, Literary, Moral and Religious* (Philadelphia, 1774), 103.

4. William Smith, *An Oration* (Philadelphia, 1773), 5, 6.

5. Philip Freneau and Hugh H. Brackenridge, "The Rising Glory of America," in Fred L. Pattee (ed.), *The Poems of Philip Freneau* (Princeton, 1902), I, 71n.

6. Benjamin Rush, *An Oration Delivered February 4, 1774* (Philadelphia, 1774), 72, 74.

7. David Rittenhouse, *An Oration Delivered February 24, 1775, Before the American Philosophical Society* (Philadelphia, 1775), 3.

boast." [8] Elkanah Watson predicted a comparable increase and warned European statesmen against attempting to oppose the inevitable course of history. "Their efforts," he declared, "will be as vain as presumptuous, and they will prove as powerless as an attempt to check the flowing of the tide. Their schemes will, in fact, be an effort to arrest the decrees of the almighty, who has evidently raised up this nation to become a lamp to guide degraded and oppressed humanity, and to direct other nations, even the nation of our oppressors, to liberty and happiness." [9]

The Declaration of Independence required that the vision of American destiny be defended against those of "our enemies" who had suggested, "that in righteous judgment for our wickedness, it would be well to leave us to that independency which we seemed to affect, and to suffer us to sink down to so many Ouran-Outans of the wood, lost to the light of science which, from the other side of the Atlantic, had just begun to break upon us." [10] One answer, often heard, was that "in Europe Science reigns no more, Their souls are fetter'd with tyrannic power." [11] In America, independence "unfetters and expands the human mind, and prepares it for the impression of the most exalted virtues, as well as the reception of the most important science." [12]

Dr. David Ramsay, Charleston patriot, gave an address on the occasion of the second anniversary of the independence of the United States in which he was able to discover numerous reasons for anticipating that independence would encourage the growth of science. "Every circumstance," he said, "concurs to make it probable, that the arts and sciences will be cultivated, extended, and improved in independent America. They require a fresh soil, and always flourish most in new countries." Monarchies, he felt, might disguise simple truth, but "in republics, mankind appear as they really are without any false colouring . . . Large empires are less favourable to true philosophy, than small independent states." What he meant by philosophy or science was made clear when he remarked, "A large volume of the

8. John Warren, Draft of Address [Oct. 12, 1775?], Warren Papers, Massachusetts Historical Society.

9. Winslow C. Watson, *Men and Times of the Revolution; or Memoirs of Elkanah Watson* (New York, 1857), 78.

10. *United States Magazine*, 1 (1779), 3.

11. [Samuel Dexter], *The Progress of Science* ([n.p.], 1780), 3.

12. Samuel Cooper, *A Sermon* [Boston, 1780], 17-18.

Book of Nature, yet unread, is open before us, and invites our attentive perusal.... We stand on the shoulders of our predecessors, with respect to the arts that depend upon experiment and observation. The face of our country, intersected by rivers, or covered by woods and swamps, give[s] ample scope for the improvement of mechanicks, mathematics, and natural philosophy." [13]

At points, particularly in his peroration, Ramsay revealed the flood of emotion that accompanied his thoughts. "May we not hope," he asked, "as soon as this contest is ended, that the exalted spirits of our politicians and warriors will engage in the enlargement of public happiness, by cultivating the arts of peace, [and] promoting useful knowledge, with an ardor equal to that which first roused them to bleed in the cause of liberty and their country?" He went on to predict, "The arts and sciences, which languished under the low prospects of subjection, will ... raise their drooping heads.... Even now, amidst the tumults of war, literary institutions are forming all over the continent, which must light up such a blaze of knowledge, as cannot fail to burn, and catch, and spread, until it has finally illuminated with the rays of science the most distant retreats of ignorance and barbarity." [14]

Many orators and penmen were ready to expand upon the theme that the freedom and liberty associated with independence would be beneficial to the growth of science. The Reverend Dr. Samuel Cooper, of Massachusetts, said that as the arts and sciences "delight in liberty, they are particularly friendly to free States." [15] John Gardiner, son of a staunch loyalist, pointed out, "The introduction and progress of *freedom* have generally attended the introduction and progress of *letters* and *science*. In despotick governments the people are mostly illiterate, rude, and uncivilized; but in states where CIVIL LIBERTY hath been cherished, the human mind hath generally proceeded in improvement,—learning and knowledge have prevailed, and the arts and sciences have flourished." [16]

13. David Ramsay, [An Oration on the Advantages of American Independence], *United States Magazine*, 1 (1779), 53.
14. *Ibid.*, 23, 25, 54.
15. Cooper, *Sermon*, 40.
16. John Gardiner, *An Oration delivered July 4, 1785* (Boston, 1785), 10; *Boston Gazette*, May 9, 1785.

Francis Bailey, editor of the *United States Magazine*, introduced the question of the relationship of the war to science when he wrote, "Liberty is of so noble and energetic quality, as even from the bosom of a war to call forth the powers of human genius, in every course of literary fame and improvement." [17] Joel Barlow was a little less certain that even liberty could overcome the deleterious effect of combat. In private correspondence, he reflected, "Literary accomplishments will not be so much noticed till some time after the settlement of peace, and the people become more refined. More blustering characters must bear sway at present, and the hardy veterans must retire from the field before the philosopher can retire to the closet." [18] When peace was finally attained, Benjamin Rush admitted that the war years had been "unfavorable to improvements of every kind in science," although he yielded to no one in his expectations of the permanent advantages that would follow from the Revolution.[19]

There was general agreement that science did not bloom in the midst of war, but a few men suggested a significant historical parallel which seemed to indicate that some wars had intellectually stimulating results. Dr. Benjamin Waterhouse, who had spent much of the Revolution studying in Europe, referred particularly to the English civil wars when he asserted, "The arts and sciences commonly flourish immediately after civil wars and commotions." [20] With more specific relevance for science, David Ramsay noted that the Royal Society, the great nursemaid of science in the colonies, had been founded "immediately after the termination of the civil wars in England." [21] To link the astonishing intellectual productivity of that period of English history with aspirations for American achievement after its Revolution was to proclaim the most exalted goals for science.

Among the literati, anticipation of a flowering in science to follow the Revolution became almost universal, but quiet, earnest Thomas Bond pointed out that there was a reciprocal relationship, too. As

17. *United States Magazine*, 1 (1779), 4.
18. Barlow to Noah Webster, Jan. 30, 1779, in Charles B. Todd, *Life and Letters of Joel Barlow* (New York, 1886), 18-19.
19. Rush to John C. Lettsom, Nov. 15, 1783, L. H. Butterfield (ed.), *Letters of Benjamin Rush* (Princeton, 1951), I, 312.
20. B[enjamin] Waterhouse, *Synopsis of a Course of Lectures, on the Theory and Practice of Medicine* (Boston, 1786), 2.
21. Ramsay, [Oration], *United States Magazine*, 1 (1779), 24.

liberty and independence encouraged the growth of science, so science was the essential support of liberty and independence. Where literature and science were unknown, he declared, "Liberty is a wilderness, —and, where they are neglected, she retreats in Disgust. Point out the Nation which has not Science, or that which has abandoned it and I will point out to you Savages or Slaves." It would not be enough to sit back and await the blooming and blossoming of science; active promotion of science was imperative. "Science and true genius," he went on, "have a native modesty, which ever accompanies them ... the public Neglect of them would be a national Dishonour and a Reflection on the Gratitude of *America*. We cannot pay too great Attention to national Character. The *European* World will soon make a closer Enquiry into our Disposition and Conduct, than merely whether we are Patriots and Soldiers. The Fame of *America* must rest on a broader Basis than that of Arms alone." "Military Fame may be won or lost in an Hour, but the Fame of Science is a Character of Duration." [22]

Dispassionate analysis of America's actual accomplishments and her reasoned prospects in the realm of science was difficult amid the highly charged emotion of the Revolution—particularly for the more doctrinaire children of the Enlightenment. Joel Barlow probably believed his own pronouncement: "The present is an age of philosophy, and America, the empire of reason. Here neither the pageantry of courts nor the glooms of superstition have dazzled or beclouded the mind." [23] In such an atmosphere, it was even possible to become convinced to a certainty that the millennium was "actually to commence in the territories of the United States." [24] Yet, the mobbing in Philadelphia of an old woman accused of being a witch and her resultant death in Anno Domini 1787 should have convinced the prophets of glory that the millennium was not yet that close.[25] Reality was not quite the same thing as the visions of enthusiasts.

In their more thoughtful moments, many of those who sang the praises of American virtues and science realized that the United

22. Thomas Bond, *Anniversary Oration Delivered May 21st* (Philadelphia, 1782), 30-33, 10.
23. Joel Barlow, *An Oration Delivered ... July 4th, 1787* (Hartford, [1787]), 19.
24. *New Haven Gazette and Connecticut Magazine*, 1 (1786), 1.
25. *Columbian Magazine*, 1 (1786-87), 558.

States had many serious obstacles to be overcome before science could be enthroned in America and the nation's destiny attained. In a moment of disillusionment, even Benjamin Rush confessed, "Philosophy does not here, as in England, walk abroad in silver slippers; the physicians (who are the most general repositories of science) are chained down by the drudgery of their professions so as to be precluded from exploring our woods and mountains. Besides, there are not men of learning enough in America as yet, to furnish the stimulus of literary fame to difficult and laborious literary pursuits. I have felt the force of this passion; Alas! my friend, I have found it in our country to be nothing but 'avarice of air.' " [26]

Thomas Jefferson went to France in 1785, sharing the great vision as fully as any man ever did. He did not abandon it during his stay in that country either, but he did come to recognize that European scientists were well ahead of anyone in America. Characteristically, however, he was able to view this unpleasant discovery in optimistic terms, concentrating upon one aspect of American society in which it excelled. He wrote with feeling, "In science, the mass of people [of Europe] is two centuries behind ours, their literati, half a dozen years before us. Books, really good, acquire just reputation in that time, and so become known to us and communicate to us all their advances in knowledge. Is not this delay compensated by our being placed out of the reach of that swarm of nonsense which issues daily from a thousand presses, and perishes almost in issuing?" [27]

The higher educational level of even the least favored Americans was a real advantage they enjoyed over Europe and one that was frequently mentioned in connection with potentialities in science. Americans were often told they could not "boast of so many learned philosophers as some older states in Europe" but that the people at large were "more intelligent, inquisitive for knowledge, and more willing to embrace new opinions, founded on reason and experience than any older countries." [28] John Gardiner declared, "In no part of the habitable globe is learning and true *useful* knowledge so uni-

26. Rush to Lettsom, Oct. 26, 1786, Thomas J. Pettigrew, *Memoirs of the Life and Writings of the Late John Coakley Lettsom* (London, 1817), II, 428.
27. Jefferson to Bellini, Sept. 30, 1785, Julian P. Boyd (ed.), *The Papers of Thomas Jefferson* (Princeton, 1950–), VIII, 569.
28. *New Haven Gazette and Connecticut Magazine*, 3 (1788), 27.

versally disseminated as in *our native country*. Who hath seen a native adult that cannot write? who known a native of the age of puberty that cannot read the bible?" [29] Yet dissemination of knowledge was not the same thing as its advancement even though it might provide a good foundation for future accomplishment.

It provided no kind of an answer to the direct charge of Abbé Guillaume Raynal: "It is astonishing that America has not yet produced a good poet, an able mathematician, a man of genius in a single art, or a single science." [30] This slur sent the prophets of glory scurrying to make up lists with which to refute Raynal. In deep passion, Jefferson replied, "In war we have produced a Washington, ... In physics we have produced a Franklin, than whom no one of the present age has made more important discoveries, nor has enriched philosophy with more, or more ingenious, solutions of the phenomena of nature. We have supposed Mr. Rittenhouse second to no astronomer living; that in genius he must be the first, because he is self taught. As an artist he had exhibited as great a proof of mechanical genius as the world has ever produced. He has not indeed made a world; but he has by imitation approached nearer its Maker than any man who has lived from the creation to this day." [31] Ezra Stiles even maintained that Americans had made as capital contributions to science in "the last half century, as in all europe," citing Godfrey's quadrant, Muirson's mercurial inoculation, Franklin's lightning rods, Eliot's production of iron from black sand, and Winthrop's calculation of the quantity of matter in comets.[32]

As hard as they tried to present America's best face, few of these men could hide from themselves the poverty of the American intellectual environment when compared with that of Europe. It was not merely that there was only one scientist of first rank in the country or that America's libraries, museums, and philosophical societies were inferior to what might be found in Paris and London, but even the basic instruments required for research in many of the sciences could

29. Gardiner, *Oration*, 21.
30. Guillaume Thomas Raynal, *Histoire Philosophique et Politique* (Maestricht, 1774), VII, 92.
31. [Thomas Jefferson], *Notes on the State of Virginia* ([Paris], 1782 [1785]), 119-20.
32. Ezra Stiles, *The United States Elevated to Glory and Honour* (Worcester, 1785), 50.

not be obtained.[33] Before the war, men had bent their efforts toward enriching this environment; now, the full-blown vision of American destiny drew them on to greater and greater endeavors. Criticism of American deficiencies did not depress them; it only increased the energy with which they sought to repair those deficiencies.

Many of the architects of the political Revolution displayed the zest David Ramsay had asked for in attempting to remold America in accord with their vision. Rush was thinking in terms of a thorough-going reformation of American life when he remarked in 1786, "The American war is over: but this is far from being the case with the American revolution." [34] Literate activity soon reached an unprecedented level. By 1785, new enterprises were multiplying on every hand in education, publishing, invention, and the erection of learned societies. Some there were who sought to launch a cultural renaissance at one stroke, but most of the activity was limited to the elimination of specific deficiencies in the American scene.

A general cultural flowering appeared to be in progress. In education, the most striking activity was evident at the collegiate level with the founding of seven new colleges during the Confederation period.[35] All of them were denominational foundations with only a general concern for science, intensified at times when interested individuals joined their faculties. The newest element in the collegiate picture was the interest demonstrated by the state governments: in almost converting William and Mary and the College of Philadelphia into state universities, and in laying the foundations, at least on paper, for the University of Georgia in 1785 and the University of North Carolina in 1789. Despite the new colleges, academic science remained very largely a story of activity within the colleges which had been founded in the colonial period.[36] As for elementary education, it took many years before the level of late colonial attainment was restored, let alone surpassed.

33. American Academy, *Memoirs*, 1 (1785), ix.
34. Nestor [Benjamin Rush], "On the Defects of the Confederation," *American Museum*, 1 (1787), 9.
35. Samuel Miller, *A Brief Retrospect of the Eighteenth Century* (New York, 1803), II, 492-506.
36. Theodore Hornberger, *Scientific Thought in the American Colleges, 1628-1800* (Austin, 1945), 14.

After the worst of the fighting was over, the presses began to supply the schools and colleges with textbooks which by that time were in very short supply. Most of them—English, Latin, and French grammars, arithmetic books, and spellers—were nothing but reprints of English works. A few Americanized editions did appear, and then a few books written specifically for the newly independent Americans.[37] Jedidiah Morse's *American Geography* was almost by definition a celebration of the United States. Nicholas Pike prefaced his new arithmetic with the statement, "As the United States are now an independent Nation, it was judged that a System might be calculated more suitable to our Meridian, than those heretofore published." [38] Most passionate of the textbook writers was Noah Webster who announced, "America must be as independent in *Literature* as she is in *politics*—as famous for *arts* as for *arms*," and then set out singlehandedly to effect the intellectual revolution.[39]

Between 1783 and 1785, the three parts of Webster's *Gramatical Institute of the English Language* were published, comprising a speller, a grammar, and a reader. It attempted, and in some measure effected, a reform of the language in the direction of simplification and Americanization. Webster's aim was not new, for an American Language Society to seek the same sort of reform had been urged as early as 1774, but his ability to execute the design was remarkable. By stressing American and patriotic themes, his books—particularly the spellers —became one of the greatest forces in the land serving to inculcate the deep and idealistic patriotism of the Revolution in the generations succeeding. Among the principles that crept into his efforts was his support of the desire to enthrone science: "Next to the sacred writ-

37. See for example, Thomas Sheridan, *A Rhetorical Grammar of the English Language* (Philadelphia, 1783); John Ash, *Grammatical Institutes* (Worcester, 1785); Rev. [Robert] Ross, *The American Latin Grammar* (5th ed., Providence, 1780); John James Bachmair, *A German Grammar* (Philadelphia, 1788); Thomas Sarjeant, *Elementary Principles of Arithmetic* (Philadelphia, 1788); Benjamin Workman, *A Treatise of Arithmetic . . . by John Gough* (Philadelphia, 1788).

38. Jedidiah Morse, *The American Geography; Or a View of the Present Situation of the United States of America* (Elizabeth Town, 1789); Nicholas Pike, *A New and Complete System of Arithmetic, Composed for the Use of the Citizens of the United States* (Newbury-Port, 1788), preface, not paged.

39. Webster to John Canfield, Jan. 6, 1783, Harry Warfel (ed.), *Letters of Noah Webster* (New York, [1953]), 3.

ings," he wrote, "those books which teach us the principles of science and lay the basis on which all our future improvements must be built, best deserve the patronage of the public." [40]

Publishing media multiplied at such a rapid rate that some ninety newspapers were established by 1790, including dailies in Philadelphia, New York, and Charleston. Magazines not only became more numerous than ever before, but for the first time some of them became reasonably self-sustaining. The most important of these were the *Columbian Magazine* and the *American Museum*, each of which had been founded at Philadelphia with the help of the recent Irish immigrant, Matthew Carey. They were so successful in gaining subscriptions throughout the states that when Noah Webster tried in 1787 to make his *American Magazine* "a federal publication," he was unable to break into the field.[41] The several New England magazines —the *Boston Magazine*, Isaiah Thomas' *Worcester Magazine* and his *Massachusetts Magazine*—had a more local appeal. Although a New Englander, the Reverend Jeremy Belknap gave important encouragement to the Philadelphia and New York journals. As the new magazines demonstrated a capacity for sustained life, the newspapers ceased to function as magazines except for such publications as the *New Haven Gazette and Connecticut Magazine* which was really a magazine in newspaper form. Scientific essays were frequently printed by the magazines but rarely now by the newspapers.[42]

Book publishing, too, attained flourishing proportions. There was some substance to the charge that "we proceed faster in the number of our Books than in the excellence of our Execution," but more original titles were published now than in any comparable earlier period and some of them were of permanent value.[43] Histories of the separate states, geographies, treatises on religion, government, and medicine,

40. Harry Warfel, *Noah Webster, Schoolmaster to America* (New York, 1936), 59, 71, 80, 83, 85; Webster to Franklin, [n.d.], Franklin Papers, XL, 139, American Philosophical Society; [Proposed Academy of Language], Colonial Society of Massachusetts, *Publications*, 14 (1911-13), 263-64.
41. Webster to Belknap, Feb. 9, 1799, *Belknap Papers*, III, Massachusetts Historical Society, *Collections*, 6th ser., 4 (1891), 385.
42. William Spotswood to Belknap, April 16, 1787, Oct. 9, 1788, *ibid.*, 331, 421; Belknap to Matthew Carey, Feb. 2, 1787, Earl L. Bradsher, *Matthew Carey* (New York, 1912), 4.
43. William Bentley, *Diary* (Salem, 1905), I, 172.

and even works on natural history and physics appeared. There was an effort in belles lettres to write great epic poems that in their message would celebrate the glorious destiny of the United States and in the excellence of their form become lasting monuments to American literature. The English writer, Thomas Day, was not unfair when he remarked that epics became tiresome and insipid without the great genius which Joel Barlow, David Humphreys, and Timothy Dwight just did not have. Nevertheless, these Hartford wits were recognized even in Philadelphia as excelling in literature—despite the concurrent work of Philip Freneau.[44]

One of the most obvious means of encouraging original literary work was to provide the protection of copyright laws, but the Revolution did not encourage government intervention of this sort on the part of the Confederation Congress. Thomas Paine was apparently thinking in terms of the Congress in 1782 when he urged the necessity of laws "to prevent depredations on literary property." [45] No action was taken, however, until Noah Webster turned his efforts to the task and then it was altogether a matter of state action. In January 1783, the first general copyright law in America was passed by Connecticut as a direct result of Webster's campaign. That was just the beginning. Webster made personal applications to the legislatures of the individual states, won a Congressional recommendation that states take such action, and finally persuaded all of the states except Vermont to enact copyright laws. The prevailing pattern was to grant the author the exclusive benefit of his labors for fourteen years—in some cases patents upon inventions were established in the same law. Until 1790, when the new government under the Constitution took over responsibility, this method of controlling literary property prevailed.[46]

Although some of the libraries of the colonial period ceased to function during the war, many were revived afterward and others added. The book trade was busiest in the Middle states in both foreign imports

44. Thomas Day to Price, April 8, 1786, "Letters of Richard Price," Massachusetts Historical Society, *Proceedings*, 2d ser., 17 (1903), 340; Rittenhouse to Jefferson, Nov. 8 [1787?], *Jefferson Papers*, Massachusetts Historical Society, *Collections*, 7th ser., 1 (1900), 35.

45. Thomas Paine, *Letter Addressed to the Abbé Raynal* (Philadelphia, 1782), iiin.

46. Warfel, *Webster*, 58; *Public Records of Connecticut*, V, 13-15; David Ramsay, *The History of South-Carolina* (Charleston, 1809), II, 378.

and domestic publications, feeblest in the Southern states, but everywhere private libraries grew. Strenuous efforts were made to stock the shelves of the new college libraries.[47] The most unusual beginning for a new library was the wartime capture by an American privateer of the library of the Irish chemist, Dr. Richard Kirwan, who later gave his blessing to the use of his books as the foundation of the Philosophical Library of Salem.[48] Renewed enthusiasm led to the expansion of the value and utility of the oldest of the subscription libraries, the Library Company of Philadelphia. Work was begun on a new building designed for the purpose by William Thornton and its collections still further expanded. The principle of the subscription library was followed in the establishment of many new institutions in the various states although the Charleston Library Society and the Society Library of New York experienced difficulty in recovering their colonial vitality.[49]

It was only after the war that museums attained a separate institutional status. There had been many museums in the colonial period but they were no more than cabinets of natural history or collections of curiosities belonging to a college, a library society, or a philosophical society—often assembled in fits of absent-mindedness. Unlike the libraries, they did not serve useful scientific purposes. Charles Willson Peale's museum established in 1784 was the first institution of its kind to aid in the dissemination and even the advancement of knowledge.

Yet Peale's museum was itself an outgrowth of an increasing interest in effective collections, particularly in the city of Philadelphia. This city, which flowered so strikingly in many ways during the Confederation period, boasted numerous collections, private and institutional. The college, the Library Company, the hospital, the Carpenter's Company, and the Philosophical Society all possessed cabinets of some sort. There was a collection of Indian relics at Fort St. David's, and the clockmaker, Robert Leslie, made a collection of models, drawings, and descriptions of implements used in foreign countries in manu-

47. Bentley, *Diary*, I, 219.

48. *Ibid.*, 151; Frances D. Robotti, *Chronicles of Old Salem* (Salem, 1948), 43.

49. Richard P. McCormick, *Experiment in Independence* (New Brunswick, 1950), 61; see Charles Evans, *American Bibliography* (Chicago, 1903-34), VI, 428.

factures and the useful arts.[50] The English Whig, Samuel Vaughan, brought to America what Andrew Ellicott valued as "the Best Philosophical Aperatus in the United States and a great variety of Petrifactious Fossils." [51] More important was the wax anatomical museum opened by Dr. Abraham Chovet as early as 1774 which, when completed, displayed the internal structure of the human body at different periods of life and revealed the effects of disease upon various organs. It was a collection of real value which the Marquis de Chastellux judged "superior to those of Bologna, but inferior to the preparations of Mademoiselle Bieron; the wax having always a certain lustre which makes them less like nature." [52]

A more general museum was that of the Swiss miniaturist, Pierre Eugene du Simitière, who had been collecting curiosities for some years when, in 1782, he began to charge admission to visitors as Chovet had finally done. The American Museum, as Du Simitière called his collection, included a great miscellany of books, engravings, water color sketches, coins, Indian and African relics, pieces of wood, specimens preserved in alcohol, and fossils. The unrelated diversity was fatiguing to the observer. Ebenezer Hazard reported to his friend Jeremy Belknap that Du Simitière was "possessed of no extraordinary genius"; he was "a mere collector of curiosities." When he died in 1784, however, the American Philosophical Society and the Library Company both became very interested in purchasing his museum. The Library Company succeeded in obtaining most of his manuscripts, but it appears that the bulk of his natural history collections went to Charles Willson Peale.[53]

50. Whitfield J. Bell, Jr., Science and Humanity in Philadelphia, 1775-1790 (Doctoral Dissertation, University of Pennsylvania, 1947), 49, 165.
51. Catherine Van C. Mathews, Andrew Ellicott, His Life and Letters (New York, 1908), 52-53.
52. Pennsylvania Packet, July 1, 1790; Marquis de Chastellux, Travels in North-America in the Years 1780-81-82 (New York, 1827), 113.
53. American Philosophical Society Minutes, Nov. 5, 1784, Nov. 12, 1784 [III], n.p.; Minutes of the Directors of the Library Company of Philadelphia, March 3, 1785, March 19, 1785, II, 244; Minutes of the General Assembly of Pennsylvania, Ninth, 154, 158; Pennsylvania Journal, July 2, 1783, April 17, 1787; Chastellux, Travels, 111; John and Samuel C. Hall, Memoirs of Matthew Clarkson... and of Gerardus Clarkson ([Philadelphia], 1890), 76; William J. Potts, "Du Simitière, Artist, Antiquary, and Naturalist, Projector of the First American Museum," Pennsylvania Magazine of History and Biography, 13 (1889), 351.

Like Du Simitière, Peale was a painter, although he concentrated upon canvases rather than miniatures. In 1784, before Du Simitière's museum was sold, Peale began to think about attaching a museum of natural history to the gallery of paintings of Revolutionary heroes which he was already exhibiting. He was encouraged in this design by Robert Patterson of the University of Pennsylvania, who viewed it as a device to advance science, and by a business man who foresaw popular encouragement. When Peale finally turned his attention to the museum, he offered the public both the paintings of American heroes and a "Collection of preserved Beasts, Birds, Fish, Reptiles, Insects, Fossils, Minerals, [and] Petrifactions." Located in the nation's capital, he was able to receive donations from foreign diplomats, citizens of other states, and merchants trading with distant parts of the world. In time he gathered the largest collection of natural curiosities in the country, he discovered new and successful methods of taxidermy, and he developed an excellent technique of presenting animals in their natural habitat. Although he charged admission to his museum, he always directed it with the intention of advancing science, his principal model being the Musée National at Paris. In later years, he attracted about the museum a remarkable group of men who ultimately became important naturalists in their own right.[54]

The museum, the libraries, the books and magazines, and the educational institutions all contributed toward improving the intellectual atmosphere in which science might be expected to flourish, but it was upon the academy or learned society that those who sought to advance science in the United States placed their major hopes. The American patriots were very conscious of the great role played by the European academies in the progress of the scientific revolution. They voiced the hope that in the United States, similar bodies, "by con-

54. Max Meisel, *A Bibliography of American Natural History* (Brooklyn, 1924-29), II, 57; Charles C. Sellers, "Charles Willson Peale and the American Philosophical Society," American Philosophical Society, *Year Book, 1944* (Philadelphia, 1945), 68; Charles C. Sellers, *Charles Willson Peale* (Philadelphia, 1947), I, 238-56; William P. and Julia P. Cutler, *The Life, Journals and Correspondence of Rev. Manasseh Cutler* (Cincinnati, 1888), I, 259-62; Jedidiah Morse, *The American Gazeteer* (Boston, 1797), not paged; *Pennsylvania Packet,* May 27, 1785, July 29, 1788, Dec. 11, 1788, Feb. 16, 1790, and July 10, 1790.

centering in a proper focus the scattered rays of science, may aid and invigorate the intellectual: benefitting by their productions, not only the communities in which they are respectively instituted, but America and the World in general." [55] Even before independence had been declared, the Continental Congress recommended that each colony establish a "society for the encouragement of agriculture, arts, manufactures, and commerce, and ... maintain a correspondence between such societies, that the rich and numerous natural advantages of this country, for supporting its inhabitants, may not be neglected." [56] This envisaged a particularly utilitarian adaptation of the learned society which, despite its manifest advantages in forwarding the war effort, proved impossible to accomplish until more tranquil times. The congressional resolution had been introduced by John Adams not only because of the utility he saw in it, but also to hold "up to the view of the nation, the air of independence." [57]

John Adams' attention to the establishment of a more general learned society at Boston did finally bear fruit. He had first become interested in this project while attending the sessions of the Continental Congress in Philadelphia and had decided that Boston must have a learned society comparable to the American Philosophical Society. Observation of natural history collections in this country and in Europe further sharpened this resolution. More than a touch of envy influenced John Adams, particularly when, in France, he heard "much conversation concerning the Philosophical Society in Philadelphia and their Volume of transactions, which was considered as a laudable Institution and an honour to our Country." [58] As soon as he returned to America in 1779, he broached the question of a Boston philosophical society to the Reverend Dr. Samuel Cooper who, although initially dubious, responded to Adams' assurances that Massachusetts had

55. James Bowdoin, *A Philosophical Discourse, Addressed to the American Academy of Arts and Sciences* (Boston, 1780), 7.

56. *Journals of the Continental Congress*, IV, 224.

57. Edmund C. Burnett, "The Continental Congress and Agricultural Supplies," *Agricultural History*, II (1928), 112.

58. Adams to Waterhouse, Aug. 7, 1805, Worthington C. Ford (ed.), *Statesman and Friend, Correspondence of John Adams with Benjamin Waterhouse* (Boston, 1927), 26; John Adams to Abigail Adams, Aug. 3, 1776, *Letters of John Adams Addressed to his Wife*, ed. by Charles F. Adams (Boston, 1841), I, 145.

enough men of learning to make it effective. Adams cleverly pointed out that the proposed society would draw its strength from Harvard, complementing the college rather than competing with it. Cooper, convinced, went on to win support for the project, aided particularly by Benjamin Guild, Harvard tutor, who brought back from a Philadelphia visit several ideas for the proposed society.[59]

The legislature of Massachusetts willingly passed an act incorporating the American Academy of Arts and Sciences in 1780. The clause Adams was busily inserting in the new state constitution to encourage such societies never had to be used.[60] The title of the institution was chosen so that it would clearly distinguish it from the American Philosophical Society, but at the same time it was intended as an honor to France.[61] This, indeed, was the period when the Philadelphia society was busying itself in electing French members and in attempting to find various ways of honoring the French. Actually, the American Academy was modeled more upon the Royal Society of London, the London Society of Arts, and the American Philosophical Society than upon the Académie des Sciences.[62]

The war-born academy associated itself with other societies "for promoting useful knowledge," announcing that it hoped "to promote most branches of knowledge advantageous to a community." [63] At the same time, the patriotic impulse was very strong among the founders of the academy who were convinced that in establishing the society, they were "animated by the generous principles, which liberty and independence inspire." The academy hoped to attract papers in basic science, although it was careful to announce that the papers published in its *Memoirs* were not all to be considered contributions

59. Adams (ed.), *Works of John Adams,* IV, 259, 260n; Albert Mathews, [no title], Colonial Society of Massachusetts, *Publications,* 14 (1911-13), 260; Franklin B. Dexter (ed.), *The Literary Diary of Ezra Stiles* (New Haven, 1901), II, 385.

60. Adams (ed.), *Works of John Adams,* IV, 260n.

61. Mathews in Colonial Society of Massachusetts, *Publications,* 14 (1911-13), 261; William Gordon, *The History of the Rise, Progress, and Establishment of the Independence of the United States* (London, 1788), III, 398; Manasseh Cutler to Jonathan Stokes, Oct. 30, 1786, Cutler, *Life of Manasseh Cutler,* II, 267.

62. Bowdoin, *Discourse,* 7.

63. American Academy, *Memoirs,* 1 (1785), [iii], iv.

to knowledge. The dissemination of knowledge as well as its advancement was an objective.[64]

With sixty resident members named in its very charter, the academy had little difficulty in developing a working organization. James Bowdoin, philosophical friend of Franklin and distinguished patriot soon to be elected governor of Massachusetts, was chosen first president. The vice-presidency went to Adams' friend Samuel Cooper and upon his death to the Reverend Joseph Willard, president of Harvard. Most of the other offices were assigned to members of the Harvard faculty, although Harvard had only minority representation on the ten-man council which included several political and civic leaders, some with no notable scientific interests. The academy's meetings were held quarterly.[65]

Several nonresident members were elected, most of the Americans being New Englanders although the most outstanding patriots of the American Philosophical Society were given representation too. A very distinguished group of European scientists was chosen to membership, D'Alembert, Euler, La Lande, and Priestley among them. Court de Gebelin was enthusiastic about the new academy, even offering to put it in correspondence with a recently established French society of sciences, letters, and arts.[66] Richard Price accepted membership with great gladness, for to him the academy promised verification of his hope "of seeing the United States distinguish'd as seats of science, liberty, and virtue." [67] The leaders of the academy were particularly solicitous of the support of Benjamin Franklin and were not long in receiving it. He was anxious to be an active member even though then residing in France and promised to send "them from time to time some of the best Publications" that appeared in Europe. Franklin also recalled that Boston was his native city, a fact he never forgot despite his long association with Philadelphia. Franklin, Price, and other nonresident members were able to give the academy substantial aid by

64. *Ibid.*, ix.
65. *Ibid.*, v, xx-xxii; *Boston Gazette*, Jan. 5, 1784, Jan. 19, 1784.
66. American Academy, *Memoirs*, 1 (1785), xxii; Court de Gebelin to Franklin, May 6, 1781, Franklin Papers, XXII, 7, American Philosophical Society.
67. Price to Bowdoin, Oct. 25, 1785, *Bowdoin and Temple Papers*, II, 78.

putting it in communication with European learned societies and men of science.[68]

The most urgent project of the academy's early years was the publication of its first volume of *Memoirs*, a feat it accomplished in 1785. Better than any other development, this volume revealed that the academy did not compete with Harvard or challenge the college's intellectual dominance of the whole region. More than half of the *Memoirs* consisted of papers written or solicited by Harvard professors. President Joseph Willard and the Reverend Samuel Williams, Winthrop's successor as Hollis Professor of Mathematics and Natural Philosophy, wrote 206 of the volume's 568 pages alone. The contents were divided into three sections: astronomy, natural philosophy and natural history, and medicine. The men who prepared the volume were perhaps unduly apologetic over the fact that it contained little in the way of contributions to basic science, for the journal was well received even though it did not sell well. At a time when all American publications were viewed askance in Britain, the *Monthly Review* declared, "[This volume may] be considered as a proof, that philosophical pursuits are carried on with vigour in the American States." [69]

With the publication of its first volume of *Memoirs*, the American Academy was accepted in America and in Europe as one of the two effective learned societies in the United States. In 1788, James Bowdoin was the first American to be elected to fellowship in the Royal Society since the Revolution—elected not because of his personal scientific accomplishments as much as in recognition of his office as president of the academy.[70] A less desirable compliment was paid the academy by America's postwar crop of magazines, especially the *Columbian Magazine* and the *American Museum*, which made a practice of extracting articles from the *Memoirs* as well as from the *Transactions* of the American Philosophical Society. Although this was a familiar procedure in England, it drew a mild rebuke from Benjamin

68. Samuel Cooper to Franklin, Feb. 1, 1781, Franklin Papers, XXI, 43, American Philosophical Society; Bowdoin to Franklin, Jan. 11, 1781; *ibid.*, 18; Willard to Franklin, Feb. 9, 1781, *ibid.*, 55; Franklin to Cooper, May 15, 1781, Bache Collection, American Philosophical Society.

69. *Monthly Review*, 79 (1788), 393.

70. Bowdoin to Charles P. Layard, May 10, 1789, *Bowdoin and Temple Papers*, II, 193; Stearns, "Colonial Fellows of the Royal Society," 267-68.

Franklin and a sharp protest from Jeremy Belknap, who felt that the academy would have had better success in selling its *Memoirs* except for this pirating. Belknap queried, "Were they not communications made to regularly instituted philosophical bodies, and by them *preserved in their repositories,* and printed at their expense, which expense unless defrayed by the sale of the books must lie as a dead weight on them." [71] When the *London Medical Journal* reprinted five articles from the *Memoirs,* there was no disposition to protest. At the same time, the American Academy and the American Philosophical Society both directed papers not published in their journals to the general magazines. [72]

The American Academy did not compete with Harvard, but even its founders were conscious that it represented something of a challenge to the older American Philosophical Society. James Bowdoin hoped that the Boston and Philadelphia institutions might "both together resembling some copious river, whose branches, after refreshing the neighboring region, unite their waters for the fertilizing a more extensive country." [73] It was the larger patriotic aspirations of each society that the founders counted on to reduce the unpleasantness of any competition. Sharply conscious of the great American experiment, Joseph Willard looked forward to harmony and a noble emulation worthy of the newly independent nation. The founders of the academy did not focus their attention upon surpassing the American Philosophical Society but upon demonstrating the truth of the sentiment freely expressed in their motto, "Arts and sciences flourish best in free States." [74]

The American Philosophical Society was experiencing considerable difficulty in developing its own programs when it was confronted by the success of the Boston academy. Although revived in 1779, the Philadelphia society had not regained its colonial vitality; indeed in 1782 attendance at its meetings sank below that for any single year of the colonial period. [75] After that low point, it slowly gained strength

71. Belknap to Carey, May 17, 1788, *Belknap Papers,* III, 402.
72. *London Medical Journal,* 7 (1786), 287-321.
73. Bowdoin, *Discourse,* 7.
74. American Academy, *Memoirs,* 1 (1785), xix; Willard to Ewing, Nov. 20, 1780, Ms. Communications to American Philosophical Society, Mathematics and Astronomy, I, 14.
75. American Philosophical Society Minutes, *passim.*

but made only the slightest progress toward the publication of the much discussed second volume of *Transactions*. With the restoration of ties with England that followed the end of the war, the society again began to elect more members from the former mother country than from any other foreign nation. Continental ties continued, however, and with Franklin's help, better judgment was shown in the election of foreign as well as domestic members. Sterile meetings still plagued the society when the publication of the American Academy's *Memoirs* led Jeremy Belknap to suggest that that event might "give them a Jog for," he said, "they will not like to be rivalled by the New Englanders, especially as they think themselves before us in point of improvements. In some respects there is a foundation for this opinion, and the most candid New England men must subscribe to it." [76] The *Memoirs* certainly provided a spur, but a strenuous effort to reinvigorate the society had been under way for some time when that publication appeared.

The principal instrument of this effort was Judge Francis Hopkinson, a clever, little man "of a Thousand whims" and some very considerable talents.[77] Satirist, poet, spare-time musician, and painter, Hopkinson had a deep devotion to science although his own work was limited to strange little inventions: a candle shield, a spring block, a water clock, and a lubricant containing rubber. Yet, where more profound thinkers were content to let matters take their course, Hopkinson was not. Late in 1783, he soundly spanked the members of the society for failing to throw off their state of lethargy, and warned them that if they did not reform, they would "unavoidably sink into contempt." [78] He demanded that the society acquire its own building, reform its administration, dispense with consideration of matters not related to science, and plan at least one scientific experiment for every meeting. He was given strong support by David Rittenhouse and the English Whig merchant, Samuel Vaughan, who applied his own great zeal to the same end. Hopkinson was opposed by the Uni-

76. Belknap to Cutler, Nov. 18, 1785, Cutler, *Life of Manasseh Cutler*, II, 234.
77. Hopkinson to Jefferson, Jan. 4, 1784, Boyd (ed.), *Papers of Jefferson*, VI, 445.
78. Francis Hopkinson, "An Address to the American Philosophical Society," *Miscellaneous Essays and Occasional Writings* (Philadelphia, 1792), I, 362.

versity of Pennsylvania faculty members who were offended at his effort to reduce science to a matter of enthusiasm. He told them, "The language of nature is not written in Hebrew or Greek; the understanding thereof is not involved in the contemptible quirks of logic, nor wrapt in the visionary clouds of metaphysical hypothesis. The great book of nature is open to all—all may read therein." [79] His was a phase of the faith of the Revolution—however unrealistic—and it was from this faith that he drew the strength to carry his fight to success. He was fighting for a principle which gave him great confidence. Even as he held Franklin up as one who had made important discoveries without benefit of liberal education, he rapped him on the knuckles with a reprimand: "[The society] expect their President will now and then favour them with his Notice." [80]

The publication of the second volume of *Transactions* in 1786 was one evidence that a new spirit had been generated in the society. It was an inferior book compared with the first volume which had made its reputation on the comprehensive treatment of the transit of Venus and had generally shown more sense of direction. No society programs were reported in the second volume because there had been none. Some of the papers included were significant but they were often obscured by articles that had little to recommend them.

Yet, the restrained reception that met this volume was not so much a reaction to its intrinsic character as it was a reflection of the great changes that had taken place since the publication of the first volume. The New Englanders were able to make points of comparison with their own *Memoirs* which permitted them to conclude that the *Transactions* was not in all respects superior.[81] Mild approval greeted the volume on the Continent, in Scotland, and in the liberal circles of England, but in some quarters bitter rancor revealed itself. The English *Critical Review*, displaying a "cool abruptness, seemingly bordering on contempt," came to the conclusion that the volume was full of "empty speculation and disputed theories." Only those papers signed with the magical name "Benjamin Franklin" were exempted from condescending attack. The *Critical Review* clearly felt that in

79. *Ibid.*, 364-65.
80. Hopkinson to Franklin, May 24, 1784, Franklin Papers, XXXI, 185, American Philosophical Society.
81. Cutler to Belknap, Oct. 11, 1786, *Belknap Papers*, III, 318.

the realm of science the Americans ought to be content to remain in a colonial status, and that they "might be more usefully employed in accumulating facts than in constructing systems." [82]

In addition to its volume of *Transactions*, the society took some other forward steps. It resumed the colonial tradition of annual orations with a series of three in 1780, 1781, and 1782 and another series in 1786, 1787, and 1788. The first two orations, delivered by the radical Whigs, Timothy Matlack and Owen Biddle, and the third by Thomas Bond were notable for their exhortations to advance science for the welfare and greater glory of the United States.[83] In the latter series Benjamin Rush's *Enquiry into the Influence of Physical Causes on the Moral Faculty* and Samuel Stanhope Smith's *Essay on the Causes of the Variety of Complexion and Figure in the Human Species* both sought to present something more in the nature of a contribution to knowledge. Dr. John Foulke brought the orations to a close with an address on longevity, the only one not published.[84] All of these orations served the important function of keeping alive a strong interest in the promotion of science among the large lay audiences which heard them.

In 1786, an important step was taken in the establishment of the Magellanic Premium. Both the American Philosophical Society and the American Academy had been interested in instituting premium programs but neither was able to do so until the unforseen donation to the Philadelphia society by John Hyacinth de Magellan, a Portuguese member of the Royal Society. He donated 200 guineas to permit

82. See Brooke Hindle, Rise of the American Philosophical Society, 1766-1787 (Doctoral Dissertation, University of Pennsylvania, 1949), 224-26; *Critical Review*, 64 (1787), 241-47, 435-41.

83. Timothy Matlack, *An Oration Delivered March 16, 1780, Before ... the American Philosophical Society* (Philadelphia, 1780); Owen Biddle, *An Oration Delivered the Second of March, 1781 at the Request of the American Philosophical Society* (Philadelphia, 1781); Thomas Bond, *Anniversary Oration Delivered May 21st Before the American Philosophical Society* (Philadelphia, 1782).

84. Benjamin Rush, *An Oration Delivered before the American Philosophical Society ... on the 27th of February, 1786* (Philadelphia, 1786); Samuel S. Smith, *An Essay on the Causes of the Variety of Complexion and Figure in the Human Species* (Philadelphia, 1787); John Foulke, "Oration Pronounced ... before the American Philosophical Society Feb. [1788]," Historical Society of Pennsylvania.

the society to grant a premium periodically "to the Author of the best discovery, or most useful improvement relating to Navigation, or to Natural Philosophy, *mere Natural History* only excepted." Magellan's motivation in establishing this premium may have been in part to obtain membership in the society. More likely is the possibility that he wanted to show his esteem for the new republic by helping to promote science within it. The one requirement he insisted upon in connection with his premium was that the entire society—not merely the council—adjudicate the award; this a reflection of his opposition to "Aristocratic governments." [85]

None of these accomplishments served to restore the society to its former effectiveness. The one thing that overcame internal dissension and introduced genuine enthusiasm into the society was the building of Philosophical Hall. This was not in itself a project that could directly advance science but somehow it became a symbol—an emotional factor that restored morale to a high point. The desire for a building of its own was a heritage from the colonial period. The society had purchased a building lot from Francis Hopkinson in 1784 and than had abandoned that plan in favor of building on Independence Square, where the state legislature proved willing to grant a lot for the purpose. A building fund inaugurated in 1785 began to receive encouraging subscriptions not only from Philadelphians but also from residents of other sections. One of the largest individual pledges came from Henry Laurens of South Carolina. The building plans were drawn up by Samuel Vaughan who also supervised construction. Before very much of the building could be completed, however, financial difficulties multiplied to such a degree that the question of abandoning the entire project was raised. When various alternatives failed to answer the needs, the society reluctantly considered a proposal that the partly finished building be sold to the Library Company of Philadelphia. [86]

It was at this point that Benjamin Franklin dramatically intervened. Returning to the United States in 1786, Franklin was able to play a much more direct role in the development of science within the

85. Magellan to American Philosophical Society, Sept. 17, 1785, July 7, 1786, Archives, American Philosophical Society; American Philosophical Society Minutes, Jan. 20, 1785, Oct. 6, 1786, Nov. 7, 1786, [III], not paged.
86. Hindle, Rise of the American Philosophical Society, 238-42.

country than he had for the past thirty years. He was particularly concerned that the society of which he was president should endure and prosper. When he learned of the proposed sale of the building, he invited the society to come to his home for their next, crucial meeting. Setting the stage for a delightful display of psychology and diplomacy, he refused, as presiding officer, to permit the decision on the society's building to come to an immediate vote. Instead, he opened the meeting by asking that two letters he had received be read: one from Patrick Wilson, the Glasgow astronomer, and the other from an American with an idea for preventing chimney fires. In thus subtly recalling to the members the purpose of their association, Franklin no doubt revealed his feeling on the matter. The motion to give up the building was subsequently defeated, and in a flurry of emotion the society directed the building committee "immediately to proceed and with all convenient dispatch to have the Walls carried up, and covered in—." A few weeks later, Franklin solved the financial crisis by subscribing a second £100 to the building fund and by lending the society £500. Philosophical Hall became a reality, a symbol of a new sense of confidence, and a monument to Benjamin Franklin.[87]

The societies in Philadelphia and Boston were the only ones that attained a national significance in this period, but others were attempted. As soon as the Harvard tutor, Benjamin Guild, brought Ezra Stiles plans of the projected American Academy of Arts and Sciences, Stiles began to talk of his own plans for a Connecticut academy. He had been fabricating formulas for learned societies since Stamp Act days but it was not until Guild's visit that he took active steps to realize his dream, now that he was president of Yale College. In November 1780, he sent the Reverend Nathan Strong of Hartford his draft of a charter for a proposed Connecticut academy which Strong then sought to push through the legislature. The bill of incorporation failed to be approved by that body because of its "Jealousy least the Academy should become constitutionally connected with [the] College."[88] This antagonism caused the defeat of a second attempt to secure a charter in 1783 despite the fact that Stiles took special care

87. *Ibid.*, 242-44; American Philosophical Society Minutes, Sept. 18, 1787, [IV], 7.

88. Dexter (ed.), *Diary of Stiles*, II, 385, 486; Edward C. Herrick, "Historical Sketch of the Connecticut Academy of Arts and Sciences," *American Quarterly Register*, 13 (1840), 23-24.

in this draft to avoid any marked connection between the college and the academy.[89]

A way out of the impasse was finally found in 1786 with the formation of a voluntary association, not incorporated by the state, to be called the Connecticut Society of Arts and Sciences. In the evident hope of ultimately securing a charter, this organization bent over backward to avoid any appearance of connection with the college. Instead, it tied itself to the state legislature, holding semi-annual meetings alternately at Hartford and at New Haven when the legislature was in session in each of those towns. The lieutenant governor of the state, Oliver Wolcott, was elected first president. Stiles remained clear of the actual formulation of the society, but he was elected to membership and the following year to the post of corresponding secretary.[90]

The society had the same aspirations and ambitions that moved other philosophical societies. Through the public press, it requested the submission of astronomical observations, natural history materials, medical and chemical essays and experiments, and material relating to agriculture and manufactures.[91] It heard some papers at its meetings and arranged the publication of the Reverend Jonathan Edwards' *Observations on the Language of the Muhhekaneew Indians.* Attempts to broaden the society led to the election of some thirty honorary members in other states and countries. Further efforts to extend the society's influence were impeded by the failure to issue a volume of transactions, as Ezra Stiles discovered when he wrote to Benjamin Franklin requesting a list of the European academies with the intention of asking them for copies of their transactions. Franklin sent a copy of the *Transactions* of the American Philosophical Society by way of reply but dissuaded Stiles from asking the Europeans for their publications until he could offer a volume in exchange.[92] That time never came.

New York's attempt to form a philosophical society was still less

89. Ezra Stiles, Draft of Charter for Society, Stiles Papers; Simeon E. Baldwin, "The First Century of the Connecticut Academy of Arts and Sciences, 1799-1899," Connecticut Academy of Arts and Sciences, *Transactions*, 11 (1901-3), xv.

90. Dexter (ed.), *Diary of Stiles*, III, 262-63.

91. *New Haven Gazette and Connecticut Magazine*, 1 (1786), 354.

92. Stiles to Franklin, July 3, 1787, July 31, 1787, Franklin Papers, XXXV, 90, 97, American Philosophical Society.

successful. The New York Society for Promoting Useful Knowledge was established late in 1784, less than a year after Thomas Paine had demanded the formation of such an organization, but it is not clear that his suggestion was the precipitating cause. Paine's motivation was strongly personal. Still smarting from his recent failure to be elected to the American Philosophical Society, he was anxious to see that group eclipsed. He felt they assumed much too much in calling themselves "the '*American* Philosophical Society'" when they were primarily a Pennsylvania institution. "Pennsylvania," he deplored, "with scarcely anybody in it who knows anything of the Matter, except Mr. Rittenhouse and one or two more, is drawing to herself laurels and Honours she does not deserve." [93] If this bitterness was shared by the men who formed the New York society, there is no evidence of it.

The New York society proceeded in much the same manner as the Connecticut society in electing political figures to its top offices. Governor George Clinton became president and John Jay and James Duane, vice-presidents. Dr. Samuel Bard, a man with scientific training and interests, was elected secretary. The society elected both resident and honorary members and revealed hopes that it might publish a volume of transactions. In 1787 it was still showing some evidences of life when Samuel Bard, rather despondently, thanked the American Philosophical Society for the gift of its *Transactions*. He remarked, "[We can] only follow you at an humble distance, and I am afraid it will be a long time before we shall return your Compliment, by sending you a Vol[u]me of our Transactions—." In 1788, Brissot de Warville learned of the society but found, "They assemble rarely and they do nothing." He correctly tagged the president, Clinton, as "any other thing rather than a man of learning." Before 1789, a new slate of officers was elected with John Sloss Hobart, state supreme court justice, becoming president and Samuel Bard moving up to a vice-presidency. Just a few years later, it was not possible to find a trace of the effects of the society in the city. [94]

93. Paine to Lewis Morris, Feb. 16, 1784, Duane Mss., New-York Historical Society; *Pennsylvania Packet,* Jan. 17, 1785; De Witt Clinton, *An Introductory Discourse, Delivered before the Literary and Philosophical Society of New-York, on the Fourth of May, 1814* (New York, 1815), 13.
94. J. P. Brissot de Warville, *New Travels in the United States of America* (London, 1794), I, 133; *New York Directory and Register for the Year 1789*

Elsewhere throughout the country, similar philosophical societies were attempted but none were given sufficient support to make them effective. Rather oddly, there was nothing of the sort in Newport or in Charleston where even the Library Society had trouble getting back on its feet.[95] In New Jersey, however, both the Trenton Society for Improvement in Useful Knowledge and the New Jersey Society for Promoting Agriculture, Commerce and Arts were established. The Kentucky Society for Promoting Useful Knowledge was formed in 1787.[96] In Virginia, the Alexandria Society for the Promotion of Useful Knowledge was founded and John Page made several efforts to revive the Virginian Society for the Promotion of Usefull Knowledge at Williamsburg.[97] Few Virginians showed any interest, in part because their attention had been diverted by another, related institution.

It was John Page who, in 1778, suggested the idea of establishing an academy to a young French officer recuperating in his home from a wound, the Chevalier Alexandre Quesnay de Beaurepaire. From that beginning, Quesnay worked out the grandiose project of the Academy of the Sciences and Fine Arts of the United States of America. sophical society but it was something of each. In the high emotion It was to be more than an educational institution, more than a philo- of the Revolution, Quesnay conceived it to be "That much needed bridge over the divide which separates Europe from America." [98]

(New York, 1789), 121; Bard to Franklin, May 13, 1785, Bard to [?], July 24, 1787, Archives, American Philosophical Society; New York Literary and Philosophical Society, *Transactions*, 1 (1815), 37.

95. Thornton to Lettsom, Feb. 15, 1787, Pettigrew, *Memoirs of Lettsom*, II, 511; William M. and Mabel S. C. Smallwood, *Natural History and the American Mind* (New York, 1941), 105.

96. McCormick, *Experiment in Independence*, 62; Lewis C. Gray, *History of Agriculture in the Southern United States to 1860* (Washington, 1933), II, 783.

97. *American Museum*, 8 (1790), 85; Page to Jefferson, Dec. 9, 1780, Jefferson Papers, Library of Congress; Gov. Harrison to Rev. James Madison, Sept. 27, 1783, Boyd (ed.), *Papers of Jefferson*, VI, 508n; Jefferson to Page, Aug. 20, 1785, *ibid.*, VIII, 417; Page to Jefferson, April 28, 1785, *ibid.*, 119.

98. Richard H. Gaines, "Richmond's First Academy Projected by M. Quesnay de Beaurepaire in 1786," Virginia Historical Society, *Collections*, 11 (1891), 168; Quesnay de Beaurepaire, *Memoir Concerning the Academy of the Arts and Sciences of the United States of America* (Richmond, 1922), 16.

Travelling through the United States and back to France after the war, Quesnay rounded up surprising support. From the people of Virginia and Maryland he raised a subscription of 60,000 francs and from France he obtained the endorsement not only of the Académie des Sciences and the Académie Royale de Peinture et Sculpture but of the great scientific figures, Lavoisier, Condorcet, and La Lande.[99]

The academy would be centered at Richmond where a building to house it was begun in 1786. It would be staffed by European instructors of whom one, John Rouelle, was actually appointed mineralogist and professor of natural history and chemistry.[100] It would offer courses in physics, astronomy, natural history, chemistry, mineralogy, geography, mathematics, and anatomy—an impressive array of scientific subjects despite Quesnay's feeling that it was not in science but in the fine arts that the United States was most markedly inferior to Europe.[101] Quesnay also intended that the academy issue two series of annual publications: one, an almanac with lists of faculty, artists, students, and subscribers and a summary of the year's work; the other, a volume of memoirs, containing scholarly papers that would be distributed to the academies of Europe. When fully developed, the academy would serve the entire country as a graduate school through extension units in such cities as New York, Philadelphia, and Baltimore.

From the beginning, this institution had been beyond the capacities of the United States to support—particularly located as it was in Richmond. There were many American leaders who came to support the idea of a national university but even that more limited project could not be accomplished. Thomas Jefferson, who was seldom restrained in his estimates of American capacity, did recognize that the academy was too "extensive for the poverty of the country."[102]

99. Le Chevalier Alexandre M. Quesnay de Beaurepaire, *Mémoire, statutes, et prospectus, concernant l'Académie des Sciences et Beaux Arts des États-Unis de l'Amérique* (Paris, 1788), [xiii], 23; *Virginia Gazette*, Richmond, June 28, 1786.

100. Quesnay, *Mémoire*, 13; John Rouelle, *A Complete Treatise on the Mineral Waters of Virginia* (Philadelphia, 1792).

101. Quesnay, *Mémoire*, 18-20.

102. See Benjamin Rush, "A Plan for a Federal University," Butterfield (ed.), *Letters of Rush*, I, 491-95; Jefferson to Quesnay, Jan. 6, 1788, Philip A. Bruce, *History of the University of Virginia, 1818-1919* (New York, 1920), I, 57.

With the outbreak of the French Revolution, French support and Quesnay's leadership were both lost. That brought to an abrupt end the entire project. The academy was just one of the many noble plans to advance science in the United States that perished before they got well under way.

That this organizing activity in science was not altogether a function of the Revolution was suggested by parallel activity in other lands. Philosophical societies were being founded throughout the world from Manchester, England to Batavia, Java. Partly as a result of the London Society of Arts turning its attention to previously neglected British colonies after the American Revolution, there was significant activity in the West Indies. No philosophical societies were established there, but the botanical gardens on Jamaica and St. Vincent which had been initiated earlier at the suggestion of the Society of Arts, were further developed. The Barbados Society for the Encouragement of Arts, Manufactures, and Commerce, which perished shortly after 1787, was a wartime copy of the London Society of Arts.[103] The intellectual life of the British West Indies was feeble— feebler than at least one point in the French West Indies where an active intellectual group took form without any stimulus from the Society of Arts. This group, at Cap François on the island of Hispaniola, reflected a strong American influence when it called itself the Cercle des Philadelphes—in part, at least, after the city on the Delaware.[104] Between 1785 and 1789, it published a surprising number of papers on topics of the sort then engaging the American philosophical societies.

The encouragement of science, particularly through the medium of philosophical societies, was a phenomenon of the Enlightenment,

103. Sir Henry T. Wood, *A History of the Royal Society of Arts* (London, 1913), 98-99; Lowell J. Ragatz, *A Guide for the Study of British Caribbean History, 1763-1834,* American Historical Association, *Annual Report for 1930,* III (Washington, 1932), especially 242-66, 328; *Gentleman's Magazine,* 57 (1787), 564; *London Medical Journal,* 1 (1781), 354.

104. M. [Charles] Arthaud, *Discours prononce a l'ouverture de la première séance publique du Cercle des Philadelphes* (Paris, 1785), not paged; [Charles Arthaud], *Dissertation, et observations sur le tétanos* (Cap François, 1786); Arthaud to Franklin, May 13, 1788, July 27, 1789, Franklin Papers, XXXV, 131, XXXVI, 164, American Philosophical Society; American Philosophical Society Minutes, Nov. 3, 1786, March 20, 1789, [III], n.p., [IV], 96.

but in America, the Revolution gave it a special urgency and, some-times, a special American flavor. Such was the character of proposals to organize science in America on a federal basis, paralleling American political organization. Just before the founding of the American Academy of Arts and Sciences, Jeremy Belknap of New Hampshire suggested that the American Philosophical Society in Philadelphia be-come a kind of central scientific congress based upon inferior societies in each of the states. "Why," he asked, "may there not be subordinate philosophical bodies connected with a principal one, as well as sepa-rate legislatures, acting in concert by a common assembly? I am so far an enthusiast in the cause of America as to wish she may shine Mistress of the Sciences, as well as the Asylum of Liberty." [105]

The college presidents, James Madison of William and Mary and Ezra Stiles of Yale, agreed "that the Republic of Letters could gain Strength from the Union of its Members by frequent communica-tion with each other." "But," said Madison, "surely it belongs to our Colleges and Universities to lay the Foundation from which the future glory of America shall arise." As for Belknap's idea, Madison felt progress was being made: "I hope [the philosophical Society estab-lished in Philadelphia] will concenter those scientific Discoveries and Observations so that Europe shall behold America not only as a new Star in the political Horizon, but in the litterary also." [106]

It was Lewis Nicola who presented the American Philosophical So-ciety with a concrete proposal designed to implement Belknap's sug-gestion. He called for the establishment of committees in each of the states, except New Jersey and Delaware which he did not think had capital cities. These groups would make their own rules and become separate societies responsible for collecting communications and send-ing them to Philadelphia. [107] By the time the society came to consider the plan seriously, the American Academy had been formed, and ef-forts were being made to establish a society in Connecticut. It was therefore suggested that the American Philosophical Society "asso-

105. Belknap to Hazard, Feb. 4, 1780, *Belknap Papers*, I, 255.

106. Madison to Stiles, Aug. 1, 1780, *William and Mary Quarterly*, 2d ser., 7 (1927), 293.

107. Nicola to American Philosophical Society, April 7, 1780, Archives, American Philosophical Society; American Philosophical Society Minutes, Dec. 15, 1780, [III], n.p.

ciate with, or rather adopt several learned societies ... forming in each State." Wisely, nothing of this sort was attempted.[108]

The fact that accomplishment seldom matched aspiration did not depress the prophets of glory or cause them to lose faith in America's destiny. Fully at home in the ideals of the Enlightenment, they were conscious of their role in history and sensitive to the opinion of Europe. Francis Hopkinson spoke for many when he said, "The eyes of Europe are turned towards America—And with what view? Not altogether to contemplate the peculiarities of our government.... But they look towards us as a country that may be a great nursery of arts and sciences—As a country affording an extensive field of improvement in agriculture, natural history and other branches of useful knowledge." [109] In the full flood of their faith, they not only sought to make the general atmosphere more congenial to scientific development, but they specifically tried to advance several fields of learning and to apply science to affairs of life wherever it seemed possible.

108. Chastellux, *Travels*, 149.
109. Hopkinson, "Address," *Miscellaneous Essays and Occasional Writings* (Philadelphia, 1792), I, 360-61.

Chapter Thirteen

MEDICINE IN THE REPUBLIC

THERE WERE FEW more ardent patriots or sanguine prophets of the beneficent effects to be anticipated from the Revolution than such physicians as Benjamin Rush, David Ramsay, and John Warren. Their expectations were not limited to nebulous hopes for the betterment of the nation; they specifically looked for the improvement of medicine under the influence of freedom and independence. Moreover, many physicians became convinced within a few years of the end of the war that the anticipated improvements were rapidly being realized. On every hand, they could see unprecedented activity in erecting medical societies, founding and rebuilding medical schools, establishing medical journals, and seeking the improvement of the profession in a variety of ways. They were led to believe that a new era was opening.

Dr. James Potter of New Fairfield, Connecticut, declared, "No demonstration in Euclid is more certain than the rapidity with which our profession hath agreeably increased in a very few years. . . . The learning of all nations and ages is concentrated in America. . . . Therefore," he advised physicians, "strain every nerve to the last extremity, collect all your powers into one vigorous point, and push your researches with relentless impetuosity, through physics vast and ample field; . . . while you rest secure under the glorious, sacred and permanent independence of America." [1] Even the less flighty Dr. James

1. James Potter, *An Oration on the Rise and Progress of Physic in America* (Hartford, 1781), 8-11.

[280]

Tilton of Delaware could report, "I flatter myself that all things are working together for good to America." [2]

This atmosphere of optimism could not disguise the fact that war and revolution had injured medicine in the United States in ways that took time to repair. This was particularly true of American relationships with physicians and medical institutions in Europe. All contact with British medical circles had not been cut off during the war but relations were reduced to a nearly ineffective level. The return of peace brought a resumption of correspondence with a display of effort to forget the problems that had divided the Americans from the mother country. "What," wrote Benjamin Rush, "has physic to do with taxation or independence?" To William Cullen, his old Edinburgh teacher, he admitted, "One of the severest taxes paid by our profession during the war was occasioned by the want of a regular supply of books from Europe, by which means we are eight years behind you in everything." [3] Although the book trade revived and the Americans further disseminated British writings by reprinting them in their own press, it was hard to catch up. [4] Some losses could not be repaired. Dr. John Fothergill, the American physicians' most constant friend in England, had died during the war and with him, William Hunter and Daniel Solander. That brought to an end the London Medical Society and its *Medical Observations and Inquiries* which had published many papers written by American doctors. [5]

Fortunately, there were new ties to replace the old although they were never quite the same. Dr. John Coakley Lettsom, the West India-born Quaker physician who succeeded to Fothergill's practice, also was able to replace him in some measure as an advocate and patron of the Americans. He maintained an extensive correspondence with the

2. Tilton to Dr. [Alexander] Monroe, July 8, 1785, Emmet Collection, III, 962, New York Public Library.

3. Rush to Cullen, Sept. 16, 1783, L. H. Butterfield (ed.), *Letters of Benjamin Rush* (Princeton, 1951), I, 310.

4. See John Hunter, *A Treatise on Venereal Disease ... Abridged by William Currie* (Philadelphia, 1787); William Cullen, *Institutes of Medicine* (Edinburgh Printed, Boston Reprinted, 1788); W. Smellie, *An Abridgement of the Practice of Midwifery* (Boston, 1786); Edward Rigby, *An Essay on the Uterine Hæmorrhage* (3d ed., Philadelphia, 1786); F. X. Swediaur, *Venereal Complaints* (New York, 1788).

5. *Medical Observations and Inquiries*, 6 (1784), iii-v; *London Medical Journal*, 8 (1787), 217.

active physicians, Rush and Waterhouse, helping them in innumerable ways: for the one, he acted as a publishing agent in England and to the other he sent an extensive mineral collection to be used in his teaching of natural history.[6] He was a reliable friend to many American medical students who now returned to England in numbers. The new medical periodicals, the *Medical Communications* and the *London Medical Journal*, proved ready to print American essays; sometimes to reprint them unsolicited. Other journals and magazines reopened their facilities to the Americans.

Just as the return of peace brought the renewal of American ties with British medical circles, it caused a slackening of the relationships established during the war with continental medicine. This logical return to established patterns was somewhat hastened by occasional wartime experiences of American students who were unimpressed by the continental medical schools.[7] Efforts to establish relations between American and French medical groups did not seem to have any very lasting effect.[8] On the other hand, wartime relationships continued to bear such fruit as Christian Friederick Michaelis' translation of the *Transactions* of the College of Physicians of Philadelphia into German, Johann David Schöpf's publication of his *Materia Medica Americana* at Erlangen, and the appearance of Thomas Bond's essay on inoculation at Strasburg.[9] A generous attempt to maintain ties was made by the French government through the enthusiastic St. John de Crèvecoeur in offering to interested societies or to individual physicians monthly copies of the *Journal de Medicine Chirurgie et Pharmacie Militaire*—"free of any expence."[10] While he was in France, Franklin—and Jefferson after him—provided a bridge to the continent that was used by physicians as well as by others. Ties continued

6. Butterfield (ed.), *Letters of Rush*, I, 313n.

7. John Foulke to Franklin, Oct. 12, 1781, Franklin Papers, XXIII, 11, American Philosophical Society.

8. Dr. James Potter to Franklin, May 10, 1780, *ibid.*, LIV, 67.

9. Fielding H. Garrison, "The Medical and Scientific Periodicals of the 17th and 18th Centuries," *Bulletin of the Institute of the History of Medicine*, 2 (1934), 310; T. Bond, *Defense de l'inoculation et relation de progres qu'elle a fait à Philadelphia en 1758* (Strasbourg, 1784), cited by Michael Kraus, *The Atlantic Civilization* (New York, 1949), 212.

10. *Boston Gazette*, March 8, 1784.

but the Americans developed no real dependence upon continental medicine.

Continental ties were more effective for transmitting the spectacular European medical news to America than they were for transplanting basic values of French medicine. The Americans, for example, were kept well informed of the antics of Franz Anton Mesmer, discoverer of animal magnetism. Mesmer claimed to have discovered a fluid similar to magnetism which pervaded nature causing phenomena in humans similar to the phenomena of magnetism in iron. He performed a series of spectacular cures upon such prominent persons as Court de Gebelin which allegedly depended upon the control of this fluid—actually the most effective agent used by Mesmer and his followers was hypnotism. Mesmer was rejected by the medical profession in Europe—particularly at its centers in Vienna and Paris, but he won a wide popular following in France. It was not through medical channels, therefore, but by lay enthusiasts—most energetically by the Marquis de Lafayette—that animal magnetism was brought to the United States.

Mesmer failed to win the support of either Franklin or Jefferson but Lafayette became one of his initiates, paying over 100 louis d'or to learn the secrets of animal magnetism. Shortly thereafter, Lafayette, who always considered himself a living bond between two worlds, came to America to preach animal magnetism. Before leaving France, he obtained Mesmer's permission to reveal some of the secrets to George Washington. Full of enthusiasm, he had a special meeting of the American Philosophical Society called as soon as he reached Philadelphia so he could communicate his enthusiasm to the members. It was not until several months later that the receipt of the *Rapport des commissaires chargés par le roi, de l'examen du magnetisme animal* convinced most members of the "absurdity and imposition of Mesmer's doctrine." [11]

This commission had been constituted of four French doctors and five academicians, including Franklin. Although he was subjected to

11. Louis Gottshalk, *Lafayette between the American and the French Revolution*, (*1783-1789*) (Chicago, 1950), 77-80; American Philosophical Society Minutes, Nov. 12, 1784, [III], n.p.; Thomson to Jefferson, March 6, 1785, *The Papers of Charles Thomson*, New-York Historical Society, *Collections for 1878*, 11 (1879), 198.

pressure from Lafayette, who did succeed in winning the support of his grandson, Franklin signed the generally skeptical report with the rest of the members. It concluded that no evidence could be found for the existence of animal magnetism in the claims of Mesmer, Charles Deslon, or any of the practitioners. Although Jefferson was easily convinced that Mesmer was a "maniac," his influence lingered on.[12] In America, it lay behind the incredible sale of Dr. Elisha Perkins' metallic tractors in the next decade and it was at least partly responsible for the still later introduction of Mary Baker Eddy's Christian Science.

Fads came to America from Britain, too, but the traffic in ideas from the former mother country continued to be more complex and more fundamental than that from any other part of Europe. The ideas came in books, journals, newspapers, letters, the minds of returning Americans, and in the persons of Britons who made the trip for the specific purpose of diffusing ideas in the new world. There was no better emissary of the Enlightenment of this sort than the blind philosopher, Dr. Henry Moyes, who made the trip in 1784 for the purpose of making a lecture tour through the American towns.

Educated at Edinburgh, Moyes had developed an excellent technique as a professional lecturer upon scientific subjects. He became associated primarily with Manchester but he was also known to the Lunar Society of Birmingham where his argumentative disposition irked James Watt. Priestley, however, regarded him in the most favorable terms as "superior to most who see." [13] In America, Moyes continued the tradition of the colonial itinerant lecturers with the greatest success. He moved up and down the coast visiting Boston, Providence, Newport, New York, Philadelphia, Baltimore, and Charleston to offer lectures on such topics as philosophical chemistry and the philosophy of natural history. He met so much success that the gossips held he had cleared £3,000 sterling before the end of 1785. In that year, too, he was appointed professor of natural history in the medical faculty of Columbia College although he never lectured from that chair.[14] In Philadelphia, Francis Hopkinson reported that he was "a

12. Jefferson to William Smith, Feb. 19, 1791, Etting Collection, Signers of the Declaration of Independence, 50, Historical Society of Pennsylvania.
13. Hesketh Pearson, *Doctor Darwin* (Toronto, [1930]), 232.
14. Theodore Hornberger, *Scientific Thought in the American Colleges, 1628-1800* (Austin, 1945), 71; David Franks, *New York Directory* (New

great favourite of the Ladies in particular" drawing audiences as large as twelve hundred persons.[15]

One of Moyes' greatest talents was his ability to inspire men to begin projects which continued to be useful long after he had left the scene. This capacity he first demonstrated in Boston in connection with the formation of a Humane Society in that city. Humane societies had spread rapidly over Western Europe after the foundation of the first such organization in Amsterdam in 1767. They were dedicated to the rather limited aim of resuscitating drowned persons or persons whose breathing had been interrupted in some other way—such as suffocation or electrocution by lightning. In the port cities of Europe, their methods had appeared to yield very promising results. There was good reason to believe, from the number of drownings reported in the American seaports, that they would be similarly effective in America.[16]

Henry Moyes arrived in the United States at a time when the Americans were demonstrating great responsiveness to all sorts of improvement projects. At the same time, he was able to build upon an earlier attempt Benjamin Waterhouse had made to form a humane society following the loss by drowning of several young persons in Newport harbor in the summer of 1782. Moyes and Waterhouse worked together upon a plan of organization which they presented to a group of Boston gentlemen in 1785. Moyes publicized the program by causing an article of Dr. Alexander Johnson on the subject to be reprinted in the city. The formation of the Humane Society of Massachusetts followed, presided over by James Bowdoin and

---

York, 1786), 67; Robert Hunter, Jr., *Quebec to Carolina in 1785-1786* (San Marino, 1943), 122; William Darlington, *Memorials of John Bartram and Humphry Marshall* (Philadelphia, 1843), 535; Harry Warfel, *Noah Webster* (New York, 1936), 121; Douglass Robertson (ed.), *An Englishman in America, 1785: Being the Diary of Joseph Hadfield* (Toronto, 1933), 214-15; *Boston Gazette*, Nov. 28, 1785, Dec. 5, 1785; *Pennsylvania Packet*, Jan. 12, 1785, Feb. 22, 1785.

15. Hopkinson to Jefferson, March 20, 1785, Julian P. Boyd (ed.), *The Papers of Thomas Jefferson* (Princeton, 1950–), VIII, 51.

16. *Directions for Recovering Persons, who are supposed to be Dead from Drowning* (Philadelphia, [1788]), 4; John Lathrop, *A Discourse before the Humane Society in Boston* (Boston, 1787), 18; Mark A. de Wolfe Howe, *The Humane Society of the Commonwealth of Massachusetts* (Boston, 1918), 8.

supported by such active physicians as John Warren and Aaron Dexter and by clerical and civic leaders of the city.[17]

The principal function assumed by the society was the publication of methods for the treatment of drowned persons. They cautioned against rolling the victim on a barrel, urging that he be placed in a warmly covered bed with head elevated. Tobacco smoke was to be blown "up the fundament" repeatedly, the nostrils were to be tickled with a feather, and the arms and legs agitated. If the patient then showed any signs of life, he was to be given a spoonful of a warm liquid followed by a draft of liquor.[18] The society also published a report by Benjamin Franklin of a case in which a drowned person had been revived by another person's blowing air in and out of the patient's mouth. Another approach to the problem of loss of life at sea was attempted by the society when it established huts equipped with needed stores along the Massachusetts coastline where shipwrecks were likely. As all learned societies did, this one, too, resolved to hear regular addresses "upon some Medical Subject, connected with the principal objects of the Society." [19]

A similar humane society had been established in Philadelphia in 1780 without benefit of the promotion of Dr. Henry Moyes although he may possibly have had something to do with its revival in 1787. Thomas Bond had been elected president of the Philadelphia Humane Society in 1783 and other prominent physicians had served with him but it was not very active until 1787.[20] At that time, the society published in the newspapers and in a separate pamphlet directions for saving persons whose breathing had been interrupted. Its method was not the same as that of the Boston group. It followed the advice of the London Humane Society in urging that air be forced into the lungs by means of a bellows and forced out by an assistant who would press the patient's belly upwards. Other steps recommended by the London society, including the use of tobacco, electricity, and the

17. Howe, *Humane Society*, 5; Lathrop, *Discourse,* 34; Benjamin Waterhouse, *The Botanist* (Boston, 1811), 233; *Boston Magazine*, 2 (1785), 405-9.
18. *The Institution of the Humane Society of the Commonwealth of Massachusetts* (Boston, 1788), 8-11; *Boston Gazette*, Sept. 8, 1788.
19. Lathrop, *Discourse,* 7; *Boston Gazette*, June 11, 1787; *Humane Society of Massachusetts*, Jan. 8, 1788, [Broadside].
20. *Pennsylvania Journal,* Sept. 6, 1783.

agitation of arms and legs, it neglected. The Philadelphia society put its full trust in a form of artificial respiration which was very dangerous because it might burst the lungs but which might actually revive a drowned person. Years later, it went on to make lifesaving awards and to offer medals for essays on the resuscitation of people whose respiration had stopped.[21] Baltimore established a humane society in 1790 and New York in 1794.[22]

In Philadelphia, Henry Moyes left a different sort of institution as a monument to his own success in stimulating other men. The Philadelphia Dispensary resulted from a suggestion Moyes made in 1785 which he followed up with extensive support when he returned to the city the following year. The dispensary, also, was copied after European models. Supported by private subscriptions, its purpose was to supply medical care for the poor in a regularized manner. The physicians had always done a certain amount of charity work and in Philadelphia, the Pennsylvania Hospital had been a great resource to the ailing poor although it had been prevented from extending its charity by bad debts and the depreciation of paper following the war. By contrast with the hospital, the dispensary was primarily concerned with outpatient care. It was a form of free, public clinic which was not entirely free or entirely public since patients were required to obtain specific authorization for treatment from one of the subscribers. In all probability, this requirement did not operate as much of a restriction upon those who wanted to use the dispensary.[23]

At any rate, the dispensary proved a success from the moment its doors were opened in April 1786. Managed by a board of prominent citizens and attended by several capable young physicians, it treated 722 patients in the balance of 1786 and 1,647 in the following year. The dispensary even offered free dental care. Supported by the sub-

21. *Ibid.*, July 7, 1787; *Directions for Recovering Persons*, 7-9; James Mease, *Picture of Philadelphia* (Philadelphia, 1811), 240-41.

22. Howe, *Humane Society*, 7n; *Universal Asylum*, 5 (1790), 202.

23. *Universal Asylum*, 6 (1791), 118; Whitfield J. Bell, Jr., Science and Humanity in Philadelphia, 1775-1790 (Doctoral Dissertation, University of Pennsylvania, 1947), 216; *Pennsylvania Journal*, Feb. 8, 1786; *Plan of the Philadelphia Dispensary for the Medical Relief of the Poor* (Philadelphia, 1786); James Mease, *Surgical Works of the Late John Jones* (Philadelphia, 1795), 27-29.

scribers' annual gifts of a guinea apiece, it soon became a permanent and valuable institution.[24] Its success led other American cities to establish similar dispensaries, New York being the first to follow suit in 1790.[25]

The widespread medical activity that was evident in all sections of the country, differed in nature from city to city. In Boston, in the midst of many achievements, the outstanding accomplishment was the establishment of a medical school. When medical schools had been founded at Philadelphia and New York during the colonial period, Boston had not been able to get further than aspirations. The chair of anatomy and surgery at Harvard toward which Ezekiel Hersey had donated £1,000 in 1770 could not be instituted before the war.[26] After the war, when the New York and Philadelphia schools were both experiencing difficulties, Harvard at length emerged with a medical school. An attempt to revive the King's College medical school in 1784 failed. Year after year, a full medical faculty was announced but little or no teaching was done and no physicians were graduated for many more years. Philadelphia's troubles were largely political, following from the suppression of the College of Philadelphia and the establishment of the University of the State of Pennsylvania. The university's medical school functioned satisfactorily and with the restoration of the college in 1789, two medical schools served the city with considerable success. It was clear by then that Philadelphia was the strongest medical training center in the country although it was not until 1791 that the two schools were combined.[27]

The establishment of the Harvard Medical School was closely related to the Revolution itself. The nationalism of the Revolution laid a foundation in the resentment it aroused toward American dependence upon Europe for medical education. Benjamin Waterhouse expressed this very well before his own return from Leyden during the

24. College of Physicians of Philadelphia, *Transactions*, 1 (1793), 1-42; *Columbian Magazine*, 1 (1786-87), 229-35; *Pennsylvania Packet*, Jan. 1, 1788.
25. Mease, *Picture of Philadelphia*, 236; John Bard, *A Letter from Dr. John Bard...to the Author of Thoughts on the Dispensary* (New York, 1791), 2.
26. James Thacher, *American Medical Biography* (Boston, 1828), I, 31.
27. *Pennsylvania Packet*, Oct. 6, 1788, Oct. 31, 1788; Francis R. Packard, *History of Medicine in the United States* (New York, 1932), I, 370-72; Joseph Carson, *A History of the Medical Department of the University of Pennsylvania* (Philadelphia, 1869), 86-98.

war. He told Benjamin Franklin, "I am in hopes that my countrymen will in time be convinced that it is not so necessary for a man to come to Europe to learn to cure diseases of his next-door neighbor as they imagine—on the contrary, I hope they will recollect that Nature with her open volume stands courting the attention of her sons in that country as well as in Europe, with this difference only that *there* we shall trust to our own eyes, *here* too much to others.—"²⁸

A more immediate stimulus to the establishment of a medical school was the hospital of the Continental Army in Cambridge where John Warren was director—John Warren, dynamic younger brother of the doctor-patriot Joseph Warren, who had fallen on the field at the Battle of Bunker Hill. As director, Warren had unusual opportunities which he probably recognized more clearly because of his army association with such men as John Morgan. His hospital was better organized than those located in the theater of war and it provided an excellent supply of clinical subjects which—lacking a general hospital—Boston had never had before. Making the most of his situation, Warren gave a course of anatomical demonstrations to the physicians of the hospital and of Boston during the winter of 1780-1781. The success of this experiment encouraged him to present the recently organized Boston Medical Society with a full scale plan for the formation of a medical school. He suggested that Dr. Isaac Rand lecture on the theory and practice of medicine, Dr. Samuel Danforth on materia medica and chemistry, and a third person on anatomy and surgery. Rand and Danforth declined, Rand remarking in a tone of resignation, "that Warren is an artful man, and will get to windward of us all." ²⁹ In a sense, he did, for when he was asked to assume the lectures on anatomy and surgery he, of course, accepted.

The lectures he gave the following season under the auspices of the Boston Medical Society were of a different character. For one thing, they were not kept secret as the first series had been, in deference to popular opposition to dissection. Then, Warren collected fees

28. Waterhouse to Franklin, Dec. 10, 1780, Franklin Papers, XX, 127, American Philosophical Society; Thacher, *American Medical Biography*, I, 26.

29. Ephraim Eliot, "Account of the Physicians of Boston," Massachusetts Historical Society, *Proceedings*, 7 (1863-64), 181; Copy of Minutes of Massachusetts Medical Society, Nov. 30, 1781, Warren Papers, Massachusetts Historical Society; Thomas F. Harrington, *The Harvard Medical School* (New York, 1905), I, 79-80.

from the students who attended. Most important, this course was attended by President Willard and members of the Harvard Corporation. Warren had the Hersey bequest and a Harvard establishment in mind from the beginning. When the senior class at Harvard attended a third series of Warren's lectures, it became clear that Harvard was becoming interested in a medical school.[30]

It was in November 1782 that the Harvard Corporation took the first steps in the founding of a medical school. It began by appointing Warren professor of anatomy and surgery. A month later Benjamin Waterhouse was elected to fill the chair of the theory and practice of physic and in May 1783, Aaron Dexter was chosen professor of chemistry and materia medica.[31] In establishing this pattern of professorships and in the later organization of the school, the corporation investigated practices both at Philadelphia and at Edinburgh. Warren himself collected much of this information. Even by the time of its first graduation in 1788, the school had a growing reputation throughout New England and as far away as Canada.[32]

Over the question of licensing medical practitioners in Massachusetts, the new school almost immediately ran into conflict with another new institution, the Massachusetts Medical Society. This organization owed much to the Revolution for its role in associating physicians and in showing them the need for the regulation of the profession. The society's beginning lay within the Boston Medical Society, a club founded in May 1780 for the purpose of regulating local fees. The Boston Medical Society continued to meet after its members had taken the lead in a movement which resulted in the incorporation of the Massachusetts Medical Society on November 1, 1781. The charter of the state society paralleled that of the American Academy of Arts and Sciences in many respects.[33] John Warren saw the primary purpose of the society as the promotion of "medical and surgical knowledge!" but the legislature seems to have been at least equally impressed with the service it might render in regulating the

30. Harrington, *Harvard Medical School*, 80.
31. Samuel E. Morison, *Three Centuries of Harvard* (Cambridge, 1936), 169.
32. Peter de Sales La Terriere, *A Dissertation on the Puerperal Fever* (Boston, 1789), 5; Edward Warren, *The Life of John Warren* (Boston, 1874), 247.
33. *Boston Gazette*, Jan. 5, 1784; Thacher, *American Medical Biography*, I, 27; Walter L. Burrage, *A History of the Massachusetts Medical Society* ([n.p.], 1923), 19.

profession.[34] The society was specifically granted the power "to examine all Candidates for the Practice of Physic and Surgery." [35]

Against this grant, the Harvard Medical School was unable to establish the contention that its degrees ought to give recipients the right to practice medicine without any further examination. The censors of the Medical Society failed the first graduates of the Medical School they examined. John Warren then very cleverly gave the candidates a public examination in which they demonstrated themselves so well qualified that the censors were forced to pass them in a very brief re-examination.[36] After that, there were no such open clashes but Harvard degrees did not automatically convey the right to practice until 1803.[37]

The Medical Society was anxious, too, to fulfill its intention of improving medical knowledge. Toward this end it was happy to respond to letters from the Royal Academy of Medicine at Paris in 1782 and the Royal Academy of Surgery at Paris in 1783. In 1786, it worked out a much broader plan of correspondence with medical societies in London, Edinburgh, Stockholm, Copenhagen, Lyons, St. Petersburg, Vienna, Leipsig, Leyden, and Philadelphia. That program was never fully implemented but a correspondence did arise with the College of Physicians of Philadelphia, the New Jersey Medical Society, and the New York Medical Society.[38] Within Massachusetts, the attempt to stimulate activity led to the formation of district societies in some of the counties which occasionally fed significant material to the Massachusetts Medical Society.[39] Following the pattern of all learned societies, a library and a cabinet were soon collected, but the most important objective was the stimulation of the writing of medical papers with a view to ultimate publication.

The original plan of holding three meetings annually devoted to scientific questions was soon abandoned and scientific matters were

34. Quoted in Burrage, *Massachusetts Medical Society*, 38.
35. Quoted in, *ibid.*, 65.
36. Eliot, "Account of the Physicians of Boston," Massachusetts Historical Society, *Proceedings*, 7 (1863-64), 183.
37. Oscar and Mary Flug Handlin, *Commonwealth* (New York, 1947), 138.
38. *Minutes of the New Jersey Medical Society*, 56, 61; Burrage, *Massachusetts Medical Society*, 47, 48, 49.
39. *Ibid.*, 323-26; Benjamin Waterhouse, *The Rise, Progress, and Present State of Medicine* (Boston, 1792), vi; *Massachusetts Magazine*, 1 (1789), 81-84.

discussed at any of the three regular meetings. By 1786, five medical papers had been received, and by 1788 there were enough on hand to plan their publication.[40] Except for Dr. Edward A. Holyoke's record of weather and disease, the papers were case reports: operations, dissections, and attempts at differing medical treatment. Even the twelve papers collected represented a valuable pooling of experience. Papers from the otherwise active Harvard medical faculty were conspicuously missing. When it came time to publish the volume, it was decided to fill it out with extracts from European medical writers and from the publications of "the celebrated Dr. Benjamin Rush." [41]

Connecticut was full of efforts to organize the profession. The most comprehensive of these enterprises was launched by a group of physicians from Connecticut, Massachusetts, and New York who met together in 1779 to form the Medical Society of Sharon. The announced aims of this organization were to unite its members, to add life to the healing art, to "suppress quackery," and to encourage medical knowledge.[42] In pursuing these objectives, the society was driven by a strong measure of patriotism. It described itself in a letter to Benjamin Franklin seeking to open a correspondence with the Royal Medical Society of France as "the First Medical Society in the Thirteen United States of America." [43] Its president, Dr. James Potter, asserted that in reply to boasts of ancient and modern European medical accomplishments, "you may now confront them with this Medical Society." [44] Behind all of this fire, there was not enough substance to make it a lasting organization.

The Medical Society of New Haven County, founded in 1784, became more effective. Formed on a less exclusive basis than the Massachusetts Medical Society, it embraced all reputable physicians with one year of practice. It planned to meet quarterly but its members agreed to work together, assisting each other without reserve

40. Burrage, *Massachusetts Medical Society*, 350.
41. *Medical Papers Communicated to the Massachusetts Medical Society*, I (1790), 115, *passim*.
42. Connecticut Medical Society, *Proceedings of the President and Fellows ... at their Annual Convention in Middleton, October, 1792* (Hartford, 1884), v.
43. Dr. James Potter to Franklin, May 10, 1780, Franklin Papers, LIV, 67, American Philosophical Society.
44. Potter, *Oration* (1781), 14.

whenever counsel was needed. From the first, the society encouraged doctors to submit communications "on the air, seasons and climate, with such discoveries as they may make in physic, surgery, botany or chemistry." [45] Nothing but the usual case reports upon the experiences of practitioners came in, but they were received in surprising numbers. By 1788, the society was able to publish twenty-five of them in a collection entitled *Cases and Observations*—the first collection of medical papers published in the United States. They were all brief and factual. All but one were written by resident members of the society.

Both the New Haven and the Sharon societies were interested in the regulation of the medical profession as were the other local groups formed in different parts of the state. With this end in view, an effort was made in 1786 to obtain a legislative charter for a state medical society which would recognize the local societies as subordinate elements, related in a kind of federal system. The state society would help to advance knowledge and put teeth in attempts to regulate the profession. Although the act of incorporation did not pass, the New Haven society simply assumed the power of examining and licensing physicians in that area, its private approval carrying considerable weight. [46]

Despite the failure to secure a state-chartered society, the New Haven physicians retained their faith in the value of medical societies. This they expressed, along with their concept of the role played by the Revolution in stimulating medical activity, in the preface to their *Cases and Observations*:

The late war brought many ingenious and learned Physicians together from all parts of the continent, and the army formed them into a temporary society, whose unreserved communications have contributed to the improvement of medical knowledge, and the establishment of a new and important aera in the healing art. By this means the faculty have become more sensible of the importance of uniting their endeavors, and several Medical societies have been formed in different parts of the United States. Should similar institutions become general, and permanent, there is

45. Medical Society of New Haven County, *Cases and Observations* (New Haven, 1788), iii.
46. *American Magazine*, 1 (1788), 730-35; *New Haven Gazette and Connecticut Magazine*, 1 (1786), 106; George C. Groce, Jr., "Benjamin Gale," *New England Quarterly*, 10 (1937), 710.

reason to hope that Medical literature will soon be in as flourishing a state in this country, as in any part of Europe.[47]

Almost everywhere, evidence of the impulse to organize the medical profession could be found. In New York, a colonial medical society maintained some life even during the occupation; at least one paper was delivered before it in 1782. The plan of the Massachusetts Medical Society was consulted when the New York group was reorganized in 1788 with John Bard as president.[48] Baltimore's society, founded in 1788, concentrated its first efforts upon obtaining laws regulating admission to the practice of medicine.[49] The Medical Society of Delaware began a vigorous life in February, 1789.[50] Even the Medical Society of South Carolina, founded in the same year, disproved the Duke de La Rochefoucauld-Liancourt's expectation that the "indolence" of Charleston would render it ineffective. Within a few years, it was able to bring about the establishment of a humane society and a dispensary in Charleston following the lead of the northern cities.[51]

When the Medical Society of New Jersey was revived in 1781, it introduced important new procedures. It began to encourage a clinical approach to medicine by urging members to bring difficult cases before the society for advice while the course of treatment might still be altered. On some occasions, committees were appointed to visit patients who could not be examined in an open meeting. Once an operation was performed in front of the society. The New Jersey society heard papers, too, and some of them were more than case reports. Dr. Jonathan Elmer, for example, gave presidential addresses on the chemical composition of air and the chemical principles of

47. Medical Society of New Haven, *Cases and Observations*, iii-iv.
48. *An Accurate and Complete Description of Sleep in a Discourse Delivered before the Medical Society, September 1782* (New York, 1784); Sidney I. Pomerantz, *New York, An American City, 1783-1803* (New York, 1938), 404.
49. *Maryland Journal*, Baltimore, Dec. 26, 1788.
50. L. P. Bush, *The Delaware State Medical Society* (New York, 1886), 1-2; Harry B. Shafer, *The American Medical Profession, 1783-1858* (New York, 1936), 228.
51. Lee W. Ryan, *French Travelers in the Southeastern United States, 1775-1800* (Bloomington, Ind., 1939), 49n; David Ramsay, *The History of South-Carolina* (Charleston, 1809), II, 105-6; *By-Laws of the Medical Society of South-Carolina, Instituted at Charleston, Dec. 24, 1789* (Charleston, 1826).

bodies. The society discussed the causes of epilepsy, dysentery, and various fevers. Although the New Jersey group did not publish a collection of case reports, some of the most distinguished papers delivered before it appeared as magazine articles.[52]

Medical activity reached its highest level in Philadelphia which had had the first American medical school, the first dispensary, the first humane society, the first general hospital, and the first philosophical society. Its physicians were the best known in the country. As for medical societies, it boasted three by 1789. Two were connected with the dominant medical institutions in the city—the medical schools. The American Medical Society was an organization of students of the University of the State of Pennsylvania, active from 1783 to 1791. A parallel student organization attached to the medical school of the College of Philadelphia, known as the Philadelphia Medical Society, was revived with the re-establishment of the college in 1789. These societies published a few of their papers in the general magazines but did not issue any collection of papers.[53]

Much more important was the College of Physicians of Philadelphia, founded in 1787. The need for such an institution had become more and more clear as the war drew to an end. Benjamin Rush began to think about it even before John Coakley Lettsom wrote him in September 1783 to urge the establishment.[54] Its final fulfillment meant to Rush not only the consummation of a long recognized need, but more specifically, it appeared as one of the beneficent results of the American Revolution. He declared, "The late revolution which has given such a spring to the mind in objects of philosophical and moral enquiry, has at last extended itself to medicine." [55]

The College of Physicians was designed to be more than a medical

52. *Minutes of the New Jersey Medical Society*, 42, 46, 49, 52, 55, 62, 74; *Columbian Magazine*, 2 (1788), 493-97, 677-82.

53. Mease, *Picture of Philadelphia*, 302; Stephen Wickes, *History of Medicine in New Jersey* (Newark, 1879), 53; *Pennsylvania Packet*, Oct. 22, 1788; *The Act of Incorporation and Laws of the Philadelphia Medical Society* (Philadelphia, 1797), 3; *Universal Asylum*, 4 (1790), 305 ff.

54. Samuel P. Griffitts to Rush, Aug. 10, 1783, John C. Lettsom to Rush, Sept. 8, 1783, W. S. W. Ruschenberger, *An Account of the Institution and Progress of the College of Physicians of Philadelphia* (Philadelphia, 1887), 18.

55. Benjamin Rush, "A Discourse Delivered before the College of Physicians of Philadelphia, Feb. 6th, 1787," College of Physicians of Philadelphia, *Transactions*, 1 (1793), xix.

society. As such it would, of course, collect and publish medical papers. It would record meteorological observations, investigate American diseases, and make use of records such as those of the Philadelphia Dispensary on epidemics. Its function as a college was a little more subtle, for in that role it must acquire the dignity and prestige that adhered to the ancient colleges of physicians at London and Edinburgh. It would serve not only the profession but the nation as the American Philosophical Society aspired to serve in a wider sphere. Mere fellowship would convey distinction and, indeed, election as one of the twelve senior fellows of the college was an honor from the outset. Thus established, the college would be in a position to assume a quasi-public role—to give sanction to an American dispensatory and to guide the government in medical matters.[56]

Although it did not attain its charter until 1789, the college proceeded immediately to collect medical papers, to press for temperance legislation, and to solicit the grant of a lot for a botanical garden from the Pennsylvania Assembly.[57] Within a few years, it was able to publish its first volume of *Transactions*. Some of the broader, nationalist-tinged objectives announced in its charter proved more difficult to accomplish. According to the charter, the purpose of the college was "to advance the science of medicine, and thereby to lessen human misery, by investigating the diseases and remedies which are peculiar to this country; by observing the effects of different seasons, climates, and situations upon the human body; by recording the changes which are produced in diseases, by the progress of agriculture, arts, population and manners; by searching for medicine in the American woods, waters, and in the bowels of the earth; by enlarging the avenues to knowledge from the discoveries and publications of foreign countries; and by cultivating order and uniformity in the practice of physic." [58]

Even with medical societies soliciting papers in all parts of the country, there was still opportunity for the general philosophical societies to serve the profession. Some of the physicians presented

56. The division of functions is that of Rush, *ibid.*, xx-xxi; Bell, Science and Humanity in Philadelphia, 205.
57. *Minutes of the General Assembly of Pennsylvania, Twelfth*, 55, 141.
58. *The Charter, Constitution and Bye Laws of the College of Physicians of Philadelphia* (Philadelphia, 1790), 3.

their most significant papers to those bodies and saw them published in the society journals. Case reports in the *Memoirs* of the American Academy were of the same character as those published by the medical societies, often by men who were disaffected from the Massachusetts Medical Society. Thus papers by Warren and Dexter of the Harvard faculty were published in the *Memoirs*.[59] To this degree, the philosophical society journals provided a competing medium but at the same time they presented two types of papers not so well cultivated by the medical societies: the more general essay related to medicine and the analytical medical study. Still another type of essay contributed very little. John Morgan's papers suggesting prenatal influence as the cause of a mottled mulatto and presenting the possibility of spontaneous generation as the origin of a worm in a horse's eye were not even abreast of the best current knowledge.[60]

One of the finest studies of this whole period was Benjamin Rush's investigation of Dr. Hugh Martin's cancer remedy, published in the *Transactions* of the American Philosophical Society and later in a separate collection of his own papers issued in 1789 under the title *Medical Inquiries and Observations*. As an apprentice under Rush and a military surgeon during the war, Martin had had a better medical education than most of the cancer doctors. In 1782, he published a pamphlet entitled, *A Narrative of a Sovereign Specific for the Cure of Cancers*. He sent a copy to the American Philosophical Society, sought a monopoly on use of the powder from the legislature, and asked George Washington for a testimonial.[61] Washington's reply included a statement that medical cures ought not be kept secret. He refused his approval of the remedy, remarking "Certificates or recommendations from those who have been restored to health by the efficacy of your medicine would be vastly more pertinent." [62]

After Martin died in 1784, Rush was able to obtain some of the powder which he subjected to a model analysis, chemical and clinical. The chemical analysis revealed that the powder was neither the simple

59. American Academy, *Memoirs*, 1 (1785), 551-55; 2 (1793), 191-200.
60. American Philosophical Society, *Transactions*, 2 (1786), 383-91, 392-95.
61. Hugh Martin, *A Narrative of a Discovery of a Sovereign Specific for the Cure of Cancers* (Philadelphia, 1782).
62. Washington to Martin, Dec. 13, 1783, John C. Fitzpatrick (ed.), *The Writings of George Washington* (Washington, 1931-44), XXVII, 272.

vegetable compound Martin had claimed it to be, nor was it of Indian origin. Rush found that the only active ingredient it contained was arsenic—then much used in cancer treatment. The vegetable components could have served no function except possibly to dilute the arsenic. An examination of the clinical record of Martin's cures convinced Rush that it had worked in several cases of cancerous ulcers but that it always failed in cancers of the lymphatic system.[63] This was America's most prolific medical writer at his best with no flights of fancy and no metaphysics.

The question of life expectancy attracted the greatest attention from the American Academy. The Reverend Edward Wigglesworth had advice in this matter from the Reverend Richard Price while making a study of the basis for an annuity system for Congregational clergymen. Dr. Edward A. Holyoke, who also published an article in the first volume of *Memoirs,* showed more concern for the relationship of death to the diseases causing it.[64] In 1785, the American Academy began an attempt to collect "regular and uniform Bills of Mortality from the several towns within this Commonwealth." This was expected not only to furnish data for compiling tables of life expectancy but to yield "a natural history of the diseases incident to our climate." [65] By 1789 enough information had been compiled to permit Dr. Holyoke to develop a table of longevity based upon the returns of sixty-two Massachusetts and New Hampshire towns. His results supported the frequent assertion that the American environment was healthful. Indeed, Holyoke showed that the life expectancy at age sixty-five was 12.45 years—only one year less than the present life expectancy in the United States for that age! [66]

The generally expanded publication facilities of the postwar period served the physicians in many ways. While the newspapers were not so useful as they had previously been, they did occasionally publish

63. Martin to Rush, July 1, 1780, Rush Papers, Unbound, Library Co. of Philadelphia; American Philosophical Society, *Transactions,* 2 (1786), 212-17.
64. Edward Wigglesworth, Meteorological Journal, not paged, American Academy of Arts and Sciences; American Academy, *Memoirs,* 1 (1785), 546-50, 565.
65. *American Academy of Arts and Sciences,* Nov. 10, 1785, [Broadside].
66. American Academy, *Memoirs,* 2 (1793), 134; Harry Hansen (ed.), *The World Almanac and Book of Facts for 1953* (New York, 1953), 442.

medical articles.[67] The magazines, often in trouble for lack of material, were very anxious to use medical essays. Some of these were specifically written by American doctors for such publications; others were lifted from medical journals, philosophical society transactions, or collected works of leading physicians.[68] The number of separate medical titles issued by the presses increased, too. They ranged in value all the way from Jeremiah Minter's *Notable Miracle*, which described a child born after nine years in the womb, to Rush's *Medical Inquiries and Observations*.[69]

The separately published essays tended to be more reasoned and sometimes more speculative than the factual case reports of the medical societies. Often they proposed a special mode of treatment for given diseases, like Richard Bayley's pamphlet on the *Angina Trachealis*, Samuel Bard's method of treating *Uterine Hemorrhagies*, or the publications on smallpox inoculation.[70] Occasionally, as in Rush's *Influence of Physical Causes upon the Moral Faculty*, purely speculative thoughts yielded surprising insight. In that printed oration Rush began thinking along a line closely related to his ultimate development of a psychiatric approach to ailments that were not obviously physical in nature.

This volume of medical publications meant that the American reliance upon British media was less than it had been in the colonial period. American articles still appeared in British medical journals and separate titles were published in Britain, but now they were most often reprints which had been issued first in the United States.[71] American dependence upon European medicine had not declined in any funda-

67. *New York Packet*, Jan. 5, 1786; *Pennsylvania Packet*, Jan. 30, 1788.

68. See *American Museum*, 2 (1787), 459-66; *Columbian Magazine*, 1 (1786-87), 182-87; *American Magazine*, 1 (1787-88), 487-88.

69. [Jeremiah Minter], *A Notable Miracle* (3d ed., Philadelphia, 1806). The first edition was published somewhere in Virginia in 1788.

70. Richard Bayley, *Cases of the Angina Trachealis* (New York, 1781); Samuel Bard, *An Attempt to Explain and Justify the Use of Cold in Uterine Hemorrhagies* (New York, 1788); Benjamin Rush, *The New Method of Inoculating* (Philadelphia, 1786); John Morgan, *Recommendation of Inoculation According to Baron Dimsdale's Method* (Boston, 1776).

71. Charles Dilly to Rush, Feb. 26, 1787, July 5, 1786, Rush Papers, XXXI, 22, 24, Library Company of Philadelphia; Medical Society of London, *Memoirs*, 1 (1787), 65-76; *London Medical Journal*, 7 (1786), 287-321; 8 (1787), 89-104.

mental sense. Now, however, the country was much better equipped for training its own physicians, for organizing efforts to improve medicine in this country, and for pooling and publishing medical experiences and observations. Much of the postwar development proved to be of mushroom growth which died out after a time, but enough was left to provide new American roots for the development of the medical profession. When the new Constitution went into effect, new facilities were still being provided.

Men were generally pleased with the evidences of advance in medicine in the United States despite occasional indications that all was not light and glory.[72] The serious problem of obtaining cadavers to be dissected in the course of medical instruction was never solved. Resentment against stealing paupers' bodies welled up in the famous doctors' riot in New York City in 1788 which for a time seemed to threaten even the lives of city physicians. It resulted in legislation granting bodies of criminals to the medical profession, but more directly it impelled doctors to look farther afield for their corpses—to Long Island towns, for instance.[73] There were not enough executed criminals to satisfy the need. Then, the matter of smallpox inoculation seemed to show retrogression rather than advance. With some reason for their fears, to be sure, the legislatures of all but five of the states came to prohibit the practice.[74] Enlightened physicians remained convinced, however, that such black spots could all be eliminated in time.

A few men overlooked specific shortcomings to set their eyes upon a still more distant horizon that was intimately related to their rising nationalism. They looked forward to the development of a distinctly *American* system of medicine. The almost innumerable studies of climate were predicated upon the assumption—then common in Europe—that the diseases of any region were closely related to its weather.[75] Beyond that, the Reverend Nicholas Collin declared, "All

72. *New Haven Gazette and Connecticut Magazine*, 3 (1788), no. 10, not paged.
73. Thacher, *American Medical Biography*, I, 52; Jules C. Ladenheim, "The Doctors' Mob of 1788," *Journal of the History of Medicine*, 5 (1950), 42-43.
74. William Currie, *An Historical Account of the Climates and Diseases of the United States of America* (Philadelphia, 1792), 3n.
75. *Ibid., passim*; American Academy, *Memoirs*, 2 (1793), 65-92; *New Haven Gazette and Connecticut Magazine*, 3 (1788), no. 20, not paged; Mease, *Surgical Works of Jones*, 22; *American Magazine*, 1 (1787-88), 765-67.

countries have some peculiar diseases, arising from the climate, manner of living, occupations, predominant passions, and other causes." [76] It was the hope of escaping "the rules of practice and forms of prescription made in other countries" that led Dr. William Currie in 1788 to undertake an extensive survey of climate and disease throughout the United States.[77] In the midst of such enterprises, Benjamin Rush strove earnestly to erect *the* American system of medicine. He declared, "To the activity induced in my faculties by the evolution of my republican principles by the part I took in the American Revolution, I ascribe in a great measure the disorganization of my old principles of medicine." [78] Rush carried his Americanism so far in constructing new principles of medicine that one of his good friends complained of his blindness in confining his prescriptions to the " 'domestic remedies of the United States.' " [79] Patriotism, however, not infrequently obscured reason.

76. American Philosophical Society, *Transactions,* 3 (1793), iii.

77. *Universal Asylum,* 8 (1792), 121.

78. George W. Corner (ed.), *The Autobiography of Benjamin Rush* (Princeton, 1948), 39.

79. Benjamin Vaughan to Rush, April 29, 1799, Rush Papers, XVIII, 43.

Chapter Fourteen

NATURAL HISTORY IN THE REPUBLIC

EFFORTS TO FULFILL the vision of a glorious American destiny led directly to demands that natural history be studied with vigor, for it seemed to provide the key to the discovery and utilization of the natural resources of the country. Faith in the existence of abounding resources had been from the beginning an important factor in creating the vision. Thomas Bond saw lying before the Americans "a vast Country, yet unexplored, and which is supposed to abound with the greatest and richest Magazines of natural Productions, and only require their Doors to be unfolded to make it the Envy of the World, a blessing to its Inhabitants, and a common Conveniency to the Universe." [1]

With the return of peace, the powerful voice of John Coakley Lettsom was added to those of many Americans who called for work in natural history. He saluted Benjamin Rush with the statement, "The season of peace is the harvest of science," and went on to demand, "Set your men of science upon studying your own country, its nature and improvable productions." [2] To Thomas Parke, he complained, "I lament much that no active and generous steps have been taken to develop the various products of your soil. You ought to institute a Natural history; and a mineralogical Society to bring to light the treasures you possess. One should ascertain the products on

1. Thomas Bond, *Anniversary Oration Delivered May 21st* (Philadelphia, 1782), 8.
2. John C. Lettsom to Rush, Sept. 7, 1785, Rush Papers, XXVIII, 7.

the surface of the earth; the other should scrutinize subtler Riches. You possess much but you do not know your wealth. . . . perhaps," he added, "the English are not such enemies to you but there might be Individuals to aid your endeavors." [3]

The promotion of natural history did not take the form of mineralogical or natural history societies although the early chemical societies, the first of which was founded in 1789, were very nearly mineralogical in their objectives. They were much interested in the chemical analysis of American minerals.[4] Insofar as natural history did find support within societies, it was the general philosophical societies that encouraged it. For the American Academy, the Reverend Manasseh Cutler sought contributions with the declaration, "Discoveries and discriptions of the natural productions of the Country merit particular attention, as they may be improved for advancing our internal wealth and resources—an object of the highest importance at this happy period, in which independence and peace is so gloriously established." [5] The number of papers published in the *Memoirs* by men who had not previously written about natural history was evidence that some success followed the efforts of Cutler and the American Academy.

The death or removal of the most distinguished American naturalists speeded the development of new members of the natural history circle. Some of the men who now became prominent were drawn from another background altogether, but others were cut of the same cloth as Bartram the seedsman, Colden the gentleman botanizer, or Garden the medical practitioner.

Seedsmen continued to make important contributions. Humphry Marshall, for example, emerged from the war as a successful seedsman with some of the breadth of interest of his cousin, John Bartram. Indeed, it was Marshall who in 1785 published the first systematic botanical book to appear in the United States, the *Arbustrum Americanum* or the American Grove, an alphabetical catalogue of American forest trees. With the publication of his book, Marshall declared that

3. Lettsom to Parke, Aug. 1788, Gilbert Collection, III, 65, College of Physicians of Philadelphia.
4. *The Medical Repository*, 1 (1800), 68-69; Wyndham Miles, "Early American Chemical Societies," *Chymia*, 3 (1950), 95.
5. Cutler to Stiles, April 2, 1783, Isabel M. Calder (ed.), *Letters and Papers of Ezra Stiles* (New Haven, 1933), 58.

ever "Since the revolution and National Establishment of Peace in our Land," he had been anxious to contribute his "Small mite into the treasury or Collection of Usefull Knowledge." [6] The book was an important compilation from the botanist's viewpoint but, for Marshall, it served another purpose as well. At the end of the treatise, which was in large measure a catalogue of trees available in Marshall's nursery, he inserted a little advertisement, "Boxes of Seeds . . . are made up in the best manner and at a reasonable rate by the Author." European business followed as a direct result of the publication of his catalogue. [7]

When Marshall ran into trouble in his efforts to find a printer, it was Samuel Vaughan who stepped in to save the situation. Vaughan, with his great faith in the American experiment, was determined that this pioneering work of science should be published in America rather than in Europe where printing costs were less. He offered to take fifty or sixty copies for his friends here and in Europe, and, if additional subscriptions could not be obtained, to see it through publication himself. [8] Thus it was published in Philadelphia. The American Philosophical Society subscribed for forty copies and Benjamin Franklin distributed copies to his scientific friends. [9] The rest of the edition sold very slowly. This unenthusiastic public response caused Marshall to abandon his plan for a second catalogue devoted to herbaceous plants. [10]

A catalogue of both trees and herbaceous plants was published by

6. Marshall to Samuel Magaw, Jan. 7, 1786, Archives, American Philosophical Society.

7. Humphry Marshall, *Arbustrum Americanum* (Philadelphia, 1785), 170; Grimwood, Hudson, and Barrit to Marshall, Aug. 1, 1787, Humphry Marshall Correspondence, I, 41, Dreer Collection, Historical Society of Pennsylvania; First Gardener of the Duke of Saxe Weimar to Marshall, April 28, 1788, *ibid.*, I, 51.

8. Vaughan to Marshall, April 30, 1785, William Darlington, *Memorials of John Bartram and Humphry Marshall* (Philadelphia, 1843), 555.

9. American Philosophical Society Minutes, Dec. 16, 1785, [III], not paged; Rochefoucauld to Franklin, Feb. 14, 1787, Franklin Papers, XXXV, 16, American Philosophical Society; Franklin to Le Roy, March 27, 1786, Albert Henry Smyth (ed.), *The Writings of Benjamin Franklin* (New York, 1905-7), IX, 502.

10. Marshall, *Arbustrum*, ix; Joseph Crukshank to Marshall, Feb. 7, 1786, Humphry Marshall Correspondence, I, 24.

another American seedsman—this one in Paris. William Young, Bartram's old bête noire who had been appointed Queen's botanist in the colonial period, issued a pamphlet entitled *Catalogue d'Arbres Arbustes et Plantes Herbacées d'Amerique* in 1783. It appeared at a shrewdly chosen moment, while the French taste for things American was still high and when, for the first time in years, untrammeled commerce seemed possible. This catalogue contained two lists of trees, shrubs, and plants: one which Young could supply and another which he could not. From a scientific standpoint, his lists were not very satisfactory. The new varieties and species named could not well be identified because of the confused system of classification adopted. In fact, they may not have been real at all but merely designed to attract attention and trade.[11]

William Hamilton and George Logan exemplified the continuing interest of gentlemen gardeners in collecting seeds and exchanging them with Europeans but it was only in South Carolina that important writing came from such amateur botanists.[12] Thomas Walter had a plantation on the Santee. From this point, he examined the flowering plants in the vicinity of his home and classified them according to the Linnaean system. Walter had no ties with the naturalists to the north but his friend, John Fraser, took his manuscript back to England and published it there in 1788 as *Flora Caroliniana*. Fraser's herbarium also found its way to England where it was used by Frederick Pursch and later students of American botany. His book, like Marshall's *Arbustrum Americanum*, continued to be relied upon by botanists for years, while Young's little work passed into oblivion.[13]

The political results of the war were in some respects beneficial to natural history studies. The resumption of British ties that followed the return of peace was not accompanied by the casting off of the

11. Samuel N. Rhoads (ed.), *Botanica Neglecta: William Young, Jr.* (Philadelphia, 1916).

12. George Logan to [?], Nov. 27, 1783, Humphry Marshall Correspondence, I, 10; William Hamilton to Marshall, Nov. 22, 1790, *ibid.*, I, 13.

13. Henry Muhlenberg, Manuscript Note in Thomas Walter, *Flora Caroliniana* (London, 1788), viii, Library Company of Philadelphia; Frederick Pursch, *Flora Americae Septentrionalis* (London, 1814), I, xxvii; E. Millicent Sowerby, *Catalogue of the Library of Thomas Jefferson* (Washington. 1952–), I, 490.

new relationships that had been established with continental nations. Both Franklin and Jefferson made sure that a two-way traffic in ideas continued between America and the continent. Franklin served as a kind of agent for American seedsmen, putting them in touch with several Europeans who sought means of importing American seeds.[14] In the opposite direction, it was through Franklin that several important treatises reached America, especially Jan Ingenhousz's studies of the life cycle which Franklin estimated as the greatest discovery made in Europe "for some time past." [15] Jefferson continued the service not only by importing a great moose to confute Buffon but by urging that even so small a curiosity as a strange American bird be made known to the French scientists. He sent his friends in America publications and information of European developments in all fields of science.[16] James Madison, president of William and Mary, remarked, "Our means of information is confined almost entirely to you." Yet, even the general magazines reflected continued continental influences in some of the pieces on natural history that they copied.[17]

The French government itself took a most important step when it sent a botanist, André Michaux, to the United States in 1785. Ostensibly, Michaux's mission was to collect plants, shrubs, and trees—especially trees—that might be useful in France. To this end he established two gardens in which he planted seeds and cuttings found in his trips through the country. The first was located in New Jersey across the river from his New York headquarters and the second, to which he devoted his attention after April 1787, in Charleston. From this point, he explored Florida and the Carolinas and in 1789 he made a trip to the Bahamas. As a competent botanist Michaux was an important help

14. Comte de Barbançon to Franklin, Nov. 24, 1786, Franklin Papers, XXXIV, 172, American Philosophical Society; [Le Veillard] to Franklin, [Circa 1783]; *ibid.*, XLII, 128; Abbé Nolin to Franklin, Sept. 4, 1783, *ibid.*, XXIX, 128.

15. [Charles L.] L'Heritier to Franklin, May 18, 1785, *ibid.*, XXXIII, 110; Franklin to Hopkinson, March 6, 1780, Smyth (ed.), *Writings of Franklin*, VIII, 32.

16. Rev. James Madison to Jefferson, April 10, 1785, Jefferson Papers, Library of Congress; Jefferson to Stiles, July 17, 1785, Julian P. Boyd (ed.), *The Papers of Thomas Jefferson* (Princeton, 1950—), VIII, 298; Jefferson to Page, Aug. 20, 1785, *ibid.*, 419; Jefferson to Madison, Oct. 2, 1785, *ibid.*, 575.

17. *Columbian Magazine*, 1 (1786-87), 326-27.

to American naturalists. It is not clear whether he also served France in political matters during this period as he unquestionably did a few years later under the direction of the French Minister, Genêt.[18]

There was no question of political activity on the part of the Germans or the Italians who continued to make contributions to the development of American natural history. The Emperor Joseph II planned to send a naturalist to the United States and thence around the world to make a collection of plants and animals for the imperial gardens and menagerie, but apparently the plan never materialized.[19] Some of the German naturalists who came to America during the war returned before the coming of peace as did Baron Wagenheim. The Reverend Frederick V. Melsheimer, who remained, did not begin creative work in entomology until much later. Certainly the most active German was Johann David Schöpf, who toured the country in 1783 and 1784, making friends of many of the American naturalists and making observations which he wrote up after his return to Germany. His most important books were *Materia Medica Americana* and *Beyträge zur Mineralogischen Kenntnis*, both published in 1787. The second opened the field of systematic mineralogy in the United States.[20] In 1788 Schöpf published his travel journal for the years 1783 and 1784. After his return to Germany, he maintained a correspondence with Henry Muhlenberg from whom he gained much of the information included in the *Materia Medica*—without any acknowledgment.[21] From Italy Count Luigi Castiglioni made an extended trip through the United States in 1785 and 1786 at which time he met many American botanists. He devoted over two hundred

18. C. S. Sargent, "Portions of the Journal of André Michaux," American Philosophical Society, *Proceedings*, 26 (1889), 4; Reuben G. Thwaites, *Early Western Travels* (Cleveland, 1904), III, 12; W. C. Coker, "The Garden of André Michaux," Elisha Mitchell Scientific Society, *Journal*, 27 (1911), 67.

19. Franklin to [unaddressed], April 22, 1783, Smyth (ed.), *Writings of Franklin*, IX, 35.

20. Johann David Schöpf, *Beyträge zur Mineralogischen Kenntnis des Östlichen Theils von Nordamerica und seiner Gebürge* (Erlangen, 1787); Johann David Schöpf, *Materia Medica Americana Potissimum Regni Vegetabilis* (Erlangen, 1787).

21. Schöpf-Muhlenberg Correspondence, Historical Society of Pennsylvania; Herbert H. Beck, "Henry E. Muhlenberg, Botanist," Lancaster County Historical Society, *Papers*, 32 (1928), 106.

pages of his published *Viaggio* to a discussion of the most useful plants he found in the country.[22]

The most important new element in the natural history circle was the product of domestic rather than foreign developments. A new class of naturalists arose in the medical schools and the colleges—the professors of natural history or botany. In the colonial period natural history had been occasionally taught as a very small part of the course in natural philosophy. For a few years a course in botany had been offered by Dr. Adam Kuhn in the Medical School of the College of Philadelphia—when the season permitted.[23] The dropping of Kuhn's course eliminated nothing of importance, but its deletion underlined the failure of the colonial colleges to become a part of the natural history circle. During the Confederation period foundations were laid which assured that, thereafter, the colleges would play a prominent role in the development of the sciences of nature. The professor began to take his place beside the gentleman gardener and the professional seedsman.

The changing character of the naturalist was strikingly demonstrated in the contrast between William Bartram and Benjamin Smith Barton. It was recognized that the mantle of old John Bartram had fallen upon William rather than upon young John who inherited the garden. William assisted his brother John in the seed business conducted from the ancestral garden and perhaps he collaborated in the preparation of a list of "Trees, Shrubs and Herbacious Plants" printed over John's name in 1790.[24] Yet when Franklin spoke of "Mr. Bartram our celebrated Botanist of Pennsylvania," it was William he had in mind.[25] It was William whom the American Philosophical Society elected to membership in 1786. Finally, when the University of Pennsylvania decided to appoint a professor of botany, it was William to

22. Luigi Castiglioni, "Osservazioni sui vegetabili pui utili degli Stati Uniti," *Viaggio negli Stati Uniti* (Milano, 1790), II, 169-402; Howard R. Marraro, "Count Luigi Castiglioni," *Virginia Magazine of History and Biography*, 58 (1950), 474, 475.

23. *Pennsylvania Gazette*, May 5, 1768, May 11, 1769; Marion E. Brown, "Adam Kuhn: Eighteenth Century Physician and Teacher," *Journal of the History of Medicine*, 5 (1950), 177.

24. John Bartram, *Catalogue* [1790], [Broadside].

25. Franklin to Ingenhousz, April 29, 1785, Smyth (ed.), *Letters of Franklin*, IX, 314.

whom the post was offered. Upon his refusal, it went instead to Benjamin Smith Barton in 1789.[26]

Benjamin Smith Barton was an ambitious young man with a sense of destiny about him. Son of the Reverend Thomas Barton, he was a nephew of David Rittenhouse who had been much helped by the younger Barton's father in his own youth. Rittenhouse, in turn, tried to aid the son in many ways. He took him along as an assistant on the expedition to run the western Pennsylvania boundary in 1785 and on several occasions he used his influence to promote his advancement. More important, perhaps, was Barton's own acknowledgment, "Your example early emplanted in me an ardent *love* of science." [27] Whether or not this was the source of his ambition, young Barton did develop a burning desire to achieve—particularly in natural history which was certainly not Rittenhouse's own field. Rittenhouse indicated that he considered descriptive natural history of distinctly less importance than astronomy or physics by offering one such paper of his own for publication in a general magazine rather than laying it before the American Philosophical Society.[28] Barton, however, chose medicine as a profession and that led him to natural history rather than to the physical sciences.

Benjamin Smith Barton began to plan his route to fame while at Edinburgh where he went to study medicine. He made a good record as a student, winning a medal for his essay on the nature and medical properties of hyoscyamus niger and being elected president of the student medical society.[29] In 1787 he published a little work entitled *Observations on Some Parts of Natural History* in which he discussed the mounds and ancient fortifications found in the west, particularly near the Ohio and Muskingum Rivers. He concluded, on no very sound basis, that they had been built by descendants of Danish immigrants —probably the Toltecs. At the same time, he was credited with first

26. Ernest Earnest, *John and William Bartram* (Philadelphia, 1940), 164.

27. Benjamin Smith Barton, *A Memoir Concerning the Fascinating Faculty which has been Ascribed to the Rattlesnake* (Philadelphia, 1796), iv; William P. C. Barton, *A Biographical Sketch ... of ... Professor Barton*, [n.p., n.d.].

28. Rittenhouse to Jefferson, [Nov. 8, 1787?], *Jefferson Papers*, Massachusetts Historical Society, *Collections*, 7th ser., 1 (1900), 35; *Columbian Magazine*, 1 (1786-87), 284ff; William Barton, *Memoirs of the Life of David Rittenhouse* (Philadelphia, 1813), 606.

29. *Pennsylvania Journal*, May 16, 1787; *Pennsylvania Packet*, April 2, 1788.

suggesting that they might have been religious monuments rather than fortifications.[30] He became so enthusiastic that he thought of writing a *"History of America"* and opened a correspondence with men in Spain and Portugal for that purpose.[31]

A more far reaching opportunity occurred to young Barton in connection with the manuscript version of William Bartram's *Travels*, which he had seen in Philadelphia before he left for his European studies. When Barton mentioned the manuscript to Dr. Lettsom and other prominent Englishmen, there was an enthusiastic demand that it be published. Somehow, Enoch Story, Jr., who was trying to arrange for printing the *Travels* by subscription in Philadelphia, heard of Barton's activities and accused him of planning to publish the journal for his own profit.[32] He denied the accusation and offered to print the manuscript at his own expense, allowing Bartram half the profits. Barton wanted to make additions of his own, tying himself as a tail to the Bartram kite. In words reminiscent of Peter Collinson, he tried to convince William Bartram that his *Travels* ought to be made available to the Europeans. *"Natural History* and *Botany,"* he declared, "are the fashionable and favourite studies of the polite as well as the learned part of Europe: whatever regards the *Natural History* of America is particularly sought after." [33]

Bartram was not to be convinced. Although the Story subscription failed, he declined Barton's invitation with the result that his *Travels* were not published until 1791 when they appeared in Philadelphia. Additions in Barton's careless prose would have added nothing to the literary charm of the *Travels* which carried it on to edition after edition in Europe. Yet, Barton's reasoned approach to natural history and his interest in the scientific aspects of the journal might have sharpened its value as a scientific document. Benjamin Smith Barton

30. Benjamin Smith Barton, *Observations on Some Parts of Natural History* (London, [1787]), 65; Manasseh Cutler to Jeremy Belknap, March 6, 1789, William P. and Julia P. Cutler, *The Life, Journals and Correspondence of Rev. Manasseh Cutler* (Cincinnati, 1888), II, 249.
31. Barton to William Bartram, Dec. 13, 1788, Bartram Papers, Historical Society of Pennsylvania, I, 5.
32. Francis Harper, "Proposals for Publishing Bartram's *Travels,*" American Philosophical Society, *Yearbook for 1945* (Philadelphia, 1946), 78.
33. Barton to William Bartram, Aug. 26, 1787, Feb. 19, 1788, Bartram Papers, I, 3, 4.

and William Bartram were worlds apart in their approach to natural history.

When the younger man accepted the proffered post at the University of Pennsylvania, he took his place among a growing number of naturalists associated with the colleges. Barton had none of Bartram's genius but in its place he had academic learning and a high degree of organization and drive. Wandering naturalists continued to appear, but leadership in the sciences of nature was passing to men with academic roots. Most of them were physicians whose teaching was done in one of the medical schools.

Benjamin Waterhouse, for example, was one of the original pillars of the Harvard Medical School. During his student years abroad, Waterhouse had been introduced to the leading English promoters of natural history and to a taste for the study of nature. He had particularly benefited from the friendship of John Fothergill and John Coakley Lettsom—both of them Quakers, like Waterhouse. He brought back with him in 1782 ties that were unusual for a New Englander and interests that were to prove of the greatest importance in redirecting attention to natural history in his section of the country.[34] In later years Dr. Lettsom, who directed toward Philadelphia so much of his effort to awaken science, always reserved a special concern for Waterhouse and his attempt to promote natural history in Boston.

After accepting the chair of the theory and practice of physic at Harvard, Waterhouse was appointed professor of natural history at the College of Rhode Island in 1784. In fulfillment of this appointment he delivered two short series of twelve public lectures on natural history just before the college commencements of 1786 and 1787.[35] In 1788, the Harvard Corporation voted to have Waterhouse "deliver annually a course of Lectures upon Natural History." Thereafter, it was in Cambridge that he offered his course on natural history to students who obtained their parents' consent and paid the one guinea

34. Josiah Charles Trent, "The London Years of Benjamin Waterhouse," *Journal of the History of Medicine*, 1 (1946), 25-40.
35. Benjamin Waterhouse, *The Botanist* (Boston, 1811), vi; J. Walter Wilson, "The First Natural History Lectures at Brown University," *Annals of Medical History*, 4 (1942), 390-98.

charge which went to the instructor. The course proved successful and widely influential in developing a taste for natural history.[36]

In these years, Waterhouse contributed little of a creative nature to natural history, his only publication being a synopsis of his lectures, financed by a loan from the Harvard Corporation. In the effort to disseminate knowledge of natural history, he exerted himself strenuously, but he encountered a serious obstacle in the objection of parents to his attempts to teach "the youth of both sexes" the Linnaean system of sexual classification. To overcome this difficulty, Waterhouse made an attempt to "drop the Linnaean metaphor of generation; and substitute that of *nutrition*." It did not work. Sir Joseph Banks protested that it was imperative for scientists to agree upon the same technical language.[37]

Although Waterhouse often felt himself alone, there were other forces and other figures encouraging natural history in Massachusetts at the same time. From France came the offer to furnish seeds and plants from the royal garden to stock a botanical garden at Cambridge if the legislature would only establish it. For that purpose, St. John de Crèvecoeur, French consul general in New York, had in hand upwards of two hundred seeds all marked according to the Linnaean system. The college president and overseers were responsive to Crèvecoeur's wish to see "the First Botanical Garden in America" established in Cambridge. They petitioned the legislature to authorize the garden but there the matter died.[38] The college collection of natural history curiosities continued to grow but the garden remained an aspiration.[39]

Advances in natural history were more directly associated with the American Academy than with Harvard, although the two institutions remained closely interlocked. The academy even deposited its own natural history curiosities in the college museum.[40] In the act of incorporation, the first named purposes of the academy had been "to

36. Waterhouse, *The Botanist*, vi; I. Bernard Cohen, *Some Early Tools of American Science* (Cambridge, 1950), 100.
37. Waterhouse, *The Botanist*, 190.
38. Harvard Corporation Records, III, 188.
39. Cohen, *Some Early Tools*, 97-98.
40. Harvard Corporation Records, III, 199-200.

promote and encourage the knowledge of the antiquities of *America*, and of the natural history of the country, and to determine the uses to which the various natural productions of the country may be applied." [41] In his presidential address, James Bowdoin further defined the academy's interests by referring to the need for "an extensive, and well-digested body of *American* natural history. For a purpose so beneficial in itself, and so honorary to our country, it is hoped," he announced, "gentlemen [of ingenuity and observation] will favour the Academy with their descriptions and collections; and also with the result of their researches, relative to the same subject." [42]

The response was encouraging. Clergymen, doctors, gentlemen, and generals submitted accounts of curiosities of nature. They found several mineral specimens of interest: an oilstone, pigments, and sulphur-bearing rock. Under academy direction a chemical analysis was made of the minerals at Gay Head, Martha's Vineyard. Fossil finds and curious earth formations were described. Two papers engaged in the old argument about the submersion of swallows, with second and third hand accounts of swallows that had allegedly revived after being buried in mud. The mineral springs at Saratoga were described. Most of the botanical papers received were directly related to agriculture. Two were more ambitious: Winthrop Sargent sent in a simple list of English names of trees found northwest of the Ohio River and the Reverend Manasseh Cutler wrote a botanical study of unusual importance. [43]

The significance of Cutler's lengthy treatise lay in its novelty. It was the first serious effort to study the botany of New England in a systematic manner since William Douglass—and Douglass' work had been neither completed nor published. As Cutler put it, "*Canada* and the southern states, beside the attention paid to their productions by some of their own inhabitants, have been visited by eminent botanists from *Europe*. But a great part of that extensive tract of country, which lies between them, including several degrees of latitude, and exceedingly diversified in its surface and soil, seems still to remain unexplored." The causes of such a failure in a region that had long excelled

41. American Academy, *Memoirs*, 1 (1785), vii.
42. *Ibid.*, 13.
43. *Ibid.*, 312-17, 372-496; 2 (1793), 43-61, 119-27, 143-65, 170-78.

other sections in academic attainment remain difficult to understand. Cutler explained, "The almost total neglect of botanical enquiries, in this part of the country, may be imputed, in part, to this, *that Botany has never been taught in any of our Colleges,* and to the difficulties that are supposed to attend it; but principally to the mistaken opinion of its inutility in common life." [44] Although Cutler went on to stress the value of botanical studies to medicine and agriculture, his comment on the relationship of botany to the colleges was the most perceptive of his remarks. A Yale graduate himself, Cutler made essential use of the Harvard library in his study and through the American Academy he came to know the academic community. With Waterhouse, he was an important factor in giving academic roots to natural history at Harvard. [45]

His "Account" was widely praised, as it deserved to be. It is true that he worked an almost virgin field where many plants had "never been so far noticed as to receive even a trivial name." Yet, in working it he displayed great industry as well as an acquaintance with the writings not only of Linnaeus but of such lesser figures as Catesby and Dr. William Withering. He developed a fruitful correspondence with Withering's collaborator, the English botanist, Dr. Jonathan Stokes. [46] His paper so impressed Benjamin Rush and Benjamin Franklin that they had portions of it published in the *Columbian Magazine.* When Cutler visited Philadelphia in 1787, Rush had a proposition to offer him. Both the American Philosophical Society and the College of Physicians had sought to establish a botanical garden; now Rush reported, an attempt was being made to raise a fund that would convert it into reality. The Philadelphians thought of Cutler as the most eligible person to take charge of the garden and at the same time to undertake botanical lectures to the medical students at the university. [47] Such an offer was clear indication that conditions were changing. New England was no longer a desert to natural history.

44. Manasseh Cutler, "An Account of some of the vegetable Productions, naturally growing in this Part of America, botanically arranged," American Academy, *Memoirs,* I (1785), 396-97.

45. Harvard Corporation Records, III, 133.

46. Cutler, "An Account," 396, 401; Stokes to Cutler, Aug. 17, 1785, Cutler, *Life of Manasseh Cutler,* I, 259-62; Cutler to Stokes, Oct. 30, 1786, *ibid.,* 262-74.

47. *Ibid.,* 257; *Columbian Magazine,* 1 (1786-87), 379 ff., 436 ff., 469-72.

Cutler told Ezra Stiles that he planned to continue his studies of plants in other states with a view to further publication.[48] He never did. Even as he rejected the Philadelphia offer, his thoughts were on the Ohio Company and plans for real estate promotion in the West. He wrote a little promotion pamphlet about the lands in which he was interested, printed first in English and then translated into French. It contained very little in the way of natural history although it was prefaced by a statement of Thomas Hutchins testifying that the facts relating to soil and natural productions were accurate.[49] Cutler had none of the lifelong devotion to natural history of William Bartram, but the contribution he made was vital.

Elsewhere, the process of engrafting the study of natural history upon academic roots was less successful than in Philadelphia and Cambridge although at Columbia College only the comatose state of the medical school prevented such action. Dr. Henry Moyes had been appointed professor of natural history in 1785, but it was not he who first served in that capacity. That was left to Dr. Samuel Latham Mitchill, who returned to New York in 1787 with a fresh Edinburgh education and an eager, active mind. Almost immediately he began to turn out a stream of publications—most of them in the field of natural history. In 1787, he appended "Geological Remarks on the maritime Parts of the State of New York" to a medical paper on absorbent tubes. He published a magazine article, "On the Nature and Origin of Peat or Turf," and the following year another on dissecting a skunk. He continued to keep the magazines supplied with such articles. A man of his vigor and versatility was much needed by the struggling college—particularly by the medical school, but it was not until 1792 that a chair suiting Mitchill's range of talents could be created. He was then appointed "Professor of Natural History, Chemistry, Agriculture, and other Arts Depending Thereon!" What is more, in the years ahead he was able to make some contribution to each of these fields as well as to several others. Even before that ap-

48. Cutler, *Life of Manasseh Cutler*, I, 257; Franklin B. Dexter (ed.), *The Literary Diary of Ezra Stiles* (New Haven, 1901), III, 268.
49. [Manasseh Cutler], *An Explanation of the Map* (Salem, 1787), 3. The French edition was given a different and more descriptive title, *Description du Sol, Des Productions, &c. &c., de Cette Portion des Etats Unis* (Paris, 1789).

pointment, Mitchill had become the commanding figure in a scientific circle that finally arose in New York.[50]

The professors who taught natural history or botany at Pennsylvania, Harvard, and Columbia were appointed because of the medical schools attached to those institutions; the case of Henry Muhlenberg was somewhat different. Already an accomplished botanist when Johann David Schöpf visited him immediately after the war, Muhlenberg acquired an academic affiliation in 1787 upon his appointment as president of the newly established Franklin College—or as he called it "der Deutschen Höhen Schule." [51] He never taught botany there, but at the new post his own botanical researches continued to flower. He observed plants, sought new specimens, and collected botanical books. In 1785, he presented the American Philosophical Society with a manuscript specimen of a Flora Lancastriensis and in 1791 with the supplemental Index Florae Lancastriensis. These beginnings were expanded until Muhlenberg had found 1,380 species of plants in Lancaster County.[52] He went on to develop an extensive American and European correspondence and to publish a catalogue of the plants of North America—as Cutler failed to do. When he died, he could with justice be hailed as "the chief of the botanists of the U.S." [53]

The new academic roots acquired by natural history caused subtle changes. Medical students received some benefit but college undergraduates very little. American textbooks were slow in coming. Bar-

50. *Columbian Magazine*, 1 (1786-87), 581-84; *American Magazine*, 1 (1788), 302-3; Samuel Latham Mitchill, *Observations Anatomical, Physiological, and Pathological on the Absorbent Tubes of Animal Bodies. To Which Are Added Geological Remarks on the Maritime Parts of the State of New-York* (New York, 1788); Courtney Robert Hall, *A Scientist in the Early Republic, Samuel Latham Mitchill* (New York, 1934), 7; *A History of Columbia University* (New York, 1904), 75.

51. Johann David Schöpf, *Travels in the Confederation, 1783-1784*, trans. by Alfred J. Morrison (Philadelphia, 1911), II, 12; Gotthilf Hen. Muhlenberg, *Eine Rede Gehalten den 6ten Juni 1787* (Lancaster, 1788), [i].

52. American Philosophical Society Minutes, July 15, 1785, [III], not paged; Feb. 18, 1791, [IV], 168; Herbert H. Beck, "Henry E. Muhlenberg, Botanist," Lancaster County Historical Society, *Papers*, 32 (1928), 104.

53. Quoted from *Analetic Magazine* by William M. and Mabel S. C. Smallwood, *Natural History and the American Mind* (New York, 1941), 300; J. M. Maisch, *Gotthilf Heinrich Ernst Muhlenberg als Botaniker* (New York, 1886), 36; Henry Muhlenberg, *Catalogus Plantarum Americae Septentrionalis* (Lancaster, 1813).

ton's *Elements of Botany* did not appear until 1803. The most important result was the greater stability and independence of the whole group interested in natural history. From this point on, systematic cataloguing of American natural history could be as well done in America as in Europe—a development that owed much to Walter, the planter, and Marshall, the seedsman, it is true. Barton, Waterhouse, Mitchill, and Muhlenberg, however, had the greater influence upon future developments in natural history. Although some of the leading naturalists of the years ahead had no academic affiliation, their labors were often related to those who did—sometimes planned and supported by them. The day when an American field worker had to send his collections to Europe and wait to hear how they were to be catalogued was past.

In related studies—particularly geography—texts appeared more quickly after the war. Here was an established need at several educational levels to which cultural nationalism contributed an additional incentive. As the Reverend Jedidiah Morse put it, rather sharply, "We are independent of Great Britain and are no longer to look up to her for a description of our own country." [54] His first text, *Geography Made Easy*, appeared in 1784. In 1789, he published *The American Geography* and wrote *The American Universal Geography*. To Morse, geography was much more than a question of rivers, mountains, lakes, and boundaries. Among a wide variety of topics he included some descriptive natural history—particularly botany and zoology. [55] Two other geographical texts, Robert Davidson's versified *Geography Epitomized* and Thomas Greenleaf's *Geographical Gazeteer*, offered nothing in natural history. [56] None of them contributed to the advance of knowledge—they only disseminated it.

The state histories, which followed the tradition of several colonial histories of individual provinces, all showed some interest in natural history. Indeed, it was usual to divide their subject matter into two parts as Samuel Williams later emphasized in his title, *The*

54. *New Haven Gazette and Connecticut Magazine*, 2 (1787), 222.
55. Jedidiah Morse, *Geography Made Easy* (New Haven, 1784); Jedidiah Morse, *American Universal Geography* (11th ed., Boston, 1807); Massachusetts Historical Society, *Proceedings*, 53 (1919-20), 45-46.
56. [Robert Davidson], *Geography Epitomized* (Philadelphia, 1784); Thomas Greenleaf, *Geographical Gazeteer of the Towns of the Commonwealth of Massachusetts* [no title page].

Natural and Civil History of Vermont.[57] In many respects the best of the state histories was the Reverend Jeremy Belknap's *History of New Hampshire,* the first volume of which came out in 1784. Belknap was in the forefront of those who wished to see the new United States flower in every way—most of all in science. His history of one state was not a particularistic study that ran counter to his broad nationalism. As he himself declared, he was impelled to undertake the book by natural curiosity about "the country which gave him birth. When he took up his residence in New-Hampshire his enquiries were more particularly directed to that part of it." [58] Although Belknap did not write of natural history until his third volume, which appeared in 1792, one part of it, his description of the White Mountains, was published in 1786. This was based upon the expedition which in 1784 made the first attempt to measure the height of Mount Washington.[59]

Among the state histories, John Filson's *Discovery, Settlement and Present State of Kentucke* was of a somewhat different character. Kentucky was then more a real estate operation than an incipient state. Filson's book had only a brief history to relate and much of that proved to be untrustworthy. As a real estate promotion tract, however, similar to Cutler's *Explanation,* it was very successful, with several references to natural productions. The accompanying map was probably its most valuable feature and longest in use.[60] As a treatise on the rapidly developing West, it was inferior to Thomas Hutchins' *Historical Narrative and Topographical Description of Louisiana and West-Florida* which appeared in the same year, 1784.

Hutchins was by then the best informed man on the American West. A native of New Jersey, he had been caught in England by the American Revolution while serving as a captain of engineers in the British Army. Amidst the difficulties of trying to sell his commission and being thrown into jail, Hutchins published in London in 1778,

57. Samuel Williams, *The Natural and Civil History of Vermont* (Walpole, N.H., 1794); Jefferson used this classification too, Sowerby, *Catalogue,* I, [xxi].
58. Jeremy Belknap, *The History of New Hampshire* (Philadelphia and Boston, 1784-92), I, iii.
59. *Ibid.,* III; American Philosophical Society, *Transactions,* 2 (1786), 42-49.
60. John Filson, *The Discovery, Settlement and Present State of Kentucke* (Wilmington, 1784).

his *Topographical Description of Virginia, Pennsylvania, Maryland, and North Carolina*, accompanied by a map of that region. The book included a discussion of animal, vegetable, and mineral productions and of climate and soil. The clear military importance of both map and book did not prevent Hutchins' assignment as geographer to the southern army under General Nathaniel Greene, upon his return to the United States, and in 1781 to the post of geographer to the United States. His *Description of Louisiana and West-Florida* contained some discussion of natural history too, but it was primarily useful because of the directions it provided for sailing about the mouth of the Mississippi and up the river itself. Hutchins was not a systematic naturalist but in the surveys and explorations he continued to make for separate states and for the United States, he accumulated a great store of information. His knowledge was relied upon by many men, including Thomas Jefferson.[61]

As a true child of the Enlightenment, Jefferson placed intellectual achievement upon the highest level and venerated those who made creative contributions to science. He became an eager patron of science in the United States—especially of natural history to which he devoted a continuing attention. Even so, it may be doubted that he was reading his own character aright when he declared, "Nature intended me for the tranquil pursuits of science." [62] There was, however, one moment in his full life when he planned to withdraw from participation in politics, diplomacy, and government and retire to the quiet of a scholarly life. Depressed by the failures of his term as governor of Virginia, he refused appointment to the peace commission sent by Congress to Paris to end the war because, as he wrote, "I have taken my final leave of everything of that nature, I have retired to my farm, my family, and my books, from which I think nothing will evermore

61. Beverly Bond (ed.), *The Courses of the Ohio River Taken by Lt. T. Hutchins* (Cincinnati, 1942), 11; Thomas Hutchins, *A Historical Narrative and Topographical Description of Louisiana and West-Florida* (Philadelphia, 1784), iii-iv; Thomas Hutchins, *A Topographical Description of Virginia, Pennsylvania, Maryland, and North Carolina* (London, 1778); *Pennsylvania Journal*, Dec. 8, 1781.

62. Jefferson to E. I. du Pont de Nemours, March 2, 1809, Andrew A. Lipscomb and Albert E. Bergh (eds.), *The Writings of Thomas Jefferson* (Washington, 1903-4), XII, 260.

separate me." [63] Shortly afterward, he wrote his only book, the *Notes on the State of Virginia*.

The stimulus for the *Notes* came from France. One of the many attempts made by the French government to obtain reliable information about the American states was to address a series of questions to a leading person in each state. If properly answered, the questions would have yielded a profile of the natural history, the government, and the economy of the states. In 1780 and 1781, these questions were sent by the secretary of the French legation in Philadelphia, the Marquis de Barbé-Marbois, to leading men in each state: to John Sullivan in New Hampshire, to Thomas McKean in Delaware, to Joseph Jones in Virginia. Sullivan wrote out brief answers and another set of answers has been preserved for South Carolina. [64] Jones, a Virginia member of the Congress, forwarded the queries on Virginia to Jefferson while he was still governor. At the time, Jefferson was unable to devote his attention to the matter but with his withdrawal from public affairs, he found time to reply. [65]

His answers, ready by December 20, 1781, turned out to be an extensive treatise based upon notes he had been collecting for many years. The sections relating to natural history were particularly full and satisfying. As he wrote out his answers, a new idea occurred to him in keeping with the contemplative and philosophical career he now had in prospect. He wrote to Charles Thomson in Philadelphia, "In framing answers to some queries which Monsr. de Marbois sent me, it occurred to me that some of the subjects which I had then occasion to take up, might, if more fully handled, be a proper tribute to the Philosophical Society." In any case, Jefferson asked Thomson, "Perhaps also you would be so friendly as to give me some idea of the subjects which would at any time be admissible into their trans-

63. Jefferson to Edmund Randolph, Sept. 16, 1781, Boyd (ed.), *Papers of Jefferson*, VI, 118.

64. Otis G. Hammond (ed.), *Letters and Papers of Major General John Sullivan*, III, New Hampshire Historical Society, *Collections*, 15 (1939), 229-39; Marbois to McKean, Feb. 10, 1781, McKean Papers, I, 42, Historical Society of Pennsylvania; Carl Bridenbaugh, *Myths and Realities* (Baton Rouge, [1952]), 67n; Marbois Questions on Pennsylvania, Gratz Collection, Case H, Box 29, Historical Society of Pennsylvania.

65. Jefferson to Marbois, March 4, 1781, Boyd (ed.), *Papers of Jefferson*, V, 58.

actions." [66] Although Thomson hastened to assure him that the *Notes* would be "a very acceptable present to the society," Jefferson remained hesitant.[67] From France, where he published the *Notes* because printing costs were so much less, Jefferson sent copies to his friends, but not to the society. "I have sometimes thought," he wrote Francis Hopkinson, "of sending a copy of my Notes to the Philosophical Society, as a tribute due to them: but this would seem as if I considered them as worth something which I am conscious they are not." [68] Thomson, as soon as he read the manuscript copy, felt otherwise. He described the *Notes* as "a most excellent natural history, not merely of Virginia, but of North America, and possibly equal, if not superior to that of any Country yet published." [69]

Jefferson thought better of his *Notes* than his words would indicate but he had the wisdom not to claim too much for the book. He was right in not calling it *A Natural History of Virginia* as Thomson urged—for it was not. It was not a systematic study of any phase of science, although it contained much important information and some good reasoning. It was the most influential of all the state and regional studies of this period. The first two-hundred-copy edition of 1785 was followed quickly by American, English, French, and German printings—some of them altogether unauthorized. In America, numerous extracts from it were issued as magazine articles.[70] It was well received in France despite Jefferson's harsh refutation of the writings of Buffon, Raynal, and Daubenton.[71] Some circles in England received the book favorably but the prevailing attitude which caused most American products to be viewed askance in these years affected the treatment accorded the *Notes*. While Franklin's reputation was able to survive the divisive influence of war and revolution,

66. Jefferson to Thomson, Dec. 20, 1781, *ibid.*, VI, 142.
67. Thomson to Jefferson, March 9, 1782, Jefferson Papers, Library of Congress.
68. Jefferson to Hopkinson, Sept. 25, 1785, Boyd (ed.), *Papers of Jefferson*, VIII, 551.
69. Thomson to Jefferson, March 6, 1785, *The Papers of Charles Thomson*, New-York Historical Society, *Collections for 1878*, 11 (1879), 200.
70. *Ibid.*; *Columbian Magazine*, 1 (1786-87), 366-73, 407-16; 2 (1788), 75-77, 86-89, 573-75.
71. Marie Kimball, *Jefferson, War and Peace, 1776-1784* (New York, 1947), 302-5.

Jefferson's was not. The *European Magazine*, for example, reviewed Jefferson's *Notes* in a most cavalier and condescending fashion while it accepted Franklin's *Philosophical and Miscellaneous Papers* as the work of a great scientist.[72]

Jefferson's passion showed through the *Notes* most clearly at points where he felt obliged to defend American dreams of glory from aspersions that had been cast upon them. This emotion lay behind his attack upon slavery within America. It accounted, too, for his assault upon the very distasteful theory of the degeneracy of life in America most authoritatively stated by the Comte de Buffon in his *Histoire Naturelle*. Jefferson combatted this widely circulated idea by compiling tables of comparative weights of American and European animals tending to show that, if anything, animals in America were larger than similar or identical species in the Old World. Jefferson also appended to his book a paper by Charles Thomson which argued that the American Indian was not physically inferior to the white man.[73] When Jefferson got to Paris, he pursued the combat further by importing at great expense the skin and bones of an American moose which John Sullivan had obtained for the purpose in New Hampshire. Buffon may have been convinced by the size of this specimen that his theories were wrong, as Daniel Webster asserted, but he certainly never recanted publicly.[74]

It was but a short step from Jefferson's refutation of Buffon's charges and of the Abbé Raynal's assertion that the Americans had produced no creative scholars to the assertion that everything in America was superior to what could be found in the Old World. In a magazine article, Lycurgus swelled with pride: "In viewing the two great continents which divide the earth into eastern and western hemispheres, we observe this remarkable difference in every article in which we can form comparisons between them; that every object in the face of nature on the eastern continent is formed upon a small and imperfect scale; on the western continent the same objects are extended upon a larger and more liberal and elevated plan. And this seems to be the reason why this country is reserved to be the last and

72. *European Magazine*, 12 (1787), 112-16, 205-6, 273-76, 379-82.
73. [Thomas Jefferson], *Notes on the State of Virginia* ([Paris], 1782 [1785]), 367-91.
74. Edwin T. Martin, *Thomas Jefferson: Scientist* (New York, 1952), 186.

greatest theatre for the improvement of mankind."[75] The vision of glory rose superior to the assaults of Buffon and Raynal.

Even the great interest Jefferson developed in ancient bones was at its beginning related to his quest for evidence that life in America was not inferior to the old world. The "mammoth" or mastodon, he was sure, was the "largest of all terrestrial beings." He began efforts to collect old bones in 1782 and watched with pleasure the rising interest in remains of ancient animals throughout the country.[76] The American Academy published an account by Robert Annam of bones found on the banks of the Walkill, New York, during the war, some of which found their way to Du Simitière's museum in Philadelphia while others were carried back to Germany by the Hessian doctor, Michaelis.[77] Most of the new finds came from the West in the process of developing that area. Big Bone Lick continued to yield important collections of bones, but it was on the Susquehanna that David Rittenhouse discovered a tooth which Peale was glad to display in his museum. A thigh bone found in New Jersey led the American Philosophical Society to encourage the exploration of the area; without important results.[78] Ezra Stiles, Henry Muhlenberg, Caspar Wistar, and Lewis Nicola, among many others, attempted to read the riddle of the ancient bones.[79]

A parallel interest developed in attempting to understand old Indian relics and the Indians themselves. The two most important problems were the long-known inscriptions on the Dighton rock at Taunton, Massachusetts, and the mounds and fortifications being found on the Ohio. The literati continued to try to puzzle out the Dighton rock, some accepting the interpretation of Court de Gebelin that

75. *New Haven Gazette and Connecticut Magazine*, 1 (1786), 1.

76. Jefferson to James Steptoe, Nov. 26, 1782, Paul L. Ford (ed.), *The Writings of Thomas Jefferson* (New York, 1892-99), III, 62; [Jefferson], *Notes on Virginia*, 77.

77. American Academy, *Memoirs*, 2 (1793), 160-64; Whitfield J. Bell, Jr., "A Box of Old Bones," American Philosophical Society, *Proceedings*, 93 (1949), 172.

78. American Philosophical Society Minutes, Oct. 5, 1787, [IV], 13.

79. Stiles to Jefferson, June 21, 1784, Boyd (ed.), *Papers of Jefferson*, VII, 312; American Philosophical Society Minutes, March 5, 1784, [III], not paged; Samuel H. Parsons to Stiles, April 27, 1786, Franklin B. Dexter (ed.), *Extracts from the Itineraries and Other Miscellanies of Ezra Stiles, D.D., LL.D., 1755-1794* (New Haven, 1916), 549; Wallace, *The Muhlenbergs*, 256.

the inscriptions were Punic while others continued to maintain that they were Indian in origin. With respect to the Ohio mounds which excited Noah Webster, Jefferson refused to be carried away by emotion. In a rational mood, he stated with perfect justice, "It is too early to form theories on those antiquities, we must wait with patience till more facts are collected. I wish our philosophical societies would collect exact descriptions of the several monuments as yet known and insert them, naked, in their *Transactions*." [80]

The differences between the Indian, the Negro, and the white man were particularly apparent to American thinkers and demanded a scientific explanation. Jefferson vehemently rejected the idea that the American climate had produced inferior human species any more than inferior animals. He defended the equality of the Indian and looked eagerly for evidence of equal capacities in the Negro.[81] The general disposition was to assign the variation of the human species to environmental causes—particularly to climate. This was argued upon rational grounds even by those who were unable to accept any other solution because of religious dogma. The best case for the viewpoint was made by the Reverend Samuel Stanhope Smith who stated at the outset that he was "not at liberty" to accept any hypothesis of separate creation. Altogether "independently of the sacred authority of revelation," however, he proceeded to adduce facts which demonstrated the common origin of all humanity.[82]

Amidst the many problems that absorbed naturalists and philosophers, it was left to Thomas Jefferson to pursue the grandest project of all with that peculiar blend of tenacity and enthusiasm he reserved for the dreams closest to his heart. In his *Notes on Virginia* he had written almost lovingly of the great Missouri River even though "since the Treaty of Paris," it had been "no longer within our limits." [83] It was in 1783 that he made the first suggestion of the ex-

80. Jefferson to Thomson, Sept. 20, 1787, Jefferson Papers, Library of Congress.

81. Jefferson to Chastellux, June 7, 1785, Boyd (ed.), *Papers of Jefferson*, VIII, 185; Jefferson to Benjamin Banneker, Aug. 30, 1791, Ford (ed.), *Writings of Jefferson*, V, 377.

82. Samuel Stanhope Smith, *An Essay on the Causes of the Variety of Complexion and Figure in the Human Species* (Philadelphia, 1787), 1, 111; *Massachusetts Magazine*, 1 (1789), 672-73, 770-75.

83. [Jefferson], *Notes on Virginia*, 10-11.

ploration of the West that lay beyond the limits of the United States but well within the limits of the American dream of glory. He was incensed when he heard of a subscription being solicited in England for the purpose of exploring the whole country from the Mississippi to California. He remarked in disbelief, "They pretend it is only to promote knowledge." When he sounded out George Rogers Clark on his readiness to lead an American expedition "to search that country," he, certainly, was not motivated by a wholly disinterested desire to advance knowledge.[84] Had it proven possible to finance such an expedition it would have increased acquaintance with the natural history of the region—but the information would have been collected with the still nebulous destiny clearly in mind.

In 1785, Humphry Marshall sought to promote a more limited expedition to explore the West within the boundaries of the United States. Support could not even be found for this objective although the ever ready John Coakley Lettsom offered to contribute twenty guineas toward a year-long expedition. Marshall's idea was to send William Bartram and Dr. Moses Marshall, his nephew, out "to make Observations &c. upon the Natural productions and Curiosities of those Regions."[85] Through Thomas Parke and Benjamin Franklin, he sought to gain support of the American Philosophical Society, but was told "few among us seem devoted to investigate the beauties of Natural History—."[86] He thought, too, of winning Congressional support. The only result of this project was the brief botanical tour made by Moses Marshall from Pittsburgh to the back country of South Carolina in 1789.[87]

Meanwhile, Jefferson had encountered a new possibility of making his exploration a reality in the person of John Ledyard whom he met in Paris. Ledyard, born in Connecticut, was a kind of professional explorer. During the War for Independence he had accompanied Cap-

84. Jefferson to George Rogers Clark, Dec. 4, 1783, Boyd (ed.), *Papers of Jefferson*, VI, 371.

85. Lettsom to Franklin, Aug. 14, 1786, Franklin Papers, XXXIV, 122, American Philosophical Society; Marshall to Franklin, Dec. 5, 1785, *ibid.*, XXXIII, 254.

86. *Ibid.*; Parke to Marshall, March 4, 1785, Humphry Marshall Correspondence, I, 15.

87. Dr. Moses Marshall to Sir Joseph Banks, Oct. 30, 1790, Darlington, *Memorials*, 563-64.

tain Cook on his last expedition which Ledyard wrote up in book form. When Jefferson met him he was full of several projects, but the encouragement of the American Minister and the failure of one of his projects led him to undertake a fantastic expedition across the whole Eurasian continent, thence by sea to the western coast of North America, and across America from California to the East. This time Jefferson was apparently not disturbed by the support extended to Ledyard from subscriptions by such Englishmen as Sir Joseph Banks, John Hunter, and Sir James Hall. Ledyard actually reached Yakutsk in Russian Siberia, within five hundred miles of the Pacific Ocean. There he was picked up by the secret police who carried him back—a third of the way around the world—to the Polish border where he was freed. Ledyard was still promising Jefferson to undertake the American exploration from Kentucky when at Cairo, Egypt, in 1789, he died. Still other efforts were to fail before Jefferson was able to have the great exploration accomplished, but the dream which drove him was fully evident in 1789.[88]

88. Jefferson to William Carmichael, March 4, 1789, Ford (ed.), *Writings of Jefferson*, V, 75; Jefferson to Thomson, Sept. 20, 1787, Jefferson Papers, Library of Congress; Jefferson, "Autobiography," Lipscomb and Bergh (eds.), *Writings of Jefferson*, I, 101; Jared Sparks, *The Life of John Ledyard, The American Traveller* (Cambridge, 1828), 177, 227; John Ledyard, *A Journal of Captain Cook's Last Voyage to the Pacific Ocean* (Hartford, 1783); *New York Journal*, July 23, 1789.

Chapter Fifteen

Natural Philosophy in the Republic

AMERICA HAD NEVER been hospitable to the basic physical sciences —Franklin's brilliant experiments in electricity and the co-operative observations of the transit of Venus notwithstanding. There simply was not support enough to sustain the study and experiment required. The Swedish Lutheran clergyman, the Reverend Nicholas Collin, who came to the United States after the Revolution reported very perceptively on this point. In America, he wrote, an ordinary skilled workman was better paid than the regular pastors. It was not a land for scholars and scientists.[1] The Revolution did not improve matters in this respect.

Support could be found much more readily for anything promising immediate utility and even for investigations of American natural resources than for research in mathematics and natural philosophy. The patriotism of the Revolution did not offer very much encouragement. It was not that the prophets of glory intended to slight these basic sciences. David Ramsay paraphrased Newton when he spoke of standing on the shoulders of earlier scientists. He was conscious of the need for work in mathematics and natural philosophy. Yet, neither he nor the others who tried so eagerly to stimulate scientific work thought primarily in terms of the physical sciences. Their constantly recurring

1. Amandus Johnson, *The Journal and Biography of Nicholas Collin* (Philadelphia, 1936), 132.

references to the "Book of Nature" were directed most urgently to questions of natural history.[2]

Concentration upon the open book of nature created particular difficulties when the concept was applied to the physical sciences. It led to strangely unrealistic appraisals of the background required for achievement in these fields. Owen Biddle neglected altogether the need for study in announcing, "Experiments were found to be the touchstones of truth, from which the true principles of nature were to be discovered."[3] Book learning was not important—by some it was even considered harmful. Just as Francis Hopkinson misread Franklin as a man without a liberal education, so Benjamin Rush found in Rittenhouse a child of nature who had accomplished much precisely because he had not been misled by academic learning. If he had had a formal education, Rush suggested, he "might have spent his hours of study in composing syllogism, or in measuring the feet of Greek and Latin poetry."[4] This unrealistic primitivism was no help in the physical sciences where it was essential to stand on the shoulders of those who had gone before.

The American Academy of Arts and Sciences was frankly pessimistic about achievement in this realm. In its first volume of *Memoirs*, it announced, "The astronomical and mathematical papers, in this volume will, perhaps, be the least entertaining of any of the collection, and will have the smallest number of readers. . . . Few, if any of them, contain deep speculations and obstruse researches and calculations: but they are chiefly of the practical kind."[5] Individual contributors often tried to give their papers popular appeal by stressing utilitarian aspects. Many writers rationalized their efforts. The Reverend Phillips Payson explained, "The use of Astronomical observations, to promote the purposes of navigation and geography, must be evident to every person that has paid any proper attention to the subject." He was even able to tie in a patriotic motive for such work. To define the boundaries and geography of the United States, he said, "The best, and indeed the

2. David Ramsay, [An Oration], *U.S. Magazine*, 1 (1779), 53.
3. Owen Biddle, *An Oration Delivered the Second of March, 1781* (Philadelphia, 1781), 31.
4. Benjamin Rush, "An Eulogium upon David Rittenhouse," *Essays, Literary, Moral, and Philosophical* (2d ed., Philadelphia, 1806), 348.
5. American Academy, *Memoirs*, 1 (1785), viii.

only proper method is, that of astronomical observations, which; it is probable, the Supreme Council of *America* will soon adopt, now the glorious revolution is so happily compleated." [6] It was difficult to make all work in the physical sciences appear either useful or patriotic.

Natural philosophy and mathematics suffered at least as much from the dislocations of the war as did the other sciences. They suffered from internal difficulties: the closing of colleges and absorption of the teachers in the pursuits of war. Still more difficult to repair were the breaks in the ties with Europe. Internationalism had always been less prominent among the physical scientists than in the prewar natural history circle. Ties had been very largely with England and Scotland and these mended but slowly. John Adams reported from London, "There is a universal desire and endeavour to forget America, and an unanimous Resolution to read nothing which shall bring it to their Thoughts. They cannot recollect it without Pain." [7] This feeling was reflected in the antagonistic or disinterested attitude displayed by English reviews toward American science and in the personal views of some of the men of science.

Fortunately there were others who continued to look upon the Americans in a friendly way and many who considered science, in any case, above and beyond political recrimination. The Royal Society had accepted American papers during the war, and even published them. As soon as the war ended, Sir Joseph Banks hailed Franklin upon the prospect of the renewal of untrammeled communications: "My sincere congratulations on the return of peace which in whatever form she is worshiped, bad peace or good peace never fails to prove herself the Faithful nurse of Science." [8] The Reverend Richard Price remained as friendly as ever. In 1785, he suggested that the Americans had laid "the foundation there of an empire which may be the seat of liberty, science and virtue, and from whence there is reason to hope these sacred blessings will spread, till they become universal." [9] He corre-

6. Phillips Payson, "Some Select Astronomical Observations," *ibid.*, 124.

7. John Adams to Mercy Warren, Dec. 25, 1787, *Warren-Adams Letters*, II, Massachusetts Historical Society, *Collections*, 73, (1925), 301.

8. See Joseph Willard's paper, *Philosophical Transactions*, 71 (1781), 502-7; Banks to Franklin, May 28, 1783, Franklin Papers, VIII, 20, University of Pennsylvania.

9. Richard Price, *Observations on the Importance of the American Revolution* (London Printed, Trenton Reprinted, 1785), 4.

sponded with Franklin, counselled James Bowdoin, and occasionally wrote to several other Americans. His influence was less ubiquitous than that of John Coakley Lettsom but in the physical sciences he was a very effective aid.

British immigrants helped too. A small but important group of friends of America decided to transplant themselves to the United States at the end of the war. Samuel Vaughan was more a promoter of science than a scientist himself. So was John Vaughan, his son, who remained in the country permanently to become a constant support of science in Philadelphia. Another son, Samuel, Jr., sought a position in the projected United States mint on the basis of his extensive studies of metallurgy and mineralogy in Europe. Benjamin Vaughan who did not come to the country until later, corresponded on momentum, combustion, and optics.[10] William Thornton, a Whig, Quaker friend of Lettsom, migrated to the United States to apply his talents to a wide variety of activities including architecture, town planning, scientific organization, language reform, and teaching the deaf to speak. The Reverend Charles Nisbet brought to the newly formed Dickinson College much of the learning of the Scottish Enlightenment.[11]

One of the most accomplished of these figures was Walter Minto, a Scottish mathematician who had favored the American Revolution and came to this country in 1786. He had studied at the University of Edinburgh, and under Giuseppe Slop, professor of astronomy at the University of Pisa. While in Britain, he wrote a book on the planets and collaborated with Lord Buchan on the study of John Napier. Offered posts at Columbia College, Washington College in Maryland, and the College of New Jersey, he finally settled down as professor of mathematics and natural philosophy at New Jersey. There, he spread his faith in the utility of mathematics and in its ability "to brighten

10. Samuel Vaughan, [Jr.] to the Commissioners of the Treasury, Dec. 29, 1786, Misc. Mss., American Philosophical Society; Benjamin Vaughan to Franklin, Oct. 8, 1781, Jan. 9, 1782, Bache Collection, American Philosophical Society; Samuel Vaughan to James Bowdoin, March 5, 1787, *Bowdoin and Temple Papers*, II, Massachusetts Historical Society, *Collections*, 6th ser., 9 (1897), 166.

11. See particularly, Jean Lowe, An Inquiry into the Life and Ideas of Charles Nisbet (Master's Thesis, Columbia University, 1944); William Thornton Papers, Library of Congress.

and enlarge" the mind of man. His creative work, however, came to an end with his removal to America.[12]

Even so, his influence served to counter factors that seemed ready to drag mathematics even below the level of attainment of the colonial period. The mathematics that appeared in the magazines was no worse than what had been published in the colonial newspapers, however useless it was except for amusement. Some of the mathematical problems were even submitted in verse.[13] The colleges did nothing to advance mathematics but at least they helped to transmit some part of the heritage. The worst mistake was made by James Winthrop.

James Winthrop, son of Professor John Winthrop, aspired to fill his father's shoes without having either the requisite capacity or character. On Professor Winthrop's death in 1779, James was considered for his father's chair at Harvard and once again in 1789, but each time he was passed over. He served as curator of the American Academy until 1782 and as college librarian until 1787. Just before he gave up his post as librarian, he submitted to the American Academy faulty solutions to the insolvable problems of the trisection of the angle and the duplication of the cube. Although informed that the solutions were wrong, he insisted upon publishing them in the second volume of *Memoirs* where they duly appeared.[14] This was characteristic of James Winthrop. On one occasion he even showed himself ready designedly to tamper with the truth. He advised Mercy Warren to revise the treatment of Thomas Hutchinson in her *History of the American Revolution*. He asked, "Would it not be better to give him on the credit of his own party, a little undeserved praise, to procure their judgement in favor of the work?" [15]

The man Harvard chose to fill the Hollis chair of Mathematics and Natural Philosophy in 1779 produced heartaches of his own despite his

12. Walter Minto, *An Inaugural Oration on the Progress and Importance of the Mathematical Sciences* (Trenton, 1788), 30; Luther P. Eisenhart, "Walter Minto and the Earl of Buchan," American Philosophical Society, *Proceedings*, 94 (1950), 282, 287.

13. *Columbian Magazine*, 1 (1786-87), 39, 142; 2 (1788), 107-284; *Boston Magazine*, 1 (1783-84), 390, 436,·483-84.

14. Massachusetts Historical Society, *Collections*, 2d ser., 10 (2d ed., 1843), 77-80; American Academy, *Memoirs*, 2 (1793), 9-13, 14-17.

15. James Winthrop to Mercy Warren, Feb. 2, 1787, *Warren-Adams Letters*, II, 282-83; see Mercy Warren, *History of the Rise, Progress and Termination of the American Revolution* (3 vols., Boston, 1805).

real talents and earnest service. The Reverend Samuel Williams displayed a continuing interest in astronomy and many of his observations were published during his tenure of office. Williams, too, was the man who began regular meteorological observations with instruments sent for the purpose at the direction of the Elector Palatine at Mannheim. These were published in the *Ephemerides* of the Meteorological Society at Mannheim.[16] As a mathematics instructor, he seems to have been competent. It was personal failings that forced his resignation in 1788, the charge against him being forgery. He was replaced by the Reverend Samuel Webber, who in later years published a mathematics text and ascended to the presidency of Harvard.[17]

None of the academic group distinguished themselves by publication. The other mathematical essays published by the American Academy in its first two volumes of *Memoirs* were relatively elementary and the Philosophical Society published nothing at all in the field in its second volume of *Transactions*. Numerous texts were issued but most of them were no more than reprints of British books. Some admitted to being abstracts or "compends." In Philadelphia, Benjamin Workman contributed something by adapting the popular English *Arithmetic* of John Gough to "the American Youth" and correcting errors in the "European" edition.[18] Nicholas Pike, a grammar school master, published a new arithmetic text in 1788. His *New and Complete System of Arithmetic, Composed for the Use of the Citizens of the United States* contained a section on algebra and other admirable features. Pike published his book in the little town of Newburyport but he sought to capture the whole New England market by printing testimonials from the most illustrious academic figures at Dartmouth, Harvard, Yale, and Rhode Island.[19]

16. Johann David Schöpf, *Travels in the Confederation, 1783-1784*, trans. by Alfred J. Morrison (Philadelphia, 1911), I, 223; I. Bernard Cohen, *Some Early Tools of American Science* (Cambridge, 1950), 55.

17. William Bentley, *Diary* (Salem, 1905), I, 100; Samuel Webber, *System of Mathematics* (2 vols., Boston, 1801).

18. Thomas Sarjeant, *An Introduction to the Counting House* (Philadelphia, 1789), 1; Alexander M'Donald, *The Youth's Assistant* (2d ed., Litchfield, 1789), 3; Benjamin Workman, *A Treatise of Arithmetic in Theory and Practice...By John Gough. To Which are added, Many valuable Additions and Amendments; More particularly fitting the Work for the Improvement of the American Youth* (Philadelphia, 1788), v.

19. Nicholas Pike, *A New and Complete System of Arithmetic* (Newbury-Port, 1788), preface, [not paged]; Lao Genevra Simons, *Bibliography of Early*

The Americans had no background of accomplishment in mathematics, but in astronomy it was another story. As colonists they had registered their most publicized achievement in the observations of the transit of Venus. While no such event could be anticipated after the war, the Americans still enjoyed advantages in certain kinds of descriptive observation. Moreover, they were known to and welcomed by the European astronomers.

In astronomy, the Americans maintained more fruitful relations with Europe than in other physical sciences. Every volume of the Philosophical Society's *Transactions* and the American Academy's *Memoirs* contained more papers on astronomy than upon any other single subject and these journals were distributed to important Europeans. Although American papers in the *Philosophical Transactions* were rare after the war, there were far more on astronomy than on any other subject. Jacques Cassini sent Rittenhouse observations made at the Paris Observatory and Rittenhouse developed a limited correspondence with Joseph-Jèrôme de La Lande. William Herschel sent the American Philosophical Society copies of his papers.[20] Richard Price very gently showed James Bowdoin that his attempt to revive the ancient concept of a crystalline sphere was not in touch with reality or with the latest work of Herschel on nebulae.[21] Thomas Short, the English telescope builder, opened a correspondence with Franklin on a rather unexpected basis—he offered to repair and improve Franklin's telescope if he would get him some information about members of his family who lived in America.[22] As for the literate public, they were kept in touch with European astronomy—at least the more spectacular phases of it—by the popular periodicals.[23]

American Textbooks on Algebra (New York, 1936), 7-8; Bentley, *Diary*, I, 61; *American Magazine*, 1 (1788), 343.

20. William Barton, *Memoirs of The Life of David Rittenhouse* (Philadelphia, 1813), 165n, 432; Rittenhouse to Jefferson, Nov. 8, [1787?], *Jefferson Papers*, Massachusetts Historical Society, *Collections*, 7th ser., 1 (1900), 35; Herschel to Franklin, Feb. 18, 1787, Franklin Papers, XXXV, 17, American Philosophical Society.

21. Price to Bowdoin, Oct. 25, 1785, *Bowdoin and Temple Papers*, I, 78.

22. Thomas Short to Franklin, May 28, 1787, Franklin Papers, XXXV, 69, American Philosophical Society.

23. *American Magazine*, 1 (1788), 388; *New Haven Gazette and Connecticut Magazine*, 2 (1787), 293.

On the face of things, there appeared to be ample stimulus and a wide field for American achievement in astronomy. The vast possibility of increased knowledge that seemed opened by astronomical work done through the Royal Society led Franklin to declare, "I begin to be almost sorry I was born so soon." [24] He was far from discouraged by the fact that very little of it had been done in America. When he wrote to Herschel on this point, he sounded like David Ramsay or Benjamin Rush in his most effervescent mood. Half jokingly, but half seriously, he told the great English astronomer, "Had Fortune plac'd you in this part of America, your Progress in these Discoveries might have been still more rapid." He went on to explain that the clearer air afforded nearly a third more days upon which observations could be made than in England.[25]

What few separate publications in astronomy were written by the Americans were of no great value. When Nehemiah Strong's professorship at Yale was slipping away from him, he published three lectures under the title, *Astronomy Improved: or A New Theory of the Harmonious Regularity Observable in the Mechanism or Movements of the Planetary System.* This elementary account failed to present anything new and even indicated that Strong did not fully understand the old system, when in one passage he announced that if the earth did not move, the orbit of the moon "would be circular." [26] Bartholomew Burges' *Short Account of the Solar System* was written against the background of the expected return of Halley's comet in 1788 or 1789.[27] This comet commanded a great deal of attention in the newspapers and magazines in large part because of a notice from Nevil Maskelyne which circulated through the states announcing its expected return.[28]

A more important work might have been produced if Charles Mason

24. Franklin to Banks, July 27, 1783, Albert Henry Smyth (ed.), *The Writings of Benjamin Franklin* (New York, 1905-7), IX, 74.
25. Franklin to Herschel, May 18, 1787, *ibid.*, 584.
26. Nehemiah Strong, *Astronomy Improved* (New Haven, 1784), 44.
27. Bartholomew Burges, *A Short Account of the Solar System* (Boston, 1789), 11; *Boston Gazette*, Dec. 8, 1788, Sept. 7, 1789.
28. Price to Willard, Jan. 22, 1787, "Joseph Willard Letters," Massachusetts Historical Society, *Proceedings*, 43 (1909-10), 625; *Maryland Journal*, Sept. 9, 1788; *Worcester Magazine*, 3 (1787), 172-74.

had not died in 1786. After being awarded £750 for his lunar tables by the Board of Longitude in London, the English astronomer emigrated to the United States where he set to work on "an American set of Lunar and Solar tables, which should be still more accurate than anything yet published in Europe." Just before he died, he turned over the portion he had finished to John Ewing in the hope that he would complete the project, but he never did.[29]

The most valued astronomical work done in America was in the form of brief reports of observations which most often appeared in the journals of the two philosophical societies. One of the most important of these papers, published in the first volume of the American Academy's *Memoirs*, reported observations made under the direction of Samuel Williams while the war was still in progress. This was the expedition to observe an eclipse of the sun from the vantage point of Penobscot Bay in the present state of Maine. The eclipse of October 1780 was observed also at Nova Scotia, Cambridge, Beverly, Newport, and Providence, but the expedition was planned with the hope of finding a spot where the eclipse would be total.[30] In support of this objective, Harvard College and the American Academy both solicited the aid of the Commonwealth of Massachusetts. This time, the assembly was not immobilized as it had been at the time of the transit of Venus of 1769. When it replied favorably, the Board of War was directed to fit out the galley *Lincoln* to convey Williams and his party to Penobscot or any other port to the eastward. Even the fact that the spot in question was then occupied by British forces did not prove insurmountable. The officer-in-charge of the garrison permitted the party to land but he required them to depart too soon to make maximum use of their opportunity.[31] The biggest disappointment proved to be their failure, largely because of inaccurate maps, to find a spot where the eclipse was total. In one respect, this was an advantage for so close were they to the region of totality that they observed the finest line

29. Charles Mason to Franklin, Sept. 27, 1786, Franklin Papers, XXXIV, 148, American Philosophical Society; *Pennsylvania Journal*, Nov. 4, 1786.
30. American Academy, *Memoirs*, 1 (1785), 129-34, 143-55; Willard to Ewing, Nov. 20, 1780, Ms. Communications to American Philosophical Society, Mathematics and Astronomy, I, 14.
31. Cohen, *Some Early Tools*, 53.

of light on one limb of the sun—a line that broke up into little drops or beads of light. Later named Baily's beads, this phenomenon was apparently first described by Samuel Williams in his report.[32]

Many observations of occasional events were made in all parts of the country. Eclipses of sun and moon, transits of Mercury, unusual meteors, and comets were reported to one of the American societies or, more rarely, to the Royal Society. James Bowdoin's speculations attempting to explain astronomical appearances were unusual. While he fortified his argument with various astronomical texts, Bowdoin also cited the Bible. The result was not very helpful.[33] The observers contributed much more, but scattered data, irregularly recorded, were no substitute for a regular course of observations made at an established observatory. Many Americans felt that the United States ought to have an observatory similar to those at Paris and Greenwich.

Attempts to establish a permanent institution had failed just before the Revolution, but there was reason to hope that better success might be met after the war. David Rittenhouse began to plan to build an observatory of his own as early as 1781. It was to help him in this project that Timothy Matlack sought financial aid from the Pennsylvania Assembly.[34] Within the next few years, Rittenhouse built his observatory—an octagonal brick structure located on Arch Street. There he soon undertook studies of the newly discovered planet Uranus. As late as 1792 it remained the only observatory in the United States known to the French astronomer, La Lande.[35]

This was not a public observatory of the sort the Reverend John Ewing had urged in 1775 with astronomers constantly on duty. Francis Hopkinson set on foot a new attempt to establish that kind of institution by engrafting it upon Rittenhouse's observatory. Emboldened by the favorable attitude of the assembly toward Rittenhouse's construction of his own observatory, Hopkinson framed a petition to create the post of "Astronomer to the State of Pennsyl-

32. American Academy, *Memoirs*, 1 (1785), 93; Ralph S. Bates, "The American Academy of Arts and Sciences," *Scientific Monthly*, 54 (1942), 267; Book of Eclipses, American Academy of Arts and Sciences.

33. American Academy, *Memoirs*, 1 (1785), 208-33.

34. *Journals of the House of Representatives of the Commonwealth of Pennsylvania*, ed. by Michael Hillegas (Philadelphia, 1782), I, 599, 611.

35. Barton, *Memoirs of Rittenhouse*, 165n, 166n.

vania" with Rittenhouse specifically named to the office.[36] The actual petition left the salary to be assigned the state astronomer blank, but Hopkinson was determined that it must be "at least £600, probably £750 pr. An." Three weeks later, he reported, "It is fixed at £500/ann." The assembly seriously considered the proposal, but despite expectations of "great beneficial effects and national reputation" which might result, it refused to assume this obligation.[37]

The state governments and even the federal government were much more ready to pay out money to astronomers in return for specific services that they required from them. The astronomers were called upon to make the observations needed to define state boundaries satisfactorily. This work was carried forward at a rapid rate during and after the war in a widespread effort to eliminate the heritage of disputes from the colonial period. The usual practice was for each state involved to appoint two or three commissioners to meet with men appointed by the other state. The commission then proceeded to obtain some points by astronomical observations and to draw the line between by the use of transit and chain. The line was usually cleared and marked by durable monuments. In this work most of the men interested in astronomy served at one time or another, both those with academic posts and those who were unattached. David Rittenhouse, John Ewing, Simeon DeWitt, Samuel Williams, John Page, the Reverend James Madison, Thomas Hutchins, Andrew Porter, and Andrew Ellicott were the most prominent among them.

David Rittenhouse was still almost universally considered to be the best of them. When the commission met to mark off the Pennsylvania-Virginia line, John Page maneuvered so that he would be assigned to the same party as Rittenhouse and then basked in reflected glory. Almost in a daze, he wrote, "As an Astronomer I doubt when the World can produce another for if every Instrument and Book upon Earth which Astronomers use were destroyed, I am certain, from the Instruments which he made and the manner in which I saw him use

36. Hopkinson to Jefferson, Feb. 23, 1784, Julian P. Boyd (ed.), *The Papers of Thomas Jefferson* (Princeton, 1950–), VI, 556; *Minutes of the General Assembly of Pennsylvania, Eighth*, 120.

37. Hopkinson to Jefferson, Feb. 23, 1784, March 12, 1784, Boyd (ed.), *Papers of Jefferson*, VI, 556; VII, 20; *Minutes of the General Assembly of Pennsylvania, Eighth*, 143, 166.

them, that he himself without any Assistance could replace Astronomy in the present State of its Perfection." [38] Page reflected something of his own incapacity in that statement, but Andrew Ellicott, who succeeded Thomas Hutchins as geographer of the United States, never failed to praise Rittenhouse's "abilities and industry." [39] For Rittenhouse, this service was always a chore—partly because of his poor health. In 1785 he was appointed by Congress along with Thomas Hutchins and John Ewing to run the long disputed New York-Massachusetts boundary. Samuel Williams also served for Massachusetts in this connection. Engaged for only seven weeks in this service, Rittenhouse announced that with its completion, he would "bid adieu, forever, to all running of lines." [40]

Other astronomers continued to define boundaries and to calculate the data required each year by the many almanacs issued all over the country. Almanacs provided another practical use for astronomy which yielded an income to many men: to Nehemiah Strong, Andrew Ellicott, Benjamin West, and Benjamin Workman—even to Benjamin Banneker, the Negro whose abilities so impressed Jefferson. Often the younger men and those who had not attained secure posts did this work.[41]

Orreries continued to appear valuable and desirable instruments. When Jefferson was planning reforms for the College of William and Mary in 1779, he included the acquisition of a Rittenhouse orrery to be called the "Ryttenhouse." [42] It was Jefferson, too, who set on foot the plan to make a gift of a Rittenhouse orrery to the King of France as a way of honoring America's ally. At the same time, he felt that "sending both Rittenhouse and his Orrery to Europe" would be the best way to confute "those flimsy Theorists" who disparaged

38. Page to Jefferson, April 28, 1785, Boyd (ed.), *Papers of Jefferson*, VIII, 118.

39. Andrew Ellicott to his wife, Aug. 8, 1786, in *Report of the Regents of the University on the Boundaries of the State of New York* (Albany, 1874), 262.

40. *New York Journal*, Sept. 3, 1789; Rittenhouse to Ellicott and Col. Andrew Porter, Sept. 29, 1789, in Barton, *Memoirs of Rittenhouse*, 240n.

41. Joseph T. Wheeler, *The Maryland Press, 1777-1790* (Baltimore, 1938), 109, 189; Charles L. Nichols, "Notes on the Almanacs of Massachusetts," American Antiquarian Society, *Proceedings*, new ser., 22 (1912), 27.

42. Thomas Jefferson, "A Bill for the Amending the Constitution of the College of William and Mary," Paul L. Ford (ed.), *The Writings of Thomas Jefferson* (New York, 1892-99), II, 235.

American learning.[43] Rittenhouse agreed to construct such an instrument for £200 and the King expressed himself ready to accept it, but it never materialized. After the King made gifts of books to the College of William and Mary and the University of Pennsylvania, Jefferson tried to revive the plan in a more modest form. In 1786, he called for a "Lunarium for the King" but even that proved impossible to get.[44]

It was in Boston that the only orrery to materialize during the Confederation period was constructed by Joseph Pope, a watchmaker. Bartholomew Burges showed the same city an apparatus that displayed planets in their proper orbits depicted by wires but he never dignified it by calling it an orrery.[45] Pope's gear-driven mechanism, completed in 1788, displayed the motions of the solar system in a horizontal plane as was common in European orreries, rather than in a vertical plane as Rittenhouse had done. Like Rittenhouse's it boasted eccentric and inclined orbits. Its handsome mahogany frame was embellished by bronze figures of Newton, Franklin, and Bowdoin, cast by Paul Revere. The £450 3s required to purchase the orrery was obtained by Harvard from a lottery authorized by the General Court for that purpose.[46]

America was not able to keep abreast of all the scientific advances made in Europe but scientific fads continued to sweep across the ocean with the smallest time lag imaginable. The most absorbing of these was the balloon fever which began in France on June 5, 1783, when the Montgolfier brothers made their first public trial of a small hot air balloon. In November, Pilâtre de Rozier made the first human flight in a larger balloon of this type and in December, the physicist, Jean-Alexandre-Césare Charles ascended in a hydrogen balloon. Shortly thereafter, "Montgolfières" and "Charlières" began to be raised from all parts of Europe. One of the early accounts of the bal-

43. The Rev. James Madison to Jefferson, Jan. 22, 1784, Jefferson Papers, Library of Congress.

44. The Rev. James Madison to Jefferson, April 10, 1785, ibid., Jefferson to Hopkinson, Jan. 3, 1786, Andrew A. Lipscomb and Albert E. Bergh (eds.), The Writings of Thomas Jefferson (Washington, 1903-4), V, 244; Jefferson to Rittenhouse, Jan. 25, 1786, ibid., 257.

45. Boston Gazette, Feb. 16, 1789.

46. Massachusetts Magazine, 1 (1789), 36, 37; Boston Gazette, Jan. 12, 1789; Cohen, Some Early Tools, 64-65.

loon rage received by the Royal Society was forwarded by Franklin.[47] Lafayette found too that Franklin had broken the news to the Philadelphians when he informed the American Philosophical Society of the new discoveries. Franklin, Lettsom, Lafayette, and Jefferson after he reached Paris, continued to supply pamphlets and news about ballooning to their friends, particularly in Philadelphia, Boston, and Virginia.[48]

The enthusiasm quickly spread over the United States where the physicians, perhaps because of their knowledge of chemistry, were more interested in experimenting than the professors. Dr. John Foulke seems to have been the first actually to raise a small paper balloon which he sent up from the garden of the Dutch minister on May 10, 1784. This success he followed with a lecture on pneumatics delivered the following week at the University of Pennsylvania.[49] At Yale, there was a flush of interest in 1785, the high point of which was the attempt to raise an eleven-foot hot-air balloon carrying a figure of an angel, an American flag, and a motto in seven languages! This was the work of Josiah Meigs and Ezra Stiles.[50]

The construction of a balloon large enough to carry a man was another matter. Such a project was presented to the American Philosophical Society by Dr. John Morgan but the society refused to support the enterprise because of rules restricting it from giving any opinion on matters submitted to it. Many of the members, however, and all of the professors of the University of Pennsylvania, subscribed to the ensuing effort to collect the needed funds.[51] The subscription

47. [?] to Franklin, Nov. 7, 1783, Franklin Papers, VIII, 33, University of Pennsylvania; Lettsom to Parke, Feb. 28, 1784, Etting Collection, Historical Society of Pennsylvania, Scientists, I, 54.

48. American Philosophical Society Minutes, March 19, 1784, [III]; not paged; Lafayette to Secretary American Philosophical Society, Dec. 10, 1783, Archives, American Philosophical Society; Jefferson to Hopkinson, Sept. 25, 1785, Boyd (ed.), Papers of Jefferson, VIII, 551; Franklin to Bowdoin, Jan. 1, 1786, Smyth (ed.), Writings of Franklin, IX, 479.

49. Hopkinson to Jefferson, March 31, 1784, May 12, 1784, Boyd (ed.), Papers of Jefferson, VII, 57, 246; Horace M. Lippincott, "Dr. John Foulke, 1780, a Pioneer in Aeronautics," General Magazine and Historical Chronicle, 34 (1932), 528.

50. Franklin B. Dexter (ed.), The Literary Diary of Ezra Stiles (New Haven, 1901), III, 157, 161; Boston Gazette, May 23, 1785.

51. American Philosophical Society Minutes, June 11, 1784, June 19, 1784, [III], not paged; Pennsylvania Gazette, June 30, 1784.

did not succeed. The first successful ascent in America—apparently in a captive balloon—was the work of Peter Carnes who sent a thirteen-year-old boy aloft in Baltimore on June 24, 1784. Three weeks later Carnes attempted a free ascent himself in Philadelphia. Fortunately he was thrown out of the balloon while it was just ten feet from the ground, for when it got well up into the air, it burst into flames and was consumed.[52] In New York, John Decker successfully sent a thirty-foot balloon aloft but never made the ascent himself.[53]

No American succeeded in making a free balloon flight, but several of them followed the soaring imaginations of leading Europeans more readily. In Paris Franklin was confronted with letter after letter proposing new types of "aerostatic machines" which he sent along for consideration by the committee of the Académie des Sciences appointed for that purpose.[54] From England, Richard Price wrote, "the last year will, I suppose, be always distinguished as the year in which mankind began to fly in France," and Sir Joseph Banks considered it the greatest achievement since the invention of shipping.[55] Franklin himself thought that the offensive possibilities of 5,000 balloons manned by two soldiers each might act as a deterrent to war.[56]

In America, the dreamers were just as active. Philip Freneau wrote of balloons being used in commerce as well as in war and even for trips to distant planets. A "science fiction" account of a trip to the moon was printed in Litchfield, Connecticut.[57] The newspapers and magazines were full of schemes for building balloons with sails, with wings, and with rudders, and for sending them around the world or merely across the ocean. Francis Hopkinson wanted to build a balloon

52. *Maryland Journal*, June 25, 1784; *Pennsylvania Journal*, July 10, 1784, July 17, 1784, July 21, 1784, July 24, 1784; Jeremiah Milbank, Jr., *The First Century of Flight in America* (Princeton, 1943), 21.

53. Dexter (ed.), *Diary of Stiles*, III, 367; *New York Journal*, Aug. 13, 1789, Aug. 20, 1789.

54. Franklin to M. Creuze, Feb. 14, 1784, Franklin Papers, XLV, 175, American Philosophical Society; Franklin to Le Roy, Feb. 25, 1784, *ibid.*, XLV, 176.

55. Price to Franklin, April 6, 1784, *ibid.*, XXXVI, 140; Banks to Franklin, Sept. 13, 1783, *ibid.*, XXIX, 146.

56. Franklin to Ingenhousz, Jan. 16, 1784, John Bigelow (ed.), *Complete Works of Benjamin Franklin* (New York, 1887-89), VIII, 432-33.

57. *An Account of Count D'Artois and his Friend's Passage to the Moon* (Litchfield, 1785); Fred L. Pattee (ed.), *The Poems of Philip Freneau* (Princeton, 1902-7), II, 276-79.

with an oblong form like the body of a bird or fish that would be driven through the air by a hand-actuated propeller.[58] It was left to James B. Pleasants of Georgetown, Maryland, to solve the problems of the flimsy texture of balloons and their great bulk by proposing a globular balloon made of cast iron or brass from which the air had been evacuated. There is no record that he ever flew one.[59]

A close relationship of chemistry to ballooning was demonstrated not only in the filling of balloons, but in the work of Dr. John Jeffries. Jeffries, a Boston loyalist émigré was the only American who did fly at this time. With the professional French balloonist Jean Pierre Blanchard, Jeffries made the first flight across the English channel in 1785. More important was his work in obtaining specimens of the upper air which Henry Cavendish analyzed and found not appreciably different from air at sea level.[60]

For the most part, chemistry in America remained an insignificant part of general college courses in natural philosophy or a minor interest of the physicians. The Americans were kept informed of Priestley's work but they were not equally acquainted with Lavoisier. Jefferson criticized Buffon for considering chemistry mere cookery, describing it himself as "among the most useful of sciences, and big with future discoveries for the utility and safety of the human race." Yet, he showed very little sympathy with the developing chemical revolution while he was located at its center in Paris. He considered Lavoisier's reforms as "premature" and he predicted that his jargon would retard progress in the field.[61] At Harvard, however, Brissot de Warville was delighted to find that Dr. Aaron Dexter "gave lectures on the experiments of our school of chemistry. The excellent

58. *Pennsylvania Journal*, July 14, 1784; *Boston Magazine*, 1 (1783-84), 147-48, 433-36, 476-78; *Boston Gazette*, March 8, 1784; Hopkinson to Franklin, May 24, 1784, Franklin Papers, XXXVI, 185, American Philosophical Society.

59. James B. Pleasants to Franklin, Oct. 1788, Ms. Communications to American Philosophical Society, Natural Philosophy, I, 16.

60. John Jefferies to Franklin, June 25, 1785, Franklin Papers, XXXIII, 146; A. Wolf, *A History of Science, Technology, and Philosophy in the Eighteenth Century* (New York, 1939), 580.

61. Jefferson to the Rev. James Madison, July 19, 1788, E. Millicent Sowerby, *Catalogue of the Library of Thomas Jefferson* (Washington, 1952–), I, 374, 378; Charles A. Browne, "Thomas Jefferson and Agricultural Chemistry," *Scientific Monthly*, 60 (1945), 55, 57.

work of my respectable master Dr. Fourcroy... taught him the rapid strides that this science had lately made in Europe." [62] Although an attack on Priestley's infidelity was published in Philadelphia, the scientists of that city were generally held to the great defender of phlogiston by their closer ties with him and his circle. [63] In an address to the Medical Society of New Jersey, Dr. Jonathan Elmer considered the phlogiston dispute and decided more experiments were needed to establish the "new science of chemistry." [64]

Most American publication was confined to less momentous problems in chemistry. Benjamin Rush, first professor of chemistry in the country, tried to apply his "favorite study" to "Domestic and Culinary Purposes" in a series of lectures he delivered at the Young Ladies' Academy. He published a syllabus of these lectures in 1787. [65] Two of the younger physicians, Dr. Caspar Wistar in Philadelphia and Dr. Samuel Latham Mitchill in New York, conducted independent studies of evaporation at almost the same time. Wistar explained the phenomenon as a heat transfer problem without any reference to humidity or saturation. He saw in it the reverse of Franklin's explanation of condensation as the transfer of heat to the colder body and the deposit of moisture. Mitchill did not suffer so much from being close to a great man. Referring to Scottish thinkers, he spoke of the varying capacity of the air to hold water and presented a more satisfactory explanation. [66] Analyses of mineral waters continued to be received and published by the philosophical societies, the most significant

62. J. P. Brissot de Warville, *New Travels in the United States of America, Performed in 1788* (Dublin, 1792), I, 108-9.

63. John Stancliff, *An Account of the Trial of Doctor Joseph Priestley* (Philadelphia, 1784); see the series of letters from Priestley to Rush, Rush Papers, XXX, 56-71, Library Company, Philadelphia; Priestley to Franklin, Dec. 10, 1781, Bache Collection; American Philosophical Society Minutes, March 19, 1784, [III], not paged.

64. Jonathan Elmer, "On the Chemical Principles of Bodies," *Columbian Magazine*, 2 (1788), 493-97.

65. Benjamin Rush, *Syllabus of Lectures, Containing the Application of the Principles of Natural Philosophy and Chemistry, to Domestic and Culinary Purposes* (Philadelphia, 1787).

66. Caspar Wistar, "Experiments and Observations on Evaporation in Cold Air," American Philosophical Society, *Transactions*, 3 (1793), 125-33; Samuel L. Mitchill, "Experiments and Observations on the Evaporation of Water," *Columbian Magazine*, 1 (1786-87), 584-87.

of these being Dr. Samuel Tenney's report on the springs at Saratoga, New York.[67]

More fundamental thinking was done in physics than in chemistry—most of it by Benjamin Franklin and David Rittenhouse. Franklin was able to return to speculative thought during his voyage back to the United States in 1785 when he wrote his papers on "Maritime Observations" and the "Cause and Cure of Smoky Chimneys." Rittenhouse emerged from the war with a still growing reputation which was fed by a continuing series of papers he published and by his extraordinary ability to impress everyone he met with his command of learning. Chastellux was almost alone in discovering limitations; he declared that Rittenhouse was "not a mathematician of the class of the Eulers, and the D'Alemberts." This was almost a self-evident truth but one that escaped most of his admirers. Yet even Chastellux admitted that Rittenhouse was "perfectly acquainted with the motions of the heavenly bodies" and that as a mechanic he was a natural genius.[68] Throughout the states he was honored. The American Academy elected him a fellow, the College of William and Mary awarded him a master of arts degree, and the College of New Jersey conferred upon him the degree of doctor of laws. In 1787, a collection of his papers was published in London.[69] Rittenhouse had a different sort of mind from Franklin's. It was more limited and it was not served by the extensive associations throughout the world that were a part of Franklin's richness. Yet, in the 1780's, the younger man demonstrated a precision of thought that occasionally led to striking insight. Rittenhouse had a facility with mathematics and a knowledge of astronomy that Franklin lacked. He did not possess the same full measure of genius.

Franklin's initial scientific interest, electricity, still commanded wide attention, although he did very little with it. Experiments continued to be performed upon electrical eels, and the effects of lightning remained sufficiently newsworthy to find places in the journals of both the Philosophical Society and the American Academy.[70] The

67. American Academy, *Memoirs*, 2 (1793), 43-61; American Philosophical Society, *Transactions*, 2 (1786), 197-98.

68. Marquis de Chastellux, *Travels in North-America in the Years 1780-81-82* (New York, 1827), 112.

69. David Rittenhouse, *Philosophical Papers* (London, 1787).

70. American Philosophical Society, *Transactions*, 3 (1793), 119-22; American Academy, *Memoirs*, 1 (1785), 247-52, 253-59.

academy also received one paper by Hugh Maxwell which noted that certain species of trees were more frequently struck by lightning than others. If this proved to be a function of the nature of the wood rather than the shape of the trees, he suggested that it would be well not to construct buildings of that wood. He did nothing to test the idea.[71]

Much the most important paper in the field of electricity was David Rittenhouse's "Experiments in Magnetism." The actual experiments he related were not particularly novel; they showed how a steel rod could be magnetized when held in a magnetic field and struck a blow. In interpreting these results, however, Rittenhouse enunciated the molecular theory of magnetism by postulating numerous magnetic dipoles within the rod which were aligned with the north poles pointing the same direction whenever the rod was magnetized. This concept was in the air as Rittenhouse wrote but it had never been quite so clearly stated.[72] Franklin's thoughts on magnetism were directed toward an attempt to explain the cause of the magnetism of the earth.[73]

It was Franklin's earlier speculations on the nature of light that provided the jumping-off point for James Bowdoin's essays on the subject. Franklin, before the war, had advanced a wave theory of light opposed to the prevailing corpuscular theory identified with Sir Isaac Newton. Bowdoin tried to answer the objections Franklin had raised and was able to do so with some success without calling to witness many of the authorities he might have used. Price praised his ideas on the nature of light and Franklin said he was sure they would help him to understand the subject better.[74]

Franklin refused to enter a controversy with Bowdoin but he did further extend his support of the wave theory in a paper he first called "Loose Thoughts on a Universal Fluid" and later "A new and curious Theory of Light and Heat." Here he suggested that heat might also be explained in terms of the subtle fluid which was presumed to fill space and which, when in vibration, accounted for light.

71. American Academy, *Memoirs*, 2 (1793), 143-44.
72. American Philosophical Society, *Transactions*, 2 (1786), 178-80.
73. *Ibid.*, 3 (1793), 1-5; Franklin to Bowdoin, May 31, 1788, *Bowdoin and Temple Papers*, II, 190.
74. Price to Bowdoin, Oct. 25, 1785, *ibid.*, 78; American Academy, *Memoirs*, 1 (1785), 208-33; Franklin to Bowdoin, Jan. 1, 1786, Smyth (ed.), *Writings of Franklin*, IX, 479.

He went on to discuss the nature and behavior of heat, tying into this larger theory some of his ideas developed in the course of his thought upon stoves and smoky chimneys.[75]

Rittenhouse was interested in optics rather than in the basic character of light. In 1786 he thought of giving a course of public lectures in that subject but never did. In attempting to answer a problem posed to him by Francis Hopkinson, he conducted several experiments which led him to the discovery of the principles of the diffraction grating. He found that when light passed through a hair grid, it was dispersed in much the same way as in passing through a prism. He measured the angles by which rays of light were bent and he described the whole process, without analyzing it very extensively.[76] Rittenhouse also solved the optical problem of why objects seen through a compound microscope and some telescopes appeared raised when they were actually depressed or depressed when they were actually raised. The answer to this cameo-intaglio illusion lay in the reversed pattern of light and shadow caused by an inverted image— not a very profound problem but one that was answered by Rittenhouse as well as it could be.[77] All of this work was well above the level of other American papers on optics, such as those of Andrew Ellicott and Samuel Latham Mitchill on looming, a kind of mirage.[78]

When men of lower attainments than Franklin and Rittenhouse turned to basic problems of physics, the results were much less satisfactory than when such men tried to describe a plant or a thunderstorm. *A New System of Philosophy; or the Newtonian Hypothesis Examined* was written in 1783 "By an American" who showed his strong nationalism in asserting that Franklin had revealed "a genius equal to any that ever went before him." Franklin's experiments, he held, contradicted Newton's idea that all matter was originally the same. With that beginning, his essay was never able to get back

75. Franklin draft, "Loose Thoughts on a Universal Fluid," Smyth (ed.), *Writings of Franklin*, IX, 227-30; American Philosophical Society, *Transactions*, 3 (1793), 5-8.

76. American Philosophical Society, *Transactions*, 2 (1786), 201-6; Barton, *Memoirs of Rittenhouse*, 603.

77. American Philosophical Society, *Transactions*, 2 (1786), 37-42.

78. *American Magazine*, 1 (1788), 575-79; American Philosophical Society Minutes, Nov. 21, 1788, [IV], 67.

on the track.[79] Still more illiterate was Isaac Ledyard's *Essay on Matter in Five Chapters,* a rambling account published the following year, which demonstrated what could happen when the uninformed accepted the idea that the "book of nature" was open for all to read.[80] Charles Vancouver's *General Compendium; or Abstract of Chemical, Experimental, and Natural Philosophy* was not that bad. Vancouver, who was well read, sought to disseminate knowledge rather than to add anything new.[81] He completed only the first forty-eight pages of a projected four volume work. Dr. Joseph B. Ladd was educated in the general sense of the word, too, but his *Essay on Primitive, Latent and Regenerated Light* was a strange and unprofitable performance.[82]

The greatest activity was demonstrated in the invention of scientific instruments—usually designed to serve distinctly utilitarian purposes. Rittenhouse wrote Jefferson, "We have abundance of projectors and pretenders to new Discoveries ... some of them ridiculous enough, The Self moving Boat, the Steam Boat, the Mechanical Miller, the improved Ring Dial for finding the Variation of the Needle. The Surveying Compass to serve 20 other purposes, And a project for finding the Longitude by the Variation of the Magnetical Needle." [83] Hopkinson almost echoed him, adding, "Amongst these semi-lunatic Projectors I must not omit myself—I have sent in to the philosophical Society a Contrivance for the perfect Measurement of Time and I see no Reason *in Theory* why it should not answer—I am now making Experiments to maintain *the Fact.*" [84]

Interest in instruments more than once led men to study the more basic problems to which better instruments sometimes provided a key.

79. *A New System of Philosophy; or the Newtonian Hypothesis Examined* (Poughkeepsie, 1783), preface.

80. [Isaac Ledyard], *An Essay on Matter in Five Chapters* (Philadelphia, 1784); cf. Charles H. Wharton, "A Few Observations on a Late Pamphlet Entitled an Essay on Matter," in *A Reply to an Address to the Roman Catholics of the United States of America* (Philadelphia, 1785), Appendix 1-9.

81. Charles Vancouver, *A General Compendium; or Abstract of Chemical, Experimental, and Natural Philosophy* (Philadelphia, 1785), I, [v]; see Sowerby, *Catalogue,* I, 349.

82. Joseph B. Ladd, *An Essay on Primitive, Latent and Regenerated Light* (Charleston, [n.d.]), especially 12, 16.

83. Rittenhouse to Jefferson, April 14, 1787, Jefferson Papers.

84. Hopkinson to Jefferson, July 8, 1787, *ibid.*

Both Franklin and Rittenhouse designed hygrometers that led them to further speculations upon humidity, cloud formation, and weather.[85] The gift of a thermometer, barometer, hygrometer, and variation needle to Harvard by the Meteorological Society of Mannheim led to a much heightened interest in weather on the part of Samuel Williams. Williams had given thought to problems of weather before. He had correctly explained the dark day of May 1780 as the result of a forest fire when many other incorrect solutions were offered. After he began to make the observations requested from Mannheim, his activities were expanded. By writing individuals at some distance from Cambridge, he sought "to ascertain the climate, in different parts of America, by meteorological observations." He also asked information about the appearance of fruits, vegetables, and animals.[86] With this impetus, the meteorological records deposited in the American Academy increased still faster.

Williams may also have spurred the Philadelphians into action. They could hardly have been pleased that Harvard had received one of the thirty-odd sets of instruments sent out from Mannheim while none had been given to them. During the war the American Philosophical Society had tried to collect observations on the severe winter of 1779 to 1780 but had received responses only from Virginia, Maryland, and New Jersey.[87] After that, only scattered papers were submitted: plans for rain gauges and meteorological records. In 1788, two of the latest hygrometers developed by Jean André De Luc were sent to the society by Benjamin Franklin. One of them was lent to the French immigrant, Pierre Le Gaux, who proceeded to make regular

85. Franklin to Nairne, Oct. 18, 1783, Smyth (ed.), *Writings of Franklin*, IX, 109; American Philosophical Society, *Transactions*, 2 (1786), 51-56; Barton, *Memoirs of Rittenhouse*, 585-86; *Columbian Magazine*, 1 (1786-87), 301.

86. American Academy, *Memoirs*, 1 (1785), 234-46; Samuel Vaughan to Marshall, May 22, 1786, William Darlington, *Memorials of John Bartram and Humphry Marshall* (Philadelphia, 1843), 558; *Columbian Magazine*, 2 (1788), 310-13.

87. *Pennsylvania Journal*, March 22, 1780; *Virginia Gazette* (Dixon and Nicholson), April 8, 1780; American Philosophical Society Minutes, Feb. 6, 1781, March 16, 1781, [II], not paged; the Rev. James Madison to Rittenhouse, Nov. 1780, Ms. Communications to American Philosophical Society, Mathematics and Astronomy, I, 17; William C. Houston, Feb. 22,[?], *ibid.*, Natural Philosophy, I, 1.

meteorological observations which he returned to the society and published in the *Columbian Magazine*.[88]

Another field in which instruments were of the greatest importance was the still unsolved problem of the earth's magnetic field. Colonial interest in this phenomenon had not led to so imposing a collection of records as had been accumulated of weather conditions. The American Academy, shortly after its founding, recommended that observations be made of magnetic variation "in various parts of the country," especially by surveyors.[89] The response was limited to Harvard. Joseph Willard made observations at Beverly and Stephen Sewall at Cambridge, both of them tracing daily changes in variation.[90] Samuel Williams consulted the old records kept by Thomas Brattle and John Winthrop to show that the variation at Cambridge had decreased over the years from 9° 0' West in 1708 to 6° 52' West in 1783.[91] For the measurement of magnetic variation, the Harvard group had a variation needle made by Edward Nairne, an azimuth compass invented by Dr. Gordon Knight, and after 1785, the variation needle from Mannheim. Williams went on with his studies until he had collected data on variation in Massachusetts, Vermont, New Hampshire, Maine, Rhode Island, and New York.[92]

Those who tried to do more than collect observations on variation quickly got into trouble. Franklin adhered to the generally accepted idea that the source of terrestrial magnetism was a magnetized core of iron ore, but he overextended himself in suggesting that it might have received its magnetism from a field existing throughout all space.[93] The Reverend Temple Henry Croker of St. Christopher reported experiments which led him to conclude that there was no internal lodestone in the earth. The dipping of the needle, he ascribed,

88. Lewis Nicola to [?], Feb. 20, 1781, *ibid.*, 10, 10a; Meteorological Observations, Dec. 1787, *ibid.*, 15, 17; American Philosophical Society Minutes, Aug. 15, 1788, March 5, 1784, [III], not paged; *Columbian Magazine*, 1 (1786-87), 101-3.

89. American Academy, *Memoirs*, 1 (1785), 318.

90. *Ibid.*, 318-26.

91. *Ibid.*, 68; American Philosophical Society, *Transactions*, 2 (1786), 122-41.

92. Samuel Williams, *The Natural and Civil History of Vermont* (Walpole, N.H., 1794), 378.

93. Franklin to Bowdoin, May 31, 1788, *Bowdoin and Temple Papers*, II, 190-92.

instead, to a loss of gravity by the south end of the needle when it was magnetized. Rittenhouse brushed this notion aside as "founded on a Mistake." [94] The still more fantastic scheme of John Churchman was not so easy to suppress.

It was clear to everyone that if the pattern of variation could but be known precisely, it would provide a marvelous aid to navigation. John Churchman, a surveyor and mapmaker, turned his attention to magnetic variation with the intention of putting it to practical use. [95] To account for the apparent irregularities in the changing variation at any given point, Churchman postulated two satellites revolving about the earth, one around the North Pole and the other around the South Pole. Their periods of revolution would be lengthy, the northern satellite taking 463 years to complete its circuit. He explained the fact that they had never been sighted by placing them so close to the earth that they could not be seen from the lower latitudes. The committee of the American Philosophical Society that considered Churchman's theory, properly dismissed it as "groundless . . . whimsical . . . and impossible in the nature of things." [96]

That rebuff did not stop John Churchman. He went over the heads of the American philosophers. He wrote to the president and to the secretary of the Royal Society, to the secretary of the Board of Longitude, to Jefferson asking him to submit his scheme to the Académie des Sciences, and he asked Franklin to write some of his "learned friends in Europe." [97] The most considered reply came from Nevil Maskelyne. The astronomer royal said that magnetic variation was not useful for finding longitude as Churchman sought to do, especially when there were so much better methods. As for the general theory, he said, "Mr. Churchman might have been well satisfied with

94. Temple Henry Croker to Franklin, June 11, 1787, with endorsement by Rittenhouse, Ms. Communications to American Philosophical Society, Natural Philosophy, I, 14.

95. John Churchman, *An Explanation of the Magnetic Atlas, or Variation Chart, Hereunto Annexed: Projected on a Plan Entirely New* (Philadelphia, 1790), [i]; *Pennsylvania Journal*, Sept. 8, 1777.

96. Churchman to American Philosophical Society, Feb. 7, 1787, Ms. Communications to American Philosophical Society, Natural Philosophy, I, 12; American Philosophical Society Minutes, Feb. 16, 1787, [III], not paged.

97. Churchman, *Explanation*, Appendix 1-4; Jefferson to Churchman, Aug. 8, 1787, Jefferson Papers; Franklin to Maskelyne, March 29, 1787, Smyth (ed.), *Writings of Franklin*, IX, 557.

the judgement of such able men and good philosophers as Mr. Ewing and Mr. Rittenhouse." [98]

Still, Churchman persisted. He patched and repaired his theory where it seemed most vulnerable. When he learned from Maskelyne's letter that Leonard Euler had written a more esteemed discussion of terrestrial magnetism, he looked it up and penned his own reply.[99] The satellites, he decided, might not be above the earth but rolling around on its surface—or possibly below the surface. In any case, it would be necessary to send an expedition to Baffin Bay, where Churchman said the north magnetic pole was located, to find out just what the cause was. With the enthusiastic support of James Madison, he presented the first Congress to meet under the Constitution with a request that they finance such an expedition. This memorial was re-jected—not because it was a crack-brained scheme but because of deranged finances and because Congress felt that a ship's captain did not have to know the cause of variation to make use of it. Before he finally gave up, Churchman carried his great idea even to England and the continent.[100]

Anything as practical as an aid to navigation had the best chance of support but only the most tenuous relationship usually existed be-tween basic science and practicable improvements in navigation. Franklin in his "Maritime Observations" suggested one hopeful navi-gational technique derived from his study of the differing temperature of the gulf stream and surrounding water. His nephew, Jonathan Williams, went on to develop this technique into a kind of thermo-metrical piloting. The idea was theoretically satisfactory but in prac-tice only a little more useful than Churchman's scheme.[101] Attempts

98. Maskelyne to Franklin, March 3, 1788, Franklin Papers, XXXIV, 31, American Philosophical Society; *Pennsylvania Packet*, July 31, 1788.

99. Charles Blagden to Churchman, July 6, 1789, Churchman, *Explanation*, Appendix 2n.

100. Churchman, *Explanation*, Appendix 5; Irving Brant, *James Madison, Father of the Constitution, 1787-1800* (Indianapolis, 1950), 331-32; *Universal Asylum*, 9 (1792), 284; A. Day Bradley, "John Churchman, Jr. of Notting-ham," *The Bulletin of the Friends Historical Association*, 43 (1954), 20-28.

101. Jonathan Williams, *Thermometrical Navigation* (Philadelphia, 1799), 10, 99; American Philosophical Society, *Transactions*, 2 (1786), 315; American Philosophical Society Minutes, Nov. 19, 1790, [IV], 151; Lawrence C. Wroth, *Some American Contributions to the Art of Navigation* (Providence, [n.d.]), 31.

to improve navigation by better instruments were on safer ground. Better quadrants, improved ring dials, and other instruments were invented by men, most of whom made no pretense of advancing basic scientific knowledge.[102] Basic science was still not very practical and, if it had been, there were not many in America to apply it properly.

102. George Wall, Jr., *A Description, With Instructions for the Use of a Newly Invented Surveying Instrument, Called the Trigonometer* (Philadelphia, 1788); Arthur Grier to Franklin, April 24, 1789, Ms. Communications to American Philosophical Society, Mathematics and Astronomy, I, 22; Daniel Byrnes to Franklin, Aug. 15, 1788, *ibid.*, 19; American Philosophical Society Minutes, Dec. 9, 1784, [III], not paged.

Chapter Sixteen

THEORY AND PRACTICE

No INFORMED PERSON would have asserted that the basic sciences flourished in the United States as productively as they did in the nations of Western Europe. Looking back in 1803 upon American achievements, the Reverend Samuel Miller had little difficulty in singling out the causes. "Genius," he said, "must be nourished by patronage, as well as strengthened by culture." American colleges and libraries were defective. Few individuals could find the requisite leisure for profound study: there were no rich fellowships in the universities, no large ecclesiastical benefices, and the academic chairs carried such small salaries that they presented "little temptation to the scholar." The state offered small additional help. "Besides," Miller reflected, "the spirit of our people is *commercial*. It has been said, and perhaps with some justice, that the *love of gain* peculiarly characterizes the inhabitants of the United States. The tendency of this spirit to discourage literature is obvious." [1]

Even admitting this situation, not everyone agreed that its results had been bad. The characteristic attitude of the 1780's was that the United States might not be distinguished for its basic science and its fine arts but that in applied science and the practical arts the record was very good. Leman Thomas Rede published an analysis in 1789 that was remarkably close to the American view—for an Englishman. With some vehemence Rede declared, "North America may want

1. Samuel Miller, *A Brief Retrospect of the Eighteenth Century* (New York, 1803), II, 406-7.

some of the fopperies of literature. She boasts not those dignified literati, who in Europe obtain adulation from the learned parasite, and applause from the uninformed multitude, for pursuits that terminate in no addition to the real elegancies or conveniences of living. She may, however, claim the possession of all useful learning." The emphasis upon utility was clearly revealed by the American taste in books. Rede found, "Whatever is useful, sells; but publications on subjects merely speculative, and rather curious than important, and generally such on arts and sciences, as are voluminous and expensive, lie on the bookseller's hands." [2]

There was a strong anti-intellectual current in this emphasis upon the utilitarian and it was not confined to the barely literate. Frequent charges were made that the colleges erred not in neglecting to attend to fundamental learning as Samuel Miller charged, but in devoting too much time to useless sciences. College graduates were told in one magazine, "You are not to live in the sun, nor moon, nor to ride upon the tails of a comet.... A *few* astronomers are enough for an age." [3] Francis Hopkinson was strongly conditioned by the vision of the open book of nature in his estimate of the educational system. He told Franklin that too much time was spent upon logical and metaphysical subtleties of no use in life while the practical branches of knowledge were neglected.[4] But for the strong conviction that even basic science would ultimately prove useful, this complex of thought might have led to a reaction against science comparable to the feeling that did develop against the dead languages.

The solution was to bend science to useful purposes. This appealed to those who sought the leisure to extend fundamental learning as well as those who wanted to minimize fundamental learning in favor of practical pursuits. Both could harmonize their feelings in the pursuit of useful knowledge. This solution had been found long ago—it was memorialized in the name of the American Philosophical Society, held at Philadelphia, for Promoting Useful Knowledge—but periodically

2. [Leman T. Rede], *Biblioteca Americana* (London, 1789), 9, 18; Stuart C. Sherman, "Leman Thomas Rede's *Biblioteca Americana*," *William and Mary Quarterly*, 3d ser., 4 (1947), 340.

3. *Universal Asylum*, 5 (1790), 78-79.

4. Hopkinson to Franklin, May 24, 1784, Franklin Papers, XXXI, 185, American Philosophical Society.

it had to be restated to keep the objective clear. One such statement was received by the American Academy from Mr. A. Crocker of Somerset, England, in 1789. It so appealed to the Harvard professors and other officers of the academy that they published it in their ensuing volume of *Memoirs* along with the rest of his "Essay on Raising Apple Trees, and Making Cider." Mr. Crocker wrote, "Notwithstanding I hold all due respect for *mathematical* and *philosophical disquisitions* (two important objects of your institution) yet I conceive that the world, in general, derive a principal advantage from the *due application* of the sciences to *the common concerns of life*." [5]

There was hardly anyone to be found who would stand up to oppose the application of science to practical purposes; the big problem was how to apply it. Unquestionably, the major effort was made in medicine with new schools, new societies, and many publications giving evidence of a conviction that the healing art ought to be based upon science. Physicians, however, had long sought to establish this relationship; the best of them already had a superior knowledge of the sciences related to medicine. With the three supports of mankind, agriculture, manufactures, and commerce, there was much less to build on. The concept of useful knowledge had always implied the application of science to these pursuits but the results of colonial efforts to do so had been tenuous at best.

After the Revolution, agriculture received more attention than any other economic activity. As in the colonial period the desire to improve agriculture involved the introduction of new crops but there was no new urgency in these efforts. The old enthusiasm for silk culture was never quite recaptured. Halfhearted efforts to encourage its revival were made by both the American Academy and the American Philosophical Society but without important effect. In Connecticut, there was more activity than in the other states. There a company was formed and, on his own, Nathaniel Aspinwall raised an estimated 100,000 worms. After this success Aspinwall tried to awaken the interest of the men of learning and the legislatures in New York and Pennsylvania, without very extensive success. [6] The greatest effort to

5. American Academy, *Memoirs*, 2 (1793), 100.
6. American Philosophical Society, *Transactions*, 2 (1786), 347-66; *American Magazine*, 1 (1788), 431; *American Museum*, 2 (1787), 354-55; *New Haven Gazette and Connecticut Magazine*, 1 (1786), 176; 3 (1788), no. 19; Franklin

promote the growth of hemp was made in Massachusetts by the American Academy.[7] Elsewhere advocates appeared to urge the cultivation of grapes, ginseng, and maple sugar. Mangel wurzel, or the scarcity root, which John Coakley Lettsom urged with such persistence, was tried in some parts of America and talked about in many.[8] More profitable were the trials made by Henry Laurens in South Carolina with varieties of rice from China, Sumatra, Italy, and Egypt.[9]

It was in improved techniques for raising established crops that a heightened interest was demonstrated. This showed itself in articles on means of protecting fruit trees from pests, techniques of grafting, courses of rotation of crops, new seed drills, and useful fertilizers.[10] Much of the information upon which this advice was based came from England where there was an increasing desire to apply science to the techniques of agriculture. The *Columbian Magazine* reprinted one pertinent article by Dr. Anthony Fothergill of the Bath Agricultural Society "On the Application of Chemistry to Agriculture, and the Rural Economy." Fothergill discussed the properties of different types of soil which permitted one to retain water better than another, but he offered no handbook of agricultural chemistry. Instead, he entered a plea for study that would lead to the application of science to agriculture.[11] Humphry Marshall also projected a plan of study in a paper he called "Observations of the propriety of applying Botanical knowledge to Agriculture, feeding cattle, &c." [12] Those

---

B. Dexter (ed.), *The Literary Diary of Ezra Stiles* (New Haven, 1901), III, 229, 351; [James Swan], *National Arithmetick* (Boston, [1786]), 31; *An Essay on the Culture of Silk* (Philadelphia, 1790); *Worcester Magazine*, 1 (1786), 57.

7. *Worcester Magazine*, 3 (1787), 34-35, 44-46; *Boston Gazette*, March 30, 1789.

8. *Pennsylvania Packet*, Aug. 12, 1788; *Boston Gazette*, March 23, 1789; *Columbian Magazine*, 2 (1788), 543; *American Museum*, 2 (1787), 448; *New Haven Gazette and Connecticut Magazine*, 3 (1788), no. 24, no. 31.

9. Henry Laurens to John Vaughan, July 23, 1789, Misc. Mss., American Philosophical Society.

10. *Pennsylvania Packet*, Dec. 27, 1788, Jan. 30, 1789; *American Museum*, 2 (1787), 397-98; *American Magazine*, 1 (1788), 4-5; *Columbian Magazine*, 2 (1787), 632, 733, 826; *Worcester Magazine*, 1 (1786), 56; 2 (1786-87), 323; 3 (1787), 58, 93-94, 105-6, 254-55; *Massachusetts Magazine*, 1 (1789), 72, 84-87, 300-1; American Academy, *Memoirs*, 1 (1785), 388-89.

11. *Columbian Magazine*, 2 (1788), 754-57.

12. Timothy Pickering to Marshall, Feb. 15, 1786, William Darlington, *Memorials of John Bartram and Humphry Marshall* (Philadelphia, 1843), 558.

who were competent to judge understood that a great deal of work had to be done before scientific knowledge could be used in actual farming.

It had been the hope of the founders of America's two learned societies that those organizations would be able to assist in the application of science to life, particularly to agriculture. That was Manasseh Cutler's idea in compiling his botanical catalogue.[13] That was the intention of the Pennsylvania Assembly when it made a grant of £150 to the American Philosophical Society "for the purpose of encouraging agriculture and commerce, by enabling that learned body to obtain such discoveries as have been made in *Europe* and other countries.[14] That, too, was the purpose of the legislature of Massachusetts when in incorporating the American Academy it declared, "The Arts and Sciences are the foundation and support of agriculture, manufactures, and commerce." [15] In the case of neither society were the expectations fulfilled.

The American Academy collected, and in its first volume of *Memoirs* published, a number of papers upon agricultural subjects. Early in 1786, it moved toward a more positive program by appointing a committee on agriculture chosen so that all parts of the commonwealth were represented.[16] The committee sought not only to stimulate experiments and to collect papers but to encourage better agricultural techniques among the farmers. To accomplish this, they followed the example of the English county agricultural societies in offering a list of premiums for essays, experiments, and achievements in farming. Essays which they reviewed favorably they published in the newspapers and magazines, Isaiah Thomas's *Worcester Magazine* being particularly responsive to their communications.[17]

One of the most important agricultural accomplishments of the American Academy was to encourage the Reverend Samuel Deane in his study of the subject. Deane, a Congregational clergyman of Dedham, Massachusetts, had carried out numerous agricultural experiments and had a broad interest in useful sciences. One of his

13. American Academy, *Memoirs*, 1 (1785), 398-99.
14. *Minutes of the General Assembly of Pennsylvania, Eighth*, 86.
15. American Academy, *Memoirs*, 1 (1785), iv.
16. *Massachusetts Centinel*, Feb. 18, 1786.
17. *Ibid.*, June 7, 1786; *American Museum*, 2 (1787), 355-56.

papers was published in the Academy's first volume of *Memoirs* and he sent them another "on Indian Corn." [18] When he decided in 1788 to publish a book on farming, the agricultural committee helped his campaign for subscriptions by recommending it "to the Publick as deserving their Encouragement and Subscription." The book appeared in 1790 as *The New England Farmer; or Georgical Dictionary*. No mere compendium or digest of English writings on agriculture, it was based upon Deane's own experience and a familiarity with American as well as European work. It had many of the merits of Jared Eliot's earlier essays but was more comprehensive in scope. [19]

Even before the agricultural committee's program was well under way, James Swan criticized it on the grounds that the academy could not give the required attention to agriculture because its aim was "universal investigation." That, he said, was the reason it had been unable to raise more than $300 for its premium program. In its stead, he proposed a specialized society dedicated wholly to the promotion of agriculture and financed by a large proportion of the dirt farmers in Massachusetts. [20]

The dirt farmer with his distrust of book learning was not yet ready for that kind of an organization, but in Maryland, another man took a more practicable route. John Beale Bordley was a gentleman farmer with an estate on Wye Island in the Chesapeake whose interest in the study of agriculture was "first elicited" by Jethro Tull's *Horse Hoeing Husbandry*, one of the pivotal writings of the agricultural revolution. Thereafter, he read extensively, depending much upon the books of Arthur Young, who was probably the greatest English publicist of the new agriculture. [21] Bordley not only read; he tried experiments in his own fields, using the methods advocated by Young and then varying them. In 1784, he published *A Summary View of the Courses of Crops, in the Husbandry of England and Maryland* in which he revealed some of the results of his experiments. He found that the productivity of

18. *Boston Gazette*, Dec. 8, 1783; *Massachusetts Centinel*, Feb. 18, 1786; American Academy, *Memoirs*, 1 (1785), 378-79.

19. *Boston Gazette*, Sept. 29, 1780; Samuel Deane, *The New England Farmer; or Georgical Dictionary* (Worcester, 1790); *Universal Asylum*, 6 (1791), 175-78.

20. [Swan], *National Arithmetick*, 34.

21. [John Beale Bordley], *Sketches on Rotations of Crops* (Philadelphia, 1797), 66.

Maryland farms on the eastern shore was very much lower, acre for acre, than the productivity of English farms. An equivalent field of 289 acres, he estimated, would yield an income of £170 in Maryland; in England using the common method of rotation of crops it would yield £400. If the improved "Norfolk" system were used, it would yield £870. He advocated, however, a modification of the Norfolk pattern of crop rotation that would make much use of clover, but none of Lord Townshend's much vaunted turnips which did not appear to do so well in America as in England.[22]

Conversant with the county agricultural societies of England, Bordley attempted to form his Maryland neighbors into such a group. It did not work. The scattered nature of their holdings and their own indifference to the promotion of agricultural knowledge did not provide the needed support. Instead, he went to Philadelphia where his project was received with enthusiasm by a well-to-do group of Philadelphians, "only a few of whom were actually engaged in husbandry."[23] On February 11, 1785, the first meeting of the Philadelphia Society for Promoting Agriculture was held for the purpose of "promoting Improvements" in agriculture "within the States of America."[24]

A significant element of the early membership of this society was made up of former members of the old American Society who had drifted away from the American Philosophical Society after the union in 1769. Among them, were Charles Thomson, George Morgan, and Samuel Powel. Indeed, Powel was elected president after Bordley declined the office to accept the vice-presidency. Charles Thomson published the thirty-eight-page *Notes on Farming*, a kind of guide to the new agriculture, "chiefly collected from Mr. Young's Farmer's

22. [John Beale Bordley], *A Summary View of the Courses of Crops, in the Husbandry of England and Maryland* (Philadelphia, 1784), 12; Rodney C. Loehr, "The Influence of English Agriculture on American Agriculture, 1775-1825," *Agricultural History*, 11 (1937), 14.

23. James Mease, *Picture of Philadelphia* (Philadelphia, 1811), 266; Olive M. Gambrill, "John Beale Bordley and the Early Years of the Philadelphia Agricultural Society," *Pennsylvania Magazine of History and Biography*, 66 (1942), 415; Elizabeth B. Gibson, *Biographical Sketches of the Bordley Family* (Philadelphia, 1865), 113-14.

24. Minutes of the Philadelphia Society for Promoting Agriculture, Feb. 11, 1785, not paged, Veterinary School, University of Pennsylvania.

Tour through England." [25] George Morgan, who carried out many experiments on his own farm, proved to be the most active essayist in the society. To all of these men, the agricultural society was a more congenial instrument for applying knowledge to "American improvement" than the American Philosophical Society. The Philosophical Society, in turn, lost much of its interest in agriculture and referred to the agricultural society one request from Baron Burgsdorf of Tiegel for information about American farming. [26]

The Society for Promoting Agriculture admitted from the beginning that it looked to England for guidance. Its members were not misled, as Timothy Matlack had been in 1781 by his patriotic zeal, into believing that the French were responsible for the advances of the agricultural revolution. They felt very keenly that the United States must apply to its own farming the agricultural knowledge that had accumulated in England. In the society's first published address, it was announced, "The Husbandry of this country, and of England, were fifty years ago both imperfect, and perhaps nearly alike;—here it has ever since remained nearly stationary, there it has been continually advancing." [27] The goal was to disseminate English techniques but not until they had been tested and perhaps improved.

The society recognized the need for agricultural improvement throughout the entire country but not by its own direct action. It would foster the "*Establishment of other Societies* or Offices of Correspondence in the Principal Places in the Country." [28] In its announcement that members of other societies might attend meetings of the Philadelphia society when in town, there was a hint of the desire to assume leadership in the movement. The premiums offered, most of them for experiments with new techniques of farming, were not restricted to residents of the area. In an effort to reach the dirt

25. [Charles Thomson], *Notes on Farming* (New York, 1787), 3.

26. American Philosophical Society, *Transactions*, 2 (1786), 199-200; 3 (1793), 226-30; American Philosophical Society Minutes, March 16, 1787, [III], not paged; Minutes Philadelphia Society for Promoting Agriculture, April 4, 1786, 43.

27. *An Address from the Philadelphia Society for Promoting Agriculture* ([n.p.], 1785), 3; cf. Timothy Matlack, *An Oration Delivered March 16, 1780, Before ... the American Philosophical Society* (Philadelphia, 1780).

28. Minutes Philadelphia Society for Promoting Agriculture, March 7, 1786, Jan. 10, 1786, not paged.

farmer, newspapers in which the society's announcements and essays appeared were chosen on the basis of their circulation among the country people. The magazines then picked up the essays and made them known to their readers.[29] Early in 1786, suggestions were made that the society establish cattle fairs and a model farm at which the best farming techniques would be demonstrated.[30] With such a program the country people would have been reached and their practice modified.

As it was, the society was most successful in reaching the gentleman farmers. The first prize awarded for an essay on a farm yard went to George Morgan, who was already awake to the need for improved farming techniques.[31] George Washington, however, was pleased with the society's beginning and wished "most sincerely that every State in the Union would institute similar ones." It was through Washington that Arthur Young forwarded to the society the early volumes of his *Annals of Agriculture*.[32] The society claimed credit for extending the use of gypsum and was responsible for effectively publicizing the use of clover. In investigation, experiment, and the collection of the best knowledge, the society kept well abreast of the time.[33]

Even if the society had reached the average farmer, how beneficial its advice would have been is open to question. The experiments of Bordley and George Logan endeavored to test the applicability of the English system to the American environment, but they did not pay enough attention to the fact that intensive agriculture was not economically feasible in America where land resources were so extensive. There was some point to increasing the yield per acre in eastern Pennsylvania or eastern Massachusetts where land values were high

29. *Ibid.*, March 22, 1785, *New Haven Gazette and Connecticut Magazine*, 1 (1786), 43; *Pennsylvania Packet*, May 1, 1788, Sept. 10, 1788; *Columbian Magazine*, 1 (1786-87), 431, 128-30; 2 (1788), 99-101; *American Museum*, 1 (1787), 529.

30. Minutes Philadelphia Society for Promoting Agriculture, March 7, 1786, Jan. 10, 1786.

31. *Ibid.*, Feb. 7, 1786.

32. Washington to James Warren, Oct. 7, 1785, John C. Fitzpatrick (ed.), *The Writings of George Washington* (Washington, 1931-44), XXVIII, 291; Washington to Arthur Young, Nov. 1, 1787, *ibid.*, XXIX, 297.

33. Percy W. Bidwell and John I. Falconer, *History of Agriculture in the Northern United States, 1620-1860* (Washington, 1925), 186.

but even there the important consideration was yield per farm. Bordley recognized that he could not promise farmers immediate economic benefit when he stated that the society hoped to reach men "whose support should not altogether depend on the produce of their farms." [34] The farmers who showed reluctance to adopt the new techniques had some sound reasoning on their side.

It was in Charleston, South Carolina, that the first similar organization was established. The South Carolina Society for Promoting and Improving Agriculture and Other Rural Concerns was established on August 24, 1785, by a group of prominent planters, comparable in their wealth and social standing to the men who formed the Philadelphia society. [35] The circumstances made it clear that the inspiration came from Philadelphia. Three out of the four South Carolinians elected to the Philadelphia society became active in the Charleston society. The time interval between the formation of the two organizations was such as to suggest influence. The South Carolina society very quickly opened a correspondence with the northern group. Moreover, the inaugural address of Thomas Heyward, Jr., as president of the South Carolina society, bore unmistakable marks of the influence of the first *Address* of the Philadelphia society. [36]

Behind the South Carolina society lay an important heritage. Nearly all of the members had agricultural interests and experience, although they were planters rather than dirt farmers. Some of them had been long at work attempting to improve agriculture; Aaron Loocock, a member of the society's executive "Committee," had published a book on the cultivation of madder in 1775. [37] South Carolina had pioneered in the introduction of agricultural fairs in 1723 and in 1740 it had

34. J[ohn] B. Bordley, *Essays and Notes on Husbandry and Rural Affairs* (Philadelphia, 1799), v.
35. *State Gazette of South Carolina*, Charleston, Aug. 29, 1785; C. Irvine Walker, *History of the Agricultural Society of South Carolina* [n.p., circa 1919], 3.
36. Walker, *Agricultural Society of South Carolina*, 3, 4; Walter H. Mills, The Relation between the Philadelphia Society for Promoting Agriculture . . . and the Agricultural Society of South Carolina, 1936, 4, 68, Veterinary School, University of Pennsylvania; Philadelphia Society for Promoting Agriculture, *Memoirs*, 6 [1939], 12.
37. Walker, *Agricultural Society of South Carolina*, 8; Aaron Loocock, *Some Observations and Directions for the Culture of Madder* (Charleston, 1775).

seen the foundation of the Winyaw Indigo Society for improving the cultivation of indigo.[38]

As Heyward described its objectives in his address, the agricultural society sought to stimulate experiments in the new techniques of agriculture. He recommended that each planter "select a small part of his grounds in order to make experiments on it by various methods," and keep "a regular journal" of every occurrence.[39] The results should be written up and submitted to the society so that it could publish a collection of essays from time to time. The Philadelphia society aspired to issue a volume of memoirs too, but it was not able to do so although it did publish several essays in the public press. In 1788, the South Carolina society issued a collection of such papers under the title *Letters and Observations on Agriculture &c. Addressed to or made by the South-Carolina Society for Promoting Agriculture, and Rural Concerns.* It received donations of various seeds and cuttings for experiment, some of them from Thomas Jefferson. The ambitious project of a model or experimental farm was not realized at this time.[40]

Elsewhere the lead of these two societies was not immediately or precisely followed. As the agricultural committee of the American Academy served a similar office in Massachusetts, in New York the Society for Promoting Useful Knowledge sought to fill the need.[41] Local agricultural societies at the county level mushroomed elsewhere. All of them were conducted on the model of the learned society and fell short of their maximum usefulness because they collected information on the best agricultural techniques more successfully than they disseminated knowledge to the farmer.[42]

In some respects the agricultural societies came to be regarded as quasi-official bodies. This was notably true of the Philadelphia society located at the seat of federal and state government. In 1788, the

38. David Ramsay, *The History of South-Carolina* (Charleston, 1809), II, 224; Edgar W. Knight (ed.), *A Documentary History of Education in the South before 1860* (Chapel Hill, 1949–), I, 276.

39. Walker, *Agricultural Society of South Carolina*, 4-5.

40. William Drayton to Jefferson, May 22, 1787, Chalmers S. Murray, *This Our Land* (Charleston, 1949), 51.

41. *New York Journal*, Feb. 1, 1787.

42. Carl R. Woodward, *The Development of Agriculture in New Jersey, 1640-1880* (New Brunswick, 1927), 62-63; Rodney H. True, "The Early Development of Agricultural Societies in the United States," American Historical Association, *Annual Report for 1920* (Washington, 1925), 299.

French Consul Barbé-Marbois sent the society a series of forty-four questions on American agriculture in similar form to his earlier, more general queries that had led to Jefferson's *Notes on Virginia*. He had received the questions from the Abbé Henri Tessier, director of the experimental farm at Rambouillet and member of the Académie des Sciences. Apparently the society sent copies of the questions to different men throughout the country asking them to supply the information on crops, soil, techniques, and pests for their states. Dr. James Tilton returned an excellent series of answers for the State of Delaware which was later published in the *American Museum*, but no other replies have been preserved.[43]

More valuable service was rendered by that society in the investigation of the Hessian fly. The crisis was precipitated when a proclamation of June 25, 1788, prohibited United States wheat from entering British ports because of fear that it might be infected with the Hessian fly. This insect ravaged the wheat crops of the Middle states region from Connecticut to Maryland. Appearing first on Long Island during the war, it reached New Jersey by the summer of 1786 and Pennsylvania the following summer.[44] The Supreme Executive Council of Pennsylvania called upon the Philadelphia Society for Promoting Agriculture to "investigate and report" whether "Wheat produced from a field infected with the Fly" was "good Grain or otherwise." By that time, the press had been flooded with articles and accounts of the insect and farmers had tried shifting to rye, variants of wheat, and fresh seed in efforts to escape the blight. The society, relying heavily upon George Morgan's study of the life cycle of the fly reported that the insect "was not transmitted by seed which had been exposed to its ravages."[45]

Not everyone agreed. Morgan combatted newspaper letters which

43. *American Museum*, 5 (1791), 375; O. R. Bausman and J. A. Munroe, "James Tilton's Notes on the Agriculture of Delaware in 1788," *Agricultural History*, 20 (1946), 176, 179.

44. Phineas Bond to Lord Carmarthen, Oct. 1, 1788, "Letters of Phineas Bond," American Historical Association, *Annual Report for 1896* (Washington, 1897), 573; *Pennsylvania Journal*, Aug. 11, 1787; Asa Fitch, "The Hessian Fly," New York State Agricultural Society, *Transactions*, 6 (1846), 316-73.

45. *Pennsylvania Packet*, June 16, 1788, Sept. 10, 1788; *Pennsylvania Mercury*, June 8, 1787, Aug. 21, 1788; Minutes Philadelphia Society for Promoting Agriculture, April 2, 1788, Jan. 6, 1789, 83, 88.

held that the seed was infected.[46] In Farmington, Connecticut, James Cowles tried a disturbing experiment. He found that if he steeped seeds from infected wheat in a solution of water and bruised elder twigs, he could grow wheat free of any flies.[47] British Consul Phineas Bond quickly interpreted this information for his superiors, implying that the seed was infected by the fly and that the studies of Morgan were "imperfect and inconclusive." He further complained, "In resorting to the leading men among the philosophical and agricultural societies here, I was aware in the sentiments they expressed, the interests of the country operated as a powerful biass." [48] Bond was no friendly critic but it does seem that here patriotism may have conflicted with the best interests of science.

The question was not easily resolved. In England, Joel Barlow found Sir Joseph Banks preoccupied with the subject and full of questions.[49] Benjamin Vaughan wrote his brother John that the report of the Privy Council on the fly was so unfavorable (relying as it did upon three of Bond's letters) that there was no hope of soon lifting the ban.[50] The Philadelphia agricultural society sought the cooperation of the New York Society for Promoting Useful Knowledge and that group solicited communications in the newspapers.[51] The most important contribution to come from New York was an essay by Samuel Latham Mitchill who made more use of European publications on the fly than any of his predecessors.[52] It was still an open question which seriously worried Thomas Jefferson when he became secretary of state under the new Constitution in 1790.[53]

46. *Pennsylvania Packet*, July 1, 1788, Sept. 10, 1788.
47. *New Haven Gazette and Connecticut Magazine*, 2 (1787), 269.
48. Bond to Carmarthen, April 22, 1788, Jan. 20, 1789, "Letters of Bond," 565, 592.
49. Charles B. Todd, *Life and Letters of Joel Barlow* (New York, 1886), 80.
50. Benjamin Vaughan to John Vaughan, May 6, 1789, Archives, American Philosophical Society.
51. Minutes Philadelphia Society for Promoting Agriculture, June 5, 1787, Sept. 4, 1787, 60, 63; *New York Journal*, Feb. 1, 1787, April 18, 1787.
52. *American Magazine*, 1 (1787), 73 ff.
53. Edwin M. Betts (ed.), *Thomas Jefferson's Garden Book* (Philadelphia, 1944), 163; E. Millicent Sowerby, *Catalogue of the Library of Thomas Jefferson* (Washington, 1952–), I, 343-44; Irving Brant, *James Madison, Father of the Constitution, 1787-1800* (Indianapolis, 1950), 340.

Although Samuel Deane's *New-England Farmer* was connected with the efforts of the American Academy to promote agriculture, other general treatises appear to have been individual enterprises altogether. Charles Varlo was an English author who sought to capitalize upon the "Charming Enthusiasm" that James Warren reported "prevailing for Agriculture" by reprinting his popular manual in Philadelphia.[54] On the title page of the *New System of Husbandry* he made the appealing announcement that he would demonstrate how to make a farm of 150 acres clear £402 4s sterling each year. Lest Americans wonder how a foreigner could write authoritatively of farming in their country, he told them he had travelled in America, "that I might not be mistaken." [55] He did, in fact, handle a great variety of subjects —even to veterinary medicine—with considerable competence.

Metcalf Bowler's *Treatise on Agriculture and Practical Husbandry* was the work of a Providence merchant who had bet on the wrong side during the Revolution and had then turned to celebrate the virtues of farming. His eighty-eight-page book was purposely brief because he believed that the large expensive books on agriculture filled with illustrations of unnecessary instruments had "very little promoted the new Method of Cultivation amongst Farmers." He declared that nothing was needed but a common plow and a seed drill. The drill could be bought from Philadelphia ready made or could easily be constructed by hand.[56] His work added little that was novel but it did prove a popular presentation of the new agriculture.

The incomplete success of the agricultural reformers in the years that succeeded the Revolution has led to an underestimate of their real accomplishments. The Norfolk system of tillage was not established in America except in a very limited degree, much less the reforms in stock breeding. It would have been hard, however, for a literate man living anywhere in America not to have become acquainted with the methods of the agricultural revolution which were placed before him in his newspapers, magazines, books, and in the efforts of societies to

54. Warren to Adams, April 30, 1786, *Warren-Adams Letters*, II, 273.

55. C. Varlo, *A New System of Husbandry* (Philadelphia, 1783), Introduction, not paged; *Boston Gazette*, Oct. 11, 1784.

56. Metcalf Bowler, *A Treatise on Agriculture and Practical Husbandry* (Providence, 1786), 3; *Pennsylvania Journal*, Aug. 18, 1787; Jane Clark, "Metcalf Bowler as a British Spy," Rhode Island Historical Society, *Collections*, 23 (1930), 101-17.

give him prizes for trying them. Few but the gentlemen attempted the new techniques, but on large farm and small, agriculture flourished. John Adams said he thought the Americans had never "fed or clothed themselves more easily or more comfortably" than during the Revolutionary period.[57] Beginnings of reform were made too: seed drills were used, clover was planted, gypsum applied, and attempts at different rotation schemes made. More important, the disposition to view agriculture as susceptible to an experimental, scientific approach grew. Many readers who were not yet ready to try the new techniques would swallow whole the newspaper assertion that "Agriculture was perhaps the parent of all those sciences, arts and employments, which have since carried their heads so far above her." [58]

The ultimate effects of the industrial revolution were even greater than the effects of the agricultural revolution, but in the United States there was not the same breadth of interest or extent of enthusiasm for the encouragement of manufactures. Even Tench Coxe, the greatest publicist of manufacturing, admitted that in America agriculture must be given primary attention.[59] The renewal of prewar attempts to promote manufactures was encouraged by the economic difficulties of the Confederation period. A new campaign to buy only American manufactures was instituted for the purpose of benefiting the economy. Britain again appeared the culprit of the piece, this time because of her refusal to execute favorable trade agreements with the United States. In terms reminiscent of the past, a Boston newspaper implored, "Cease to import Goods from that imperious, insolent, debauched and intrigueing Kingdom of *Great Britain:* Abstain from British luxury, extravagance, and dissipation, or *you are an* UNDONE PEOPLE." [60]

Evidences of the movement could be found all over the country. In Boston, Hartford, and Richmond, societies pledged against the use of imported manufactures were formed.[61] The tradesmen and manu-

57. Charles F. Adams (ed.), *Works of John Adams* (Boston, 1850-56), VII, 183.

58. *Boston Gazette*, Feb. 9, 1784; Charles Leavitt, "Attempts to Improve Cattle Breeds in the United States," *Agricultural History*, 7 (1933), 52.

59. [Tench Coxe], *Observations on the Agriculture, Manufactures and Commerce of the United States* (New York, 1789), 15.

60. *Richmond Gazette*, June 19, 1788; *Boston Gazette*, Jan. 31, 1785.

61. *New York Packet*, Nov. 14, 1786; *Boston Gazette*, July 16, 1787; *Worcester Magazine*, 2 (1786-87), 437-38.

facturers of Boston even won a state law limiting the importation of manufactures. With less success they sought also to re-establish a system of committees of correspondence in the port cities to extend and coordinate the movement.[62] In Philadelphia, the emphasis was on a manufacturing society which not only set up new machinery to manufacture cotton cloth but also proceeded like a learned society in collecting information and in seeking to establish a general correspondence.[63] Similar manufacturing societies were formed in New York and Baltimore.[64] In Delaware, the people of Wilmington agreed to meet on January 1, 1789, dressed altogether in American products. In Charleston, a plan was considered for establishing a factory to supply the state with the jeans and corduroys used by slaves.[65] Everywhere, new establishments were founded for making iron and iron products, paper, glass, chemicals, and cotton and woolen cloth. In this process, state legislatures cooperated by granting monopolies, bounties, exemptions, and outright gifts of money to the enterprising.

Patriotism and the expectation that science could be applied to manufactures were elements in the creed of the promoters. Tench Coxe felt that the situation of the colonies had been unfavorable to manufactures but that the Revolution had introduced "immense advantages."[66] Another view had it that as an independent people, the Americans now required manufactures "to support their rank and influence abroad and at home."[67] As for science, Coxe pointed to David Rittenhouse. "Every combination of machinery may be expected from a country," he said, "a NATIVE SON of which, reaching this inestimable object at it's highest point, has epitomized the motions of the spheres, that roll throughout the universe." Ritten-

62. *Boston Gazette*, Oct. 10, 1785.
63. *Pennsylvania Packet*, March 20, 1788, Jan. 15, 1789; Minutes of the Pennsylvania Society for the Encouragement of Manufactures and Useful Arts, I, not paged, Historical Society of Pennsylvania; *Columbian Magazine*, 2 (1788), 172.
64. *New York Journal*, April 2, 1789, May 28, 1789.
65. *New Haven Gazette and Connecticut Magazine*, 3 (1788), no. 4, no. 30.
66. Tench Coxe, *An Address to an Assembly of the Friends of American Manufactures* (Philadelphia, 1787), 5, 11.
67. *The Plan of the Pennsylvania Society for the Encouragement of Manufactures and the Useful Arts* (Philadelphia, 1787), 3.

house, in fact, did become vice-president of the local manufacturing society.[68]

Ingenuity was continually being demonstrated by the Americans in a bewildering number of inventions and gadgets many of which had little enough relationship to the effort to promote manufactures. Machines were invented to clean wells and to clean chimneys, and a method was proposed for building chimneys so they never would have to be cleaned. Numerous improvements were made in clocks: from New York came a clock with a self-regulating pendulum, from Connecticut a self-winding clock, and from Massachusetts a clock mechanism to turn meat cooking before the fire.[69] Thomas Paine and Francis Hopkinson worked out separate improvements in candle lamps. William Thornton invented a steam cannon.[70] Benjamin Dearborn of Portsmouth, New Hampshire, invented an improved steelyard and a new engine for pumping water. From Salem, Massachusetts, came a simplified air pump and from Manheim, Pennsylvania, a pile driver. The list was almost endless but at the bottom were the inevitable flying machines and perpetual motion machines which were never quite completed.[71]

Some of the inventions were a part of the drive to promote manufactures. Pennsylvania lent Whitehead Humphreys £300 to develop his method of manufacturing steel, and in Massachusetts, the Reverend Daniel Little worked out a still different process. Little combined bar

68. Tench Coxe, "Sketches of the subject of American manufactures in 1787," *A View of the United States of America* (Philadelphia, 1794), 41.

69. *Boston Gazette*, Dec. 1, 1783; *Public Records Connecticut*, V, 237; *American Magazine*, 1 (1788), 271; *Columbian Magazine*, 1 (1786-87), 423; 2 (1788), 28-29; James Rumsey to American Philosophical Society, May 20, 1788, Ms. Communications to American Philosophical Society, Trade, 8.

70. Paine to Franklin, Dec. 31, 1785, Franklin Papers, XXXIII, 262, American Philosophical Society; *Columbian Magazine*, 1 (1786-87), 420; Francis Hopkinson, Miscellaneous Essays and Occasional Writings, II, 167-72, American Philosophical Society; Thornton to Lettsom, July 26, 1788, Thomas I. Pettigrew, *Memoirs of the Life and Writing of the Late John Coakley Lettsom* (London, 1817), II, 530.

71. American Philosophical Society Minutes, Oct. 16, 1790, [IV], 148; American Academy, *Memoirs*, 1 (1785), 497-519, 520-24; 2 (1793), 23; Dr. Ebenezer Brooks to American Philosophical Society, Ms. Communications to American Philosophical Society; Mechanics, Machinery, Engineering, etc., 1; *Pennsylvania Journal*, Jan. 8, 1784.

iron, pulverized rockweed, wood ashes, and urine in the furnace to
produce excellent steel—so he reported.[72] In connection with the mill-
ing of flour, there were many inventions and improvements, the most
widely used of which were probably those by Oliver Evans.[73] There
were also inventions of tide mills and improvements to water wheels.
Most important of all were the textile machines patterned after those
of the great English inventors. It was in November 1789 that Samuel
Slater arrived in New York from London, attracted by inducements
offered in the newspapers to textile artisans who could duplicate English machines in America. He was immediately employed by the New
York Manufacturing Society but was unimpressed by their machinery
or their motive power. In December, he got into communication with
Moses Brown of Providence. He, rather than the native American
inventors or the manufacturing societies, established the first successful
cotton textile factory at Pawtucket for the firm of Almy and Brown.[74]

Occasionally, serious attempts were made to apply science to manufacturing processes. Without much profit Miers Fisher consulted articles in Franklin's file of the *Mémoires* of the Académie des Sciences
when he was trying to improve techniques of paper manufacture. On
another occasion Franklin was able to show Benjamin Chambers, who
thought he had invented a new method of "blowing... Furnaces by
a Fall of Water," that it was a technique long known. He pulled down
three books from his library in which he was able to trace the history
of the process in several countries back through a hundred years.[75]
The manufacture of chemicals continued to demand closer attention
to science. When a new establishment began to manufacture Glauber

72. *The Statutes at Large of Pennsylvania*, comp. by James T. Mitchell and
Henry Flanders (Harrisburg, 1896-1911), XII, 235-36; American Academy,
Memoirs, 1 (1785), 527.
73. *Minutes of the General Assembly of Pennsylvania, Tenth*, 228; *Eleventh*,
172, 192; American Philosophical Society Minutes, Aug. 18, 1786, April 18,
1788, May 2, 1788, [III], not paged; Stiles to Franklin, Oct. 10, 1780, Franklin
Papers, IV, 57, University of Pennsylvania; *Pennsylvania Packet*, April 3, 1788.
74. William R. Bagnall, *Samuel Slater and the Early Development of the
Cotton Manufacture in the United States* (Middletown, Conn., 1890), 29.
75. Miers Fisher to American Philosophical Society, Aug. 5, 1788, Franklin
Papers, XXXIV, 79, American Philosophical Society; Franklin to Benjamin
Chambers, Sept. 20, 1788, Albert Henry Smyth (ed.), *The Writings of Ben-
jamin Franklin* (New York, 1905-7), IX, 664.

salts and sal ammoniac, it was a reasonable step for the owners to seek approval of their product from the American Philosophical Society.[76] One of the most direct efforts to apply science to manufacturing was the article written by the Harvard professor of chemistry, Aaron Dexter, on potash. In this, he sought to instruct manufacturers how to make the purest product in the most profitable manner. Writing to men who often had little education and no acquaintance with the terminology of chemistry, he "endeavoured to avoid prolixity, and all chemical terms." [77]

The encouragement of commerce to most merchants was a question of improving the condition of trade rather than improving techniques of sailing—with or without the help of science. Some inventions did relate to commerce but they were of minor importance: dock dredges, new sheathing material for ship's bottoms, a mechanical ship's log, and Hopkinson's spring block.[78] All of the shipbuilders were practical men who worked by rule of thumb. Joshua Humphreys was elected to membership in the American Philosophical Society but the communications he offered the society had nothing to do with building ships.[79] By far the most important attempt to study seriously ship construction and marine practices was Franklin's "Maritime Observations." Franklin pioneered in calling attention to the need for a different hull form under water from that at the water line and above. He also considered the sail form that would be least resistant to wind but would catch all of the wind possible. He recalled a Chinese practice of watertight compartmentation of ships which he recommended, cannily suggesting that the greater cost would be compensated by lower insurance rates. That Franklin was able to make such pregnant suggestions in this field was partly a result of his knowledge of the French attempt at a scientific approach to shipbuilding. Because he was not acquainted with

76. *Pennsylvania Journal*, June 28, 1786.
77. American Academy, *Memoirs*, 2 (1793), 170.
78. *American Magazine*, 1 (1787), 67, 511; *Minutes of the General Assembly of Pennsylvania, Ninth*, 42; Hopkinson, *Miscellaneous Essays*, I, 274-85; Vires acquiret cedendo to Franklin, Dec. 24, 1787, Franklin Papers, XXXV, 161, American Philosophical Society.
79. Joshua Humphreys to Rittenhouse, Jan. 3, 1789, Feb. 16, 1789, Ms. Communications to American Philosophical Society, Mechanics, Machinery, Engineering, etc., 4, 5.

such mathematical concepts as metacentric height, he went wrong in his discussion of stability.[80]

Enthusiasm for internal improvements reached a new peak. The states stepped in now to subsidize road building on a large scale. For the traveller, Christopher Colles provided a new service in his *Survey of the Roads of the United States* which included eighty-six sheets of road maps presenting accurately marked distances from point to point. These he determined by a mileometer of his own design attached to a carriage wheel.[81] Many states became interested in extending the navigation of their rivers and in building canals. Plans were pushed to the point of making surveys for canals between the Cooper and Santee Rivers in South Carolina, through the Dismal Swamp to connect North Carolina and Virginia, and from Lake Champlain to the St. Lawrence River.[82] Learning of the difficulties encountered, particularly in the swampy North Carolina region, Hugh Williamson obtained from the Society of Sciences at Haarlem information on machines and techniques the Dutch had used in canal construction. It was Jefferson who obtained a French memoir of 1785 on the building of a Panama canal—more an indication of his faith in the manifest destiny of the United States than of his interest in canals.[83] Christopher Colles applied his talents to canal planning too, with a magnificent proposal that a canal be built westward from Albany through the Mohawk Valley to Lake Ontario. It was coupled with a scheme to develop the land, Colles as engineer of the canal to receive 5 per cent of the lock

80. *Columbian Magazine*, 1 (1786-87), 434-36; American Philosophical Society, *Transactions*, 2 (1793), 294-329.

81. Washington to Jefferson, Feb. 25, 1785, Julian P. Boyd (ed.), *The Papers of Thomas Jefferson* (Princeton, 1950–), VIII, 3; Belknap, *History of New Hampshire*, III, 80; *Pennsylvania Packet*, Sept. 4, 1788; Christopher Colles, *A Survey of the Roads of the United States* ([n.p.], 1789); Wilbur C. Plummer, *The Road Policy of Pennsylvania* (Philadelphia, 1925), 39.

82. *Company for Opening Navigation between the Santee and Cooper Rivers, Rules* (Charleston, 1786); *An Act for Extending the Navigation of the Potomac* (Alexandria, [1784]); Ira Allen, *The Natural and Political History of the State of Vermont* (London, 1798), 4; *Pennsylvania Journal*, April 15, 1786; Clifford R. Hinshaw, Jr., "North Carolina Canals before 1860," *North Carolina Historical Review*, 25 (1948), 5, 15.

83. Williamson to Robert Patterson, May 24, 1784, Archives, American Philosophical Society; Memoire Sur ... un Canal dans l'Amérique Septentrionale, June 30, 1785, with endorsement by Jefferson, American Philosophical Society.

tolls and 5 per cent of the lands remaining in the hands of the canal company from the projected grant of 250,000 acres by the state.[84] Most of these programs were accomplished in the next few decades. Bridges provided a fine opportunity for the exercise of ingenuity. In 1786, a 1,503-foot bridge with a thirty-foot draw was completed over the Charles River at Boston. In Philadelphia, several plans were submitted for a bridge across the Schuylkill, one calling for four 100-foot spans framed of steel or wood.[85] It was for this spot that Thomas Paine designed his bridge to be constructed of iron bars. It would span the river with one arch. Unable to get his plan adopted in this country, Paine took a model to France where the Académie des Sciences praised it and Caron de Beaumarchais was led to urge that such a bridge be built in Paris at the Jardin Royal. Paine took the model to England, too, and there it ultimately did serve as the basis for a bridge over the Wear at Sunderland.[86]

It was during the Confederation period that the first successful steamboat was built and the first enthusiasm developed for steam navigation. Behind this accomplishment lay the unsatisfied need for boats that could move upstream against the current to make the maximum use of the nation's great river system. This need led William Henry, the Lancaster gunsmith, to suggest a wind-driven boat propelled by the force supplied from a kind of windmill rotor. He thought it would ease "the difficult return passage on the Mississippi and the Ohio."[87] About the same time, James Rumsey, Maryland innkeeper, successfully ran a mechanical pole boat upstream, propelled by the force of

84. [Christopher Colles], *Proposals for the Speedy Settlement of the Waste and Unappropriated Lands on the Western Frontiers of the State of New York* (New York, 1785), 3-5.

85. Jeremy Belknap, *The History of New Hampshire* (Philadelphia and Boston, 1784-99), III, 62; *Boston Gazette*, July 11, 1785; *Massachusetts Centinel*, June 15, 1786; *Columbian Magazine*, 1 (1786-87), 244; *Pennsylvania Packet*, Jan. 2, 1788; Minutes Philadelphia Society for Promoting Agriculture, July 3, 1787, 61.

86. Rochefoucauld to Franklin, Feb. 6, 1788, Franklin Papers, XXXVI, 16, American Philosophical Society; *Minutes of the General Assembly of Pennsylvania, Eleventh*, 166; *Thirteenth*, 25; Philip S. Foner (ed.), *The Complete Writings of Thomas Paine* (New York, 1945), II, 1026-47, 1266-68, 1411-12.

87. Johann David Schöpf, *Travels in the Confederation, 1783-1784*, trans. by Alfred J. Morrison (Philadelphia, 1911), II, 14; American Philosophical Society Minutes, March 4, 1785, [III], not paged.

the current itself. Abner Cloud invented a similar boat in Pennsylvania, and in Boston, John Mason conceived still another sort of mechanical boat.[88] The steamboat was the logical culmination of these thoughts, although in Europe it was an old idea.

The story of the steamboat quickly became that of a duel between two men each supported by his partisans: James Rumsey, the pole boat inventor, and John Fitch, a versatile craftsman from Connecticut, who had turned his hand to watch repairing, silversmithing, and cartography. Neither of them was a man of science or liberal education, but in the success of each, science and the scientific community played a significant role.

Where Fitch got his idea of a steamboat it is not possible to know, despite his circumstantial, introspective account of a vision much like the fable of Newton and the apple. He wrote that he was walking along a road near Neshaminy, Pennsylvania, one April Sunday in 1785, when a horse-drawn carriage passed him. The idea of a steam-powered, horseless carriage immediately popped into his mind although he had never heard of a steam engine—much less seen one. Soon converting his dream to a steam-driven boat, he took it to the local clergyman, the Reverend Nathaniel Irwin, for advice. Irwin rose to the occasion by taking from his shelf a volume of Benjamin Martin's *Philosophia Brittanica* in which he showed Fitch descriptions of the Newcomen and Savary steam engines. At first chagrined, Fitch, on second thought, reflected that the early engines confirmed his ideas.[89]

James Rumsey in the course of his experiments in steam propulsion referred also to accounts of earlier work done in the field. Upon one of the works of Jean T. Desaguliers, he set such great store that, in an effort to keep the ideas related to steam engines to himself, he refused to let an acquaintance handle the book.[90] He also used dictionaries or encyclopedias, perhaps even the one to which Fitch had been referred.

88. *Minutes of the General Assembly of Pennsylvania, Eighth*, 35; *Ninth*, 129; John Mason to Franklin, April 20, 1786, Franklin Papers, XXXIV, 55, American Philosophical Society.

89. John Fitch, Journal, I, 2, Library Company of Philadelphia; James T. Flexner, *Steamboats Come True* (New York, 1944), 76ff.

90. Perhaps Jean T. Desagulier's translation of W. J. 'sGravesande, *Mathematical Elements of Natural Philosophy* (4th ed., London, 1731); Englehart Cruse, *The Projector Detected* (Baltimore, 1788), 4-5.

Each man came to rely heavily upon the scientific community of Philadelphia in his efforts to make the steamboat a reality. Rumsey began his struggle by seeking the support of the great men in his own locality, particularly George Washington. He actually built and demonstrated a steamboat on the Potomac near Shepherdstown before he went to Philadelphia. Rumsey resolved that if he could not find the requisite support in Philadelphia, he would turn to South Carolina. Both Fitch and Rumsey toured the country seeking monopolies from various state legislatures. In addition, Fitch sought support from the Congress of the United States and from the Spanish minister, Don Diego de Gardoqui. It was just a few months after his first vision that Fitch set out for Philadelphia. There he called upon the Reverend John Ewing whom he had known in connection with a western land company but who was now provost of the university and one of the principal figures in the city's scientific community. By Ewing's invitation, he appeared before the American Philosophical Society to present his plans; the approval of that body would have meant much more than even Ewing's certificate suggesting that a steamboat was feasible. Rumsey went to the city on the Delaware after he had done extensive experimenting which he had written up in *A Short Treatise*. Rather than appear before the American Philosophical Society himself, he had his numerous projects presented by a member.[91]

Somehow, Fitch failed to win the support of the leading men of science. The members of the Philosophical Society had little new to suggest to him and few of them were personally ready to support the company he formed to build a boat. His biggest disappointment was Franklin. When Fitch found that Franklin had given support to a steamboat propelled by a jet of water in his "Maritime Observations" he feared the great man was trying to steal his thunder. He became more convinced when Franklin did not even reply to his letter inviting support. Incensed, Fitch interpreted Franklin's offer of alms rather than money for investment in his company as an "insult." Worst of all was the trouble he had with Franklin and Hopkinson when he

91. Fitch, Journal, I, 3, 7, 35; John C. Fitzpatrick (ed.), *The Diaries of George Washington* (New York, 1925), II, 282; James Rumsey, *A Short Treatise on the Application of Steam* (Philadelphia, 1788); American Philosophical Society Minutes, Dec. 2, 1785, Sept. 27, 1785, [III], not paged; Ella May Turner, *James Rumsey, Pioneer in Steam Navigation* (Lancaster, 1930), 116.

sought to obtain a copy of the "Maritime Observations," which he wanted to contest Arthur Donaldson's application for a Pennsylvania monopoly for operating a jet steamboat.[92] Staunch support by the recent immigrants, William Thornton and Samuel Vaughan, did not serve to disguise the fact that the bulk of the scientific community left Fitch alone. He was at least fortunate in gaining the active help of an excellent mechanic, Henry Voight, who later served creditably with Rittenhouse in the mint.

Contrasted with the neurotic Fitch, James Rumsey was tactful and capable of dealing with men effectively. A significant number of the leading members of the American Philosophical Society, with Franklin at the head of the list, joined an organization they called the Rumseian Society which was formed not only to promote his jet propelled steamboat but his mill and boiler improvements as well. Unlike Fitch's commercial company, it followed the organization of a learned society. Benjamin Rush became a strong partisan of Rumsey and Jefferson considered him "upon the whole . . . the most original and the greatest mechanical genius I have ever seen." The Rumseian Society finally raised a fund to send Rumsey to England where he could buy a Watt and Boulton engine and continue his steamboat experiments in the most favorable environment.[93]

Despite his failure to win the scientific community, it was Fitch who built the first successful steamboat. He had a boat running on the Delaware River while the Constitutional Convention was meeting in Philadelphia in August of 1787—before Rumsey's Shepherdstown experiment. Not long afterward, he ran a steamboat on a commercial schedule between Philadelphia and Burlington propelled by his rickety system of cranks and paddles. Rumsey died in England without ever having run a boat at a high enough speed to be useful. Meanwhile, another American inventor, Nathan Reed, demonstrated a steamboat in Massachusetts before the governor and leading members of the American Academy. With Reed, however, the beginning was also the

92. Fitch, Journal, I, 21.
93. *William and Mary Quarterly*, 1st ser., 24 (1915-16), 173; Turner, *Rumsey*, 120; Rumsey, *A Short Treatise*, [last page]; Joseph Barnes, *Remarks on Mr. John Fitch's Reply to Mr. James Rumsey's Pamphlet* (Philadelphia, 1788); Jefferson to Willard, March 24, 1789, Sowerby, *Catalogue*, I, 547; Rush to Lettsom, May 4, 1788, Pettigrew, *Lettsom*, 430.

end. Even Fitch's mechanically successful boat proved a commercial failure. The day of the steamboat was not yet.[94]

The same men who thought that science could be applied to the cultivation of farms, the manufacture of cloth, and the building of steamboats looked also for the application of science to human affairs. The Enlightenment recognized no fundamental difference between knowledge of physics and astronomy and knowledge of government and economics. The sciences of man were expected to yield just as precise laws as the physical sciences. This view was accepted by many of the Americans who took part in the establishment of new state and national governments during and after the Revolution.

When the French minister Turgot criticized the constitutions of the American states, he did so on the ground that they were not sufficiently scientific. They contained "an unreasonable imitation of the usages of England"; they had not enough sought "uniform principles." [95] John Adams' *Defence of the Constitutions* was an endeavor to refute the charge by applying scientific principles to the evaluation of government. "The arts and sciences, in general," he began, "during the three or four last centuries, have had a regular course of progressive improvement. . . . is it not unaccountable that the knowledge of the construction of free governments, in which the happiness of life, and even the further progress of improvement in education and society, in knowledge and virtue, are so deeply interested, should have remained at a full stand for two or three thousand years?" [96] He went on to classify and characterize many ancient and modern governments much as the botanists classified their plants.

Spurred by the same desire to apply science to government, other men formed learned societies for that purpose. The first of these was the Constitutional Society founded in Virginia in 1784 on the initiative of Philip Mazzei. This group planned to hear and discuss papers on government and even thought of electing corresponding members.

94. Essex Institute, *Historical Collections*, 1 (1859), 184; J. P. Brissot de Warville, *New Travels in the United States of America Performed in 1788* (Dublin, 1792), I, 235; William Thornton Papers, XVI (Science), Library of Congress.

95. Turgot to Price, March 22, 1788, Adams (ed.), *Works of John Adams*, IV, 279, 281.

96. John Adams, "A Defence of the Constitutions of Government of the United States of America," *ibid.*, 283.

Although many of the most important men in the state accepted membership, the society did not long continue.[97] The Society for Political Inquiries, established at Philadelphia in 1787, met with a little better success. With Franklin as president, an attempt was made to bring into the membership men of very different political opinions. Several papers were read to the society and occasionally one was published. Following the example of societies instituted for other purposes, premiums were offered for essays on announced topics. Soon after the new Constitution went into effect in the United States, this society, too, was discontinued.[98]

The most extensive attempt at a scientific approach to government was made by James Madison after the weaknesses of the Articles of Confederation had become painfully clear. From Paris Jefferson aided his efforts by sending him nearly two hundred books, many of them "treatises on the ancient or modern Federal Republics." These he used in his studies preparatory to attending the Constitutional Convention in Philadelphia. From this investigation, he wrote two papers: "Notes of Ancient and Modern Confederacies" and "Vices of the Political System of the United States." Some of the results of this research found their way into the new Constitution of the United States, but, as in the case of the steamboat, science was a minor part of that instrument. It was constructed largely by applying the extensive experience of the American states in self-government.[99]

The achievements of the Americans in applying science to useful purposes fell as far short of their aspirations as did their success in developing basic science to a level they considered worthy of their

97. *The Constitutional Society of Virginia* [pamphlet without title page found in the Libr. of Cong.]; *American Historical Review,* 32 (1927), 550-52, 792-93; Philip Mazzei to John Blair, May 12, 1785, Howard R. Marraro (trans.), "Philip Mazzei on American Political, Social, and Economic Problems," *Journal of Southern History,* 15 (1949), 376-77.
98. Minutes of the Society for Political Inquiries, Historical Society of Pennsylvania; *Pennsylvania Packet,* April 25, 1788; Samuel Vaughan, Jr. to Bowdoin, March 5, 1787, *Bowdoin and Temple Papers,* II, 165; [Tench Coxe], *An Enquiry into the Principles on which a Commercial System for the United States of America Should be Founded* ([Philadelphia], 1787).
99. Douglass Adair, "James Madison," in Willard Thorp (ed.), *The Lives of Eighteen from Princeton* (Princeton, 1946), 150-51; Gaillard Hunt (ed.), *Writings of James Madison* (New York, 1900-10), II, 361-90; Dumas Malone, *Jefferson and the Rights of Man* (New York, 1951), 87, 162.

destiny. Their magnificent accomplishment in government stood beside their failure to make steamboats a practical reality. The Americans were daunted by neither experience nor was their vision dimmed. George Washington truly described the Constitution as "an astonishing victory gained by enlightened reason over brutal force." [100] John Fitch wrote out his steamboat experiments in five folio volumes and waited for posterity to vindicate his dream. More than any other people, the Americans looked to posterity for fulfillment. If science did not yet flourish in their land as it ought, tomorrow it would. If science could not yet be extensively applied to the improvement of living, tomorrow it would. The vision still glowed brightly when the "Grand Federal Processions" crawled down the streets of the major cities of the nation celebrating the ratification of the new Constitution.

100. Washington to the Rev. Dr. Lathrop, June 22, 1788, Mark A. De Wolfe Howe, *The Humane Society of the Commonwealth of Massachusetts* (Boston, 1918), 43-44.

Chapter Seventeen

THE AMERICAN REVOLUTION AND THE
PURSUIT OF SCIENCE

THE AMERICAN REVOLUTION was one of the pivotal points of history, with meaning not only for the United States but for the entire Western world. The accompanying military struggle was not exceptional for its magnitude or for the brilliance of command demonstrated during its course. Economic and social patterns were altered only at a few points and then usually in minor ways.[1] Even in the political life of the country, continuity was more conspicuous than change. Indeed, except for the great accomplishment of independence, there were few overt changes in which the Revolution appeared revolutionary. Its significance lay deeper.

To some participants, the Revolution appeared the fruition of the age of Enlightenment and the harbinger of a better world to come. Ensuing events in Europe placed it as the herald of an "age of revolutions" which in due time would eliminate injustice, tyranny, and undeserved privilege from human society. It was possible to view the French Revolution with its sweeping consequences and the unrealized English Revolution as extensions of the American Revolution. In this sense, it was an expression of faith that the heavenly city which reli-

1. Cf. Frederick B. Tolles, "The American Revolution Considered as a Social Movement: A Re-Evaluation," *American Historical Review*, 60 (1954-55), 1-12.

gious men had looked for in the other world might, instead, be erected upon this earth.[2]

There were others who held very different views of the meaning of the Revolution, and, despite continuing study which has illuminated many facets of the movement, all differences of opinion about its nature have not been resolved—even now. Beyond question, a great deal of intellectual ferment preceded and accompanied the Revolution, but how was the thought related to the action? There is still room for debate over the extent to which this intellectual and philosophical speculation served as a motivating cause and the extent to which it was used merely as a rationale to justify action taken for other reasons.

There is no question that the Revolution was identified in the most direct manner with the patterns of thought which were the essence of the Enlightenment. The simplicity and perfection of Sir Isaac Newton's laws of motion had become the ideal of the age and influenced every field from physics to religion. The faith of the Enlightenment in the reasonableness and discernible regularity of the universe extended specifically to government and human relations. Indeed, David Hume considered "the science of man" the most important of all the sciences and the science of politics he declared directly dependent upon it.[3] The thought of the writers upon whom the Americans most relied was of a piece with this pattern. The concepts of natural law, natural rights, government by compact, and the right of revolution did not appear patently absurd to men who had experienced compacts and who discovered the necessity of exercising the right of revolution. The Americans were not only capable of comprehending these ideas, but, because of their own extensive experience in self-government, they were in a position to make positive contributions of their own. The Revolution revealed that in government the Americans were abreast of the best theory of the Enlightenment and well in advance of its best practice.

It was hardly conceivable that men who understood so well the political thought of the age and its supposed scientific character would

2. The key to this faith was supplied by Diderot in his statement, "posterity is to the philosopher what the other world is to the religious man," and brilliantly developed by Carl Becker, *Heavenly City of the Eighteenth Century Philosophers* (New Haven, 1932).

3. David Hume, *A Treatise of Human Nature* [1739] (Oxford, 1928), xx.

be content to stop there. Acceptance of prevailing political concepts rested upon and demanded acceptance of the central importance of science. The more the Americans clarified their own political position, the more it came to seem desirable to promote science in all its phases.

The high valuation placed upon science in European intellectual circles had been accepted by active and influential Americans for many years before the Revolution. As quickly as the country had become capable of it, men had sought to strengthen the support of science. This was a part of the expected course of maturation in the eighteenth century. The novel element was the cultural nationalism of the Revolution which added important dimensions to the demand for attainment in science. Patriotism and pride conceived within the framework of the Enlightenment led men who were sensitive to intellectual values to feel that scientific accomplishment was a necessary measure of national justification.

The first peak of cultural nationalism associated with the Revolution rose to the accompaniment of the protests against the Sugar Act, the Stamp Act, and the Townshend Acts. It coincided with the first successful establishment of a philosophical society in America, the founding of the first medical schools and permanent medical societies, and the cooperative observations of the transit of Venus of 1769—all supported with distinct expectations of practical advantage. The enthusiasm declined with the development of more violent political action, particularly after the Boston Tea Party.

A second and higher peak of enthusiasm began to rise before the end of the war and did not subside until the end of the Confederation period. Aspirations in the newly independent nation were for the most part directed toward the fulfillment of goals formulated before the war had broken out. In all parts of the country, it became imperative to demonstrate to the world the "favourable influence that Freedom has upon the growth of useful Sciences and Arts." [4] Organizational development and attempts to apply science were no longer so conspicuously Philadelphian because such efforts had become generally characteristic of the new nation. Separated not only from the political domination of England but also from the direct cultural ties of the

4. Address of Welcome to Benjamin Franklin, Sept. 27, 1785, American Philosophical Society Minutes, [III], not paged.

colonial period, the United States had to depend more and more upon support and stimulus from within the country. The difficulties in the way of creative scientific accomplishment were more serious than ever, but such statesmen as Washington, Franklin, Jefferson, and Adams were determined that the new nation must not fail in this essential attainment. Behind them stood ranks of men of less prominence who dedicated much of their time to the same objective. For a while, unprecedented attention was devoted to cultural matters including science. Then when the new government under the Constitution was established, the intensity of interest decreased. A straw in the wind was the correspondence between Thomas Jefferson and Francis Hopkinson which during the Confederation period was rich in science, art, music, and a wide variety of intellectual topics; as the Constitution loomed on the horizon, all this was subtly replaced by a predominant concern with political matters.[5] The objective was not abandoned but most of the leaders of the nation found their attention diverted to more urgent tasks.

These periods of enthusiasm encouraged continuing contributions to knowledge and resulted in the erection of new platforms for the support of science and in the strengthening of the old. Even so, the most basic and permanent results of the cultural nationalism of the Revolution can be altogether missed if attention is limited to positive accomplishments. Strikingly, no scientific research of the time matched the importance of Franklin's colonial experiments in electricity. Even in the realm of natural history where the Americans did become more self-reliant in the interpretation of their data, it is not clear that Western science benefited as much as American pride. The great positive accomplishment was the adjustment of the Americans to a new political and cultural status in which, even though bereft of important British supports, they proved capable of continuing scientific work at a productive level and, in some respects, of improving it. This internal life of science was not so directly responsive to emotional and patriotic motives as the promoters wished it to be.

As contrasted with the internal life of science, the manner and objectives of the pursuit of science were in considerable measure deter-

5. Most of this correspondence is found in the Jefferson Papers, Library of Congress.

mined by external factors.[6] This was the reason for the great attention devoted to organization and reorganization in science and related fields. Long accustomed to voluntary association for the attainment of specific ends, the Americans found it necessary at this time to combine more energetically than ever before. Organization was one of the most prominent features of the Revolutionary movement from the combination of merchants in opposition to English policy to the formation of new state and central governments. Necessarily the first step in promoting science would be the erection of scientific societies following European models. The foundation of medical schools and societies, new colleges, agricultural and manufacturing societies, new journals, and all the multifarious organizational activity of the time were the accepted means of approaching the objective of a creative and useful science. American as well as European patterns of promotion determined this mode of attack.

The expectation of practical advantage from the pursuit of science was another product of the social and intellectual environment. The Baconian assertion of the utility of science had still not been adequately demonstrated but the concept had been widely accepted in Europe and extensively disseminated in America where it was readily approved. The American environment strongly reinforced this European idea by providing significant support for scientific projects only when some clear utilitarian gain seemed likely. The war intensified this tendency and the separation from England further emphasized it by forcing the Americans to rely more upon their own resources.

The nationalism and emotional faith of the Revolution also crept into the pursuit of science. American political patterns were reflected in plans to form scientific societies into a federal organization, and in the studied reference to "the Republic of Letters."[7] Even the internationalism evident in the demands that science be kept superior to the distractions of national wars was a part of the ideology that accompanied the American Revolution. The demand that scientific societies be granted written charters followed the emphasis upon written po-

6. For this differentiation of internal and external factors, see Richard H. Shryock, "The Interplay of Social and Internal Factors in the History of Modern Medicine," *Scientific Monthly*, 76 (1953), 221-30.

7. Madison to Stiles, Aug. 1, 1780, *William and Mary Quarterly*, 2nd ser., 7 (1927), 293.

litical constitutions and the style of the documents was sometimes reminiscent of the great state papers of the day. The American Philosophical Society's wartime charter was typical in its unquestioning declaration of faith that, "the cultivation of useful knowledge, and the advancement of the liberal arts and sciences in any country, have the most direct tendency towards the improvement of agriculture, the enlargement of trade, the ease and comfort of life, the ornament of society, and the increase and happiness of mankind." [8]

Science was so central to the thought of the Enlightenment and it lay so directly behind the Revolutionary argument, that the men who made the American Revolution were thoroughly committed to the pursuit of science. They were convinced of its utility and confident of America's capacity to achieve in this realm. The heritage left by the Revolutionary generation contained no more important element than this faith that science would flourish in America and that it would be instrumental in advancing the wealth and happiness of the nation. Their zeal and their vision were transmitted almost intact. The unprecedented richness of modern America is a monument to the faith of the Revolutionary generation in the power and beneficence of science, just as its form of government is a monument to their faith in man's capacity to govern himself.

8. American Philosophical Society, *Transactions*, 1 (1789 ed.), xi.

Bibliographical Note

Numerous and extensive sources upon the promotion of science in early America are available. The correspondence and journals of the men who sought to promote science in America and of the scientists themselves were a fundamental recourse. At least equally important were the archives and publications of institutions on both sides of the Atlantic which nourished the growth of science in America. Occasionally, governmental records were helpful. Newspapers, magazines, and various journals yielded much information. By 1789, the number of separate titles in science and related fields published in America or written by Americans had become large. Most of them have been consulted. Whenever there were relevant, recent studies, they were used also. There are some excellent books and articles of this sort, but the most significant thing about this category of writings is its scarcity. Dozens of volumes remain to be written upon subjects within the scope of this book.

Most of the personal correspondence upon which so much dependence has been placed was used in printed form. Among the most useful collections for the early naturalists were James Edward Smith (ed.), *A Selection of the Correspondence of Linnaeus* (2 vols., London, 1821) and William Darlington, *Memorials of John Bartram and Humphry Marshall* (Philadelphia, 1843). Some smaller collections were also helpful, notably E. G. Swem (ed.), "Brothers of the Spade; Correspondence of Peter Collinson, of London, and of John Custis, of Williamsburg, Virginia, 1734-1746," American Antiquarian Society, *Proceedings*, 58 (1948), 17-190, and "Letters from Dr. William Douglass to Cadwallader Colden of New York," Massachusetts Historical Society, *Collections*, 4th ser., 2 (1854), 164-89. Important for all phases of science before the Revolution were *The Colden Letter Books*, 2 vols., New-York Historical Society, *Collections*, 9, 10 (1876, 1877) and *The Letters and Papers of Cadwallader Colden*, 9 vols., New-York Historical Society, *Collections*, 50-56, 67-68 (1917-23, 1934-35). The correspondence of Benjamin Franklin was of great value in all parts of the book. The only collection used extensively was Albert Henry Smyth (ed.), *The Writings of Benjamin Franklin* (10 vols., New York, 1905-7).

For the latter half of the book, the richest single printed collection was Julian P. Boyd (ed.), *The Papers of Thomas Jefferson* (9 vols., Princeton, 1950–) although it had to be supplemented by older collections of Jefferson's letters in the period after 1786. Of the same order of value, was Lyman H. Butterfield (ed.), *Letters of Benjamin Rush* (2 vols., Princeton, 1951). Both of these works were useful for general reference as well as for the letters they contained. Several significant collections were published by the Massachusetts Historical Society: *The Belknap Papers*, 3 vols., in its *Collections*, 5th ser., 2, 3 (1877), 6th ser., 4 (1891); *Warren-Adams Letters*, 2 vols., *ibid.*, 72, 73 (1917, 1925); *The Bowdoin and Temple Papers*, 2 vols., *ibid.*, 6th ser., 9 (1897), 7th ser., 6 (1907); "Correspondence between John Adams and Prof. John Winthrop," *ibid.*, 5th ser., 4 (1878), 289-316. Smaller groups of letters in the *Proceedings* of the Massachusetts Historical Society yielded some information: "Joseph Willard Letters, 1781-1822," 43 (1909-10), 609-46 and "Letters of Richard Price," 2d ser., 17 (1903), 262-378.

Among separately published collections was Worthington C. Ford (ed.), *Statesman and Friend, Correspondence of John Adams with Benjamin Waterhouse* (Boston, 1927). This and Charles Francis Adams (ed.), *Letters of John Adams Addressed to his Wife* (Boston, 1841) supplied the most useful of the Adams letters. Important Cutler letters were found in William P. and Julia P. Cutler, *Life, Journals, and Correspondence of Rev. Manasseh Cutler* (2 vols., Cincinnati, 1888). Thomas J. Pettigrew, *Memoirs of the Life and Writings of the Late John Coakley Lettsom* (3 vols., London, 1817) offered principally letters to Lettsom. William Barton, *Memoirs of the Life of David Rittenhouse* (Philadelphia, 1813) provided many letters not otherwise obtainable. None of the collections of that period, unfortunately, was so reliable as more recent publications such as Isabel M. Calder (ed.), *Letters and Papers of Ezra Stiles* (New Haven, 1933).

The journals used most extensively were those of Ezra Stiles: Franklin B. Dexter (ed.), *The Literary Diary of Ezra Stiles* (3 vols., New Haven, 1901) and Franklin B. Dexter (ed.), *Extracts from the Itineraries and Other Miscellanies of Ezra Stiles, D.D., LL.D., 1755-1794* (New Haven, 1916). A more conventional journal was James Thacher, *A Military Journal during the American Revolutionary War* (2d ed., Boston, 1827). The most useful of the travel journals kept by visiting foreigners were Adolph B. Benson (ed.), *Peter Kalm's Travels in North America* (2 vols., New York, 1937); Johann David Schöpf, *Travels in the Confederation, 1783-1784,* trans. by Alfred J. Morrison (2 vols., Philadelphia, 1911); J. P. Brissot de Warville, *New Travels in the United States of America Performed in 1788* (Dublin, 1792); and [François Jean] Marquis de Chastellux, *Travels in North-America in the Years 1780-81-82* (New York, 1827). George W. Corner, *The Autobiography of Benjamin Rush* (Princeton, 1948) was occasionally helpful.

Manuscript correspondence was used only for material not otherwise obtainable. The largest and finest collection of letters related to the development of science was found in the Franklin Papers and the Bache Collection at the American Philosophical Society. A very small proportion of these letters, most of them written to Franklin, has been published. A similar though smaller collection of letters from men interested in science in different parts of America and in Europe addressed to Benjamin Rush was used at the Library Company of Philadelphia. The John Bartram Papers and the Humphry Marshall Papers were probably the most fruitful manuscripts among many used at the Historical Society of Pennsylvania, fruitful because of Darlington's faulty editing or his inadequate coverage in his *Memorials*. In the Colden Collection at the New-York Historical Society, an unpublished block of papers exclusively devoted to science yielded important information. The most useful item used at the New York Public Library was the manuscript diary of William Smith of New York. At Yale University, the Papers of Ezra Stiles contained many valuable items that have not been published. The Massachusetts Historical Society was used principally for the Warren Papers, only a portion of which has been published. The Jefferson Papers at the Library of Congress have been used only for the portion not yet covered by Julian P. Boyd (ed.), *The Papers of Thomas Jefferson*.

By far the most important institutional archives used were those of the American Philosophical Society. The manuscript minutes revealed a great deal that could not be found in the calendared minutes published as *Early Proceedings of the American Philosophical Society* (Philadelphia, 1884). The minutes together with the many communications to the society and other archival material provided an unmatched record of scientific interests and achievement in the Revolutionary period. The importance of these holdings is basically a reflection of the importance of the society throughout the period 1768-1789. It has been enhanced by the failure of the American Academy of Arts and Sciences and similar organizations to preserve any records worth mentioning.

The colleges and the medical societies have been somewhat more successful in preserving their archives, but the only college archives used to any degree were those of Harvard—consulted because Harvard was the dominant intellectual institution in a very large part of New England. No medical society archives were used in manuscript form but the published *Rise, Minutes, and Proceedings, of the New Jersey Medical Society* (Newark, 1875) proved very useful. For the two organizations with the best records, good accounts were available based upon the archives: Walter L. Burrage, *A History of the Massachusetts Medical Society* (n.p., 1923) and W. S. W. Ruschenberger, *An Account of the Institution and Progress of the College of Physicians of Philadelphia* (Philadelphia, 1887).

Government archives were used only in limited measure. The best resources proved to be: *Journals of the Continental Congress*, ed. by Worthington C. Ford (34 vols., Washington, 1904-37); *The Statutes at Large; Being a Collection of all the Laws of Virginia*, ed. by William W. Hening (13 vols., Richmond, 1819-23); *Votes and Proceedings of the House of Representatives of the Province of Pennsylvania, Pennsylvania Archives*, 8th ser. (8 vols., 1931-35); *Minutes of the General Assembly of the Commonwealth of Pennsylvania* (1781-92); *Public Records of Connecticut*, ed. by James H. Trumbull and Charles J. Hoadly (15 vols., Hartford, 1850-90); *Public Records of the State of Connecticut*, I–, ed. by Charles J. Hoadly and Leonard W. Labaree (Hartford, 1894–); and *Acts and Resolves, Public and Private, of the Province of the Massachusetts Bay* (21 vols., Boston, 1869-1924).

Scientific journals were important because of the papers Americans published in them and because they were so effective in supplying information about European developments. The greatest of these, and the only one of any importance in the early colonial period was the *Philosophical Transactions* of the Royal Society. It continued to be important until the Revolution, although the *Transactions* of the American Philosophical Society, in 1771, destroyed its monopoly. After the war, and after the appearance of the first volume of *Memoirs* of the American Academy of Arts and Sciences in 1785, very few American papers were published in the *Philosophical Transactions*.

The Americans were served by British medical journals both before and after the Revolution, the most important being: *Medical Observations and Inquiries* (6 vols., 1757-84); *Medical Essays and Observations* (5 vols., 1733-42); *Medical Communications*, vols. 1, 2 (1784, 1790); and *London Medical Journal* (11 vols., 1781-90). The only such publication to appear in the United States before 1789 was the *Cases and Observations* of the New Haven Medical Society issued in 1788. Papers submitted prior to 1790 were among those included in the first volume of *Medical Papers Communicated to the Massachusetts Medical Society* of 1790 and the first

BIBLIOGRAPHICAL NOTE

volume of *Transactions* of the College of Physicians of Philadelphia published in 1793.

American newspapers and American and British general magazines were of great use. The newspapers were most valuable in the colonial period; after the Revolution their printing of scientific essays declined as the magazines became more effective. The most widely circulated English magazine of the colonial period was the *Gentleman's Magazine* (303 vols., 1731-1907). The others most used throughout the study were the *Monthly Review*, vols. 1-81 (1749-89) and the *Critical Review*, vols. 1-70 (1756-90). The American magazines to devote the greatest attention to science in the colonial period were Lewis Nicola's *American Magazine* of 1769 and the *Pennsylvania Magazine* of 1775 and 1776. In the national period the best were the *Columbian Magazine* (5 vols., 1786-90); the *American Museum* (12 vols., 1787-92); the *American Magazine* (1 vol., 1787-88); the *Worcester Magazine* (3 vols., 1786-88); and the *Massachusetts Magazine* (8 vols., 1789-96). All of the American magazines are conveniently accessible in the *American Periodical Series* of microfilms (1741-1825) issued by University Microfilms of Ann Arbor (33 reels, 1942).

No group of sources was more essential to this study than the many separate publications related to science and particularly to science in America which appeared between 1735 and 1789. The volume of books and pamphlets in the broad field of science written by Americans or by Europeans about America was very great. Despite their importance, it is not feasible to list them here; they will be found throughout the footnotes. The best collections of this material were found in the Library Company of Philadelphia, the American Philosophical Society, the Library of Congress, the New York Academy of Medicine, the New York Public Library, and Harvard College Library.

It would be even more futile to attempt to list the numerous secondary writings: histories, biographies, and special studies which contribute to an understanding of science and the promotion of science in early America. The best guide is Whitfield J. Bell, Jr., *Early American Science: Needs and Opportunities* (Williamsburg, 1955). His articles in the field and his unpublished doctoral dissertation, Science and Humanity in Philadelphia, 1775-1790 (University of Pennsylvania, 1947), are studies in depth. I. Bernard Cohen's edition of *Benjamin Franklin's Experiments* (Cambridge, 1941), his *Some Early Tools of American Science* (Cambridge, 1950), and his several articles are rich in understanding and information. Louis W. McKeehan, *Yale Science, the First Hundred Years, 1701-1801* (New York, 1947), is not as good but it is the kind of study that ought to be made in increasing numbers. Donald Fleming, *Science and Technology in Provi-*

dence, 1760-1914 (Providence, 1952), is far too brief for the period of this study, but helpful. Theodore Hornberger, *Scientific Thought in the American Colleges, 1628-1800* (Austin, 1945), is a suggestive survey. William M. and Mabel S. C. Smallwood, *Natural History and the American Mind* (New York, 1941), is a poorly digested collection of excellent material. An excellent unpublished doctoral dissertation is Winthrop Tilley, The Literature of Natural and Physical Science in the American Colonies (Brown University, 1933). Valuable articles by Raymond P. Stearns, Frederick G. Kilgour, and Josiah W. Trent are found in several periodicals. Carl Bridenbaugh's books, particularly Carl and Jessica Bridenbaugh, *Rebels and Gentlemen* (New York, 1942), contain important insights.

INDEX

INDEX

Connecticut, silk culture, 200, 203-4, 355; manufacturing, 206, 207; canals, 210; medical societies, 292-93
Connecticut Society of Arts and Sciences, 272-73
Connecticut Wits, 259
Constitution, effects, 383; application of science to, 378
Constitutional Society, 377-78
Continental Congress, 235, 245, 263
Cook, Capt. James, 221, 325-26
Cooper, Rev. Samuel, 251, 263-64, 265
Copyright legislation, 259
Coste, Dr. Jean-François, 238
Coudray, Gen. Tronson du, 241
County agricultural societies in England, 357, 359
Cowles, James, 365
Coxe, Tench, 367, 368
Crèvecoeur, St. John de, 282, 312
Critical Review, 144, 269-70
Crocker, A., 355
Croghan, George, 32-33, 184
Croker, Temple Henry, 349-50
Cullen, Dr. William, 281
Cultural flowering, 256
Cultural nationalism, 382-85
Currie, Dr. William, 301
Custis, John, 25, 30
Cutler, Rev. Manasseh, 303, 313-14, 357

D'Alembert, Jean-le Rond, 265, 344
Dalton, Tristram, 155, 157-58
Danforth, Dr. Samuel, 289
Danforth, Thomas, 148, 149
D'Anmours, Charles-François, Chevalier, 227
Dark day, 348
Darwin, Erasmus, 91
Daubenton, Louis, 321
Davidson, Robert, 317
Day, Thomas, 259
Deane, Samuel, 357-58, 366
De Brahm, William Gerard, 31-32, 176, 177, 181, 247
Decker, John, 341
Degeneracy theory, 250, 321, 322, 338-39
Degree of latitude, measurement, 175
Delaware, medical society, 294; manufacturing, 368
Delaware-Chesapeake canal, 210
Delisle, Joseph-Nicolas, 82
De Luc, Jean André, 348
De Normandie, Dr. John, 134, 187
Dentistry, 287

Derham, William, 87
Desaguliers, Jean T., 87, 374
Deslon, Charles, 284
Destiny, American sense of, 105, 116-17, 121, 125-26, 248-49
DeWitt, Simeon, 243, 337
Dexter, Dr. Aaron, Humane Society, 286; medicine, 290, 297; chemistry, 342-43, 371
Dickinson, James, 150-51
Dickinson, John, 130-31, 137
Dickinson College, 330
Diffraction grating, 346
Dighton rock, 184-85, 323-24
Dillenius, John James, 14, 17, 18, 24, 36
Dipping needle, 177, 180
Dispensaries, 287-88
Dissection, 300
Dissenters and science, 18
Dissenting academies, 86
Dixon, Jeremiah, 153, 167, 169, 175
Dobson, James, 181
Doctors' riot, New York, 300
Donaldson, Arthur, 242, 376
Dossie, Robert, 208-9
Douglass, Dr. William, background, 38; sketch, 48-50; map, 58; medical society, 60, 61; botany, 313
Dredge, 371
Dry Tortugas, charting, 176
Duane, James, 274
DuBois, Dr. Isaac, 73
Dubourg, Barbeu, 145
Dudley, Paul, 4-5, 17, 30, 60, 94
Dufay, Charles, 77
Duhamel du Monceau, Henri Louis, 17-18
Duncan, Sir William, 119
Duportail, Gen. Louis Lebeque de Presle, 241, 242
Du Simitière, Pierre Eugene, 143, 261, 323
Dutch canal influence, 372
Dwight, Timothy, 259
Dymond, Joseph, 150, 158, 183

Earthquakes, 94
Eclipses, of Jupiter's satellites, 171, 175; of sun, 335-36
Edinburgh Essays (Essays and Observations), 51
Edinburgh *Medical Observations. See Medical Essays and Observations*
Edinburgh Society of Arts and Sciences, 25, 56

INDEX